# Lecture Notes in Mathematics 1878



**Subseries:**
Ecole d'Eté de Probabilités de Saint-Flour

R. Cerf

# The Wulff Crystal in Ising and Percolation Models

Ecole d'Eté de Probabilités
de Saint-Flour XXXIV - 2004

Editor: Jean Picard

Springer

Author

Raphaël Cerf
Mathématiques, Bâtiment 425
UMR CNRS 8628
Université de Paris-Sud
91405 Orsay cedex
France
*e-mail: rcerf@math.u-psud.fr*

Editor

Jean Picard
Laboratoire de Mathématiques Appliquées
UMR CNRS 6620
Université Blaise Pascal (Clermont-Ferrand)
63177 Aubière Cedex
France
*e-mail: jean.picard@math.univ-bpclermont.fr*

Cover: Blaise Pascal (1623-1662)

Library of Congress Control Number: 2006921625

Mathematics Subject Classification (2000): 49Q20, 60F10, 82B05

ISSN print edition: 0075-8434
ISSN electronic edition: 1617-9692
ISSN Ecole d'Ete de Probabilités de St. Flour, print edition: 0721-5363

ISBN 978-3-540-30988-8 Springer Berlin Heidelberg New York



Typesetting: by the authors and SPI Publisher Services using a Springer LaTeX package

Cover design: *design & production* GmbH, Heidelberg

Printed on acid-free paper SPIN: 11601623 VA41/3100/SPI 5 4 3 2 1 0

To Katti, Benjamin and Nelly

# Foreword

Three series of lectures were given at the 34th Probability Summer School in Saint-Flour (July 6–24, 2004), by the Professors Cerf, Lyons and Slade. We have decided to publish these courses separately. This volume contains the course of Professor Cerf. We cordially thank the author for his performance at the summer school, and for the redaction of these notes.

69 participants have attended this school. 35 of them have given a short lecture. The lists of participants and of short lectures are enclosed at the end of the volume.

The Saint-Flour Probability Summer School was founded in 1971. Here are the references of Springer volumes which have been published prior to this one. All numbers refer to the *Lecture Notes in Mathematics* series, except S-50 which refers to volume 50 of the *Lecture Notes in Statistics* series.

| | | | |
|---|---|---|---|
| 1971: vol 307 | 1980: vol 929 | 1990: vol 1527 | 1997: vol 1717 |
| 1973: vol 390 | 1981: vol 976 | 1991: vol 1541 | 1998: vol 1738 |
| 1974: vol 480 | 1982: vol 1097 | 1992: vol 1581 | 1999: vol 1781 |
| 1975: vol 539 | 1983: vol 1117 | 1993: vol 1608 | 2000: vol 1816 |
| 1976: vol 598 | 1984: vol 1180 | 1994: vol 1648 | 2001: vol 1837 & 1851 |
| 1977: vol 678 | 1985/86/87: vol 1362 & S-50 | | 2002: vol 1840 |
| 1978: vol 774 | 1988: vol 1427 | 1995: vol 1690 | 2003: vol 1869 |
| 1979: vol 876 | 1989: vol 1464 | 1996: vol 1665 | 2004: vol 1878 & 1879 |

Further details can be found on the summer school web site
`http://math.univ-bpclermont.fr/stflour/`

Jean Picard, Université Blaise Pascal
Chairman of the summer school

# Preface

This text is a synthesis of recent works aiming at a mathematically rigorous justification of the phase coexistence phenomenon, starting from a microscopic model. It is intended to be self–contained. The proofs which can be found only in research papers have been included, whereas results for which the proofs can be found in classical textbooks are only quoted.

Here is a brief outline of the structure of the text. The main results on the Wulff crystal in Ising and Percolation models are presented in chapter 5. Throughout the text, I focus on three fundamental models: the Bernoulli percolation model, the FK or random cluster model and the Ising model. These models are respectively introduced in chapters 2, 3, 4. The reader interested mainly in the Wulff crystal of the Ising model can proceed directly to section 5.1. Some background on large deviations is provided in chapter 6, and the surface large deviation principles leading to the results of chapter 5 are stated in chapter 7. The associated volume large deviation principles are presented in chapter 8. Chapters 9, 10, 11, 12 contain the fundamental probabilistic estimates for the proofs. The basic geometric tools involved in the proofs are the object of chapters 13, 14, 15. The final steps of the proofs for each model are described in chapters 16, 17, 18, 19.

The simulations are performed with a one Ghz computer running under Linux. The percolation simulations were done with the program gpbond and the Ising simulations with the program gising. These programs are available under the GNU General Public License through the web page http://www.math.u-psud/˜cerf. The pictures of the droplets of oil in the water were taken in my saucepan.

I thank Jean Picard for the efficient and smooth organization of the Saint–Flour summer school. I wish to express my gratitude to Olivier Couronné, Reda Messikh, Katti Millock and Thierry Quentin de Gromard for their help and comments.

Orsay,
October 2005

Raphaël Cerf

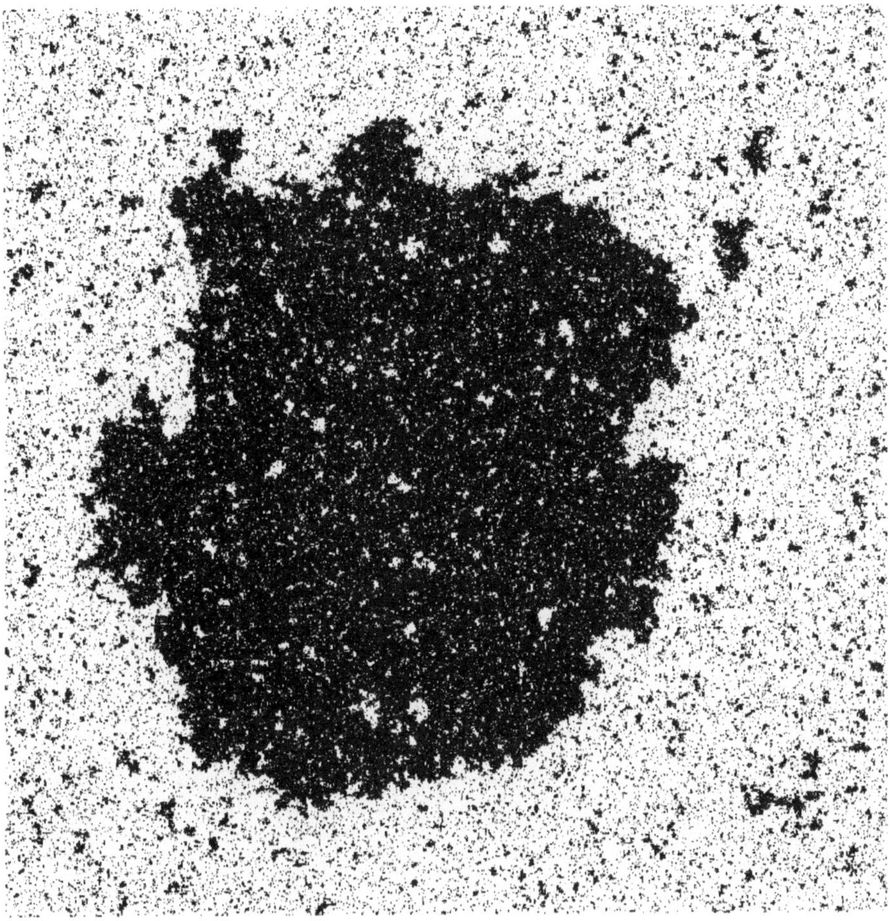

Simulation of an Ising Wulff crystal at $T = 2.22$

# Contents

## Part III Main results



## Part IV Large deviation principles



## Part V Fundamental probabilistic estimates

# Part I

# Introduction

# 1

# Phase coexistence and subadditivity

This initial chapter introduces the three core ingredients of our story: a little physical experiment of everyday life illustrating the phase coexistence phenomenon, the subadditive lemma and finally Cramér's theorem in $\mathbb{R}$. The reader interested mainly in the Wulff crystal of the Ising model can proceed directly to section 5.1 of chapter 5.

## 1.1 Water and oil

Let us consider a volume of water in absence of gravity at ordinary temperature. We start to pour a very small quantity of oil into the water. First, nothing noticeable happens on the macroscopic scale, i.e., the oil is perfectly dissolved throughout the water and the oil molecules are homogeneously spread within the water: by observing the liquid at the macroscopic level, we cannot even tell that it is a mixture of two distinct types of particles, which nevertheless have the tendency to repel each other. Let us keep pouring oil into the water. We know that the solubility of oil in water is not infinite; at some density threshold (which increases with the temperature), we obtain a solution of water saturated with oil. This solution is still a pure phase, completely homogeneous on the macroscopic level, and it realizes a perfect tradeoff between entropy and energy; we call it the water phase. Let us pour in some more oil. The excess of oil is not dissolved any more and it precipitates: macroscopic droplets of oil emerge. These droplets are not regions where there are only oil molecules, rather in these regions we observe the symmetric pure phase consisting of oil saturated with water, which we call the oil phase. The droplets are delimited by an abrupt change of the local density of water and oil molecules. We wish to understand the law governing the evolution and the shapes of these droplets.

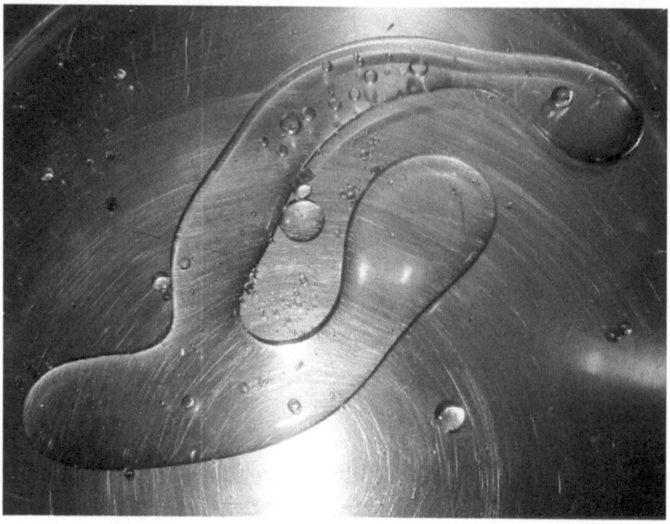

Oil in water far from the final equilibrium

The classical phenomenological theory asserts the existence of a macroscopic surface free energy $\mathcal{I}$ and that the droplets evolve so as to minimize $\mathcal{I}$. For instance, at equilibrium, in case $\mathcal{I}$ is isotropic, one observes a unique spherical droplet of the oil phase floating in the sea of the water phase. Our aim is to confirm the predictions of the phenomenological theory starting from a truly microscopic model. We wish to understand how the random forces acting at the atomic level, more precisely the probabilistic repulsive effect between the two types of particles, can induce such deterministic macroscopic effects. One of the most famous results in probability theory is the law of large numbers: If $(X_n)_{n\in\mathbb{N}}$ is a sequence of independent identically distributed random variables with mean $m$, then, with probability one,

$$\lim_{n\to\infty}\frac{1}{n}(X_1+\cdots+X_n) = m\,.$$

What we have in mind is a generalization of the law of large numbers, but in a fundamentally new context, which we could state informally as follows:

$$\lim_{\text{number of particles}\to\infty}\begin{pmatrix}\text{global effect of the random}\\ \text{microscopic repulsive forces}\end{pmatrix} = \text{single droplet}\,.$$

The limiting deterministic object is the shape of the droplet at equilibrium and the problem is intrinsically geometric; we deal with spatially dependent random variables and we leave radically and definitively the independent framework. Hence the geometry enters the problem in a decisive way, in the random interactions and in the formulation of the result itself.

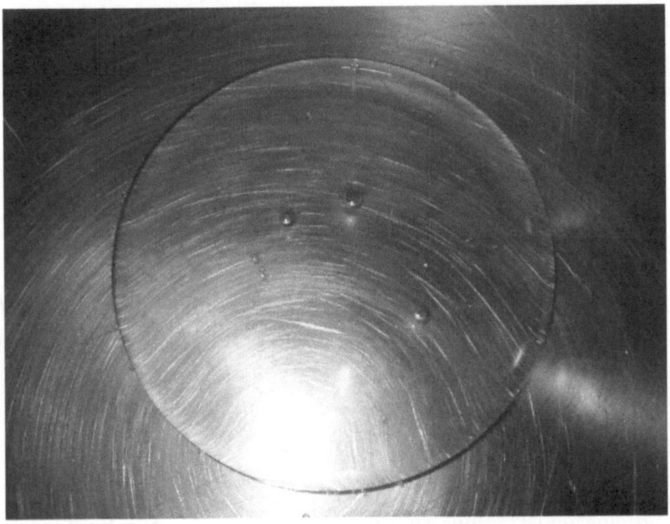

Oil in water close to the final equilibrium

## 1.2 Subadditivity

The keys to the proofs are beautiful subadditive arguments. Lanford imported subadditive ideas from statistical mechanics into Cramér's theory [94], which were further exploited by Bahadur and Zabell [14]. We cannot resist the pleasure to include at this point the statement of the subadditive lemma. This lemma should apparently be called Fekete's lemma: for historical remarks, we refer the reader to [123].

**Lemma 1.1.** *Let $f : \mathbb{N} \setminus \{0\} \to [0, \infty]$ be a subadditive map, i.e.,*

$$\forall m, n \in \mathbb{N} \setminus \{0\} \qquad f(m+n) \leq f(m) + f(n)\,.$$

*We suppose that:* $\exists N \geq 1 \quad \forall n \geq N \quad f(n) < \infty$. *Then*

$$\lim_{n \to \infty} \frac{f(n)}{n} = \inf_{n \geq 1} \frac{f(n)}{n}\,.$$

*Proof.* Let $m, n$ be integers such that $n \geq m \geq N$. Let $n = mq + r$ be the Euclidean division of $n$ by $m$. We have

$$f(n) = f(mq+r) = f(m(q-1)+m+r) \leq (q-1)f(m) + f(m+r)$$

whence

$$\frac{f(n)}{n} \leq \frac{q-1}{n} f(m) + \frac{1}{n} \max_{0 \leq k < m} f(m+k)\,.$$

The hypothesis ensures that $\max_{0\le k<m} f(m+k)$ is finite. Sending $n$ to $\infty$, we get

$$\forall m \geq N \qquad \limsup_{n\to\infty} \frac{f(n)}{n} \leq \frac{f(m)}{m}\,.$$

Taking the infimum over $m \geq N$, we get

$$\limsup_{n\to\infty} \frac{f(n)}{n} \leq \inf_{m\geq N} \frac{f(m)}{m}\,.$$

Let $k \in \{1 \cdots N-1\}$ be such that $f(k) < \infty$. Then $f(Nk) \leq Nf(k)$ whence $f(Nk)/Nk \leq f(k)/k$ and $\inf_{m\geq N} f(m)/m = \inf_{m\geq 1} f(m)/m$. That's it! □

## 1.3 Cramér's theorem

Here is the classical Cramér theorem in $\mathbb{R}$, which is the first large deviation theorem to be taught. We present a proof whose strategy is similar to what we shall do later in the more complicated Ising and percolation models. In particular, the rate function is built as a subadditive limit.

Let $(X_n)_{n\geq 1}$ be a sequence of i.i.d. random variables with values in $\mathbb{R}$. We consider the empirical mean

$$\forall n \in \mathbb{N}\setminus\{0\} \qquad \overline{S}_n = \frac{1}{n}(X_1 + \cdots + X_n)\,.$$

The Cramér transform of the law of $X_1$ is the map $I : \mathbb{R} \to [0, +\infty]$ defined by

$$\forall x \in \mathbb{R} \qquad I(x) = \sup_{y<x<z} \lim_{n\to\infty} -\frac{1}{n} \ln P(y < \overline{S}_n < z)\,.$$

The limit inside the supremum exists thanks to lemma 1.1. Indeed, for any $y < z$, any $n, m \geq 1$, using successively the convexity of the interval $]y, z[$, the independence and the stationarity of the sequence $(X_n)_{n\geq 1}$, we get

$$\begin{aligned} P(y < \overline{S}_{n+m} < z) &\geq P\Big(y < \overline{S}_n < z,\, y < \frac{1}{m}(X_{n+1} + \cdots + X_{n+m}) < z\Big) \\ &= P(y < \overline{S}_n < z)\, P(y < \overline{S}_m < z)\,. \end{aligned}$$

Therefore the sequence $(-\ln P(y < \overline{S}_n < z))_{n\geq 1}$ is subadditive. The following lemma shows that the hypothesis of the subadditive lemma 1.1 is fulfilled.

**Lemma 1.2.** *Let $U$ be an open interval of $\mathbb{R}$. Either $P(\overline{S}_n \in U) = 0$ for all $n \geq 1$ or there exists $N \geq 1$ such that $P(\overline{S}_n \in U) > 0$ for $n \geq N$.*

*Proof.* Suppose that there exists $m \geq 1$ such that $P(\overline{S}_m \in U) > 0$. A probability measure on $\mathbb{R}$ is convex regular, hence there exists a compact convex set $K$ included in $U$ such that $P(\overline{S}_m \in K) > 0$. Let $M > 0$ be such that

$P(|X_1| \leq M) > 0$. Let $n \geq m$ and let $n = mq + r$ be the Euclidean division of $n$ by $m$. If

$$\forall i \in \{0 \cdots q-1\} \qquad \frac{1}{m} \sum_{j=im+1}^{(i+1)m} X_j \in K\,,$$
$$\forall j \in \{mq+1 \cdots n\} \qquad |X_j| \leq M\,,$$

then $\overline{S}_n$ belongs to $K + [-mM/n, mM/n]$. For $n$ large enough so that the latter set is included in $U$, we have

$$P(\overline{S}_n \in U) \geq P(\overline{S}_m \in K)^q \, P(|X_1| \leq M)^r > 0\,. \qquad \square$$

By lemma 1.2, we can apply the subadditive lemma 1.1 to get the existence of the limit defining $I$.

**Theorem 1.3.** *For any $x \in \mathbb{R}$, the following limits exist:*

$$J(x) = \lim_{n\to\infty} -\frac{1}{n} \ln P(\overline{S}_n \geq x)\,, \quad K(x) = \lim_{n\to\infty} -\frac{1}{n} \ln P(\overline{S}_n \leq x)\,.$$

*The maps $J, K : \mathbb{R} \to [0, \infty]$ are convex lower semicontinuous and satisfy*

$$\forall x \in \mathbb{R} \quad J(x)K(x) = 0\,, \quad \lim_{x\to-\infty} J(x) = 0\,, \quad \lim_{x\to+\infty} K(x) = 0\,.$$

*Moreover $J + K = I$. The sequence $(\overline{S}_n)_{n\geq 1}$ satisfies the full large deviation principle with speed $n$ and governed by the convex rate function $I$, i.e., for any Borel subset $A$ of $\mathbb{R}$,*

$$-\inf\{\, I(x) : x \in \overset{o}{A} \,\} \leq \liminf_{n\to\infty} \frac{1}{n} \ln P\big(\overline{S}_n \in A\big)$$
$$\leq \limsup_{n\to\infty} \frac{1}{n} \ln P\big(\overline{S}_n \in A\big) \leq -\inf\{\, I(x) : x \in \overline{A} \,\}\,.$$

*Proof.* We deal only with the map $J$ because the arguments for the map $K$ are the same. Let $x \in \mathbb{R}$. We claim that the sequence $(-\ln P(\overline{S}_n \geq x))_{n\geq 1}$ is subadditive. Indeed, for any $n, m \geq 1$, using successively the convexity of the interval $[x, +\infty[$, the independence and the stationarity of the sequence $(X_n)_{n\geq 1}$, we get

$$P(\overline{S}_{n+m} \geq x) \geq P\Big(\frac{1}{n}(X_1 + \cdots + X_n) \geq x,\ \frac{1}{m}(X_{n+1} + \cdots + X_{n+m}) \geq x\Big)$$
$$\geq P\Big(\frac{1}{n}(X_1 + \cdots + X_n) \geq x\Big) P\Big(\frac{1}{m}(X_{n+1} + \cdots + X_{n+m}) \geq x\Big)$$
$$= P(\overline{S}_n \geq x)\, P(\overline{S}_m \geq x)\,.$$

Therefore

$$\forall m, n \geq 1 \qquad -\ln P(\overline{S}_{n+m} \geq x) \leq -\ln P(\overline{S}_n \geq x) - \ln P(\overline{S}_m \geq x)\,.$$

If $P(X_1 \geq x) = 0$, then $P(\overline{S}_n \geq x) = 0$ for any $n \geq 1$ and $J(x) = +\infty$. If $P(X_1 \geq x) > 0$, then $P(\overline{S}_n \geq x) \geq P(X_1 \geq x)^n$ and $-\ln P(\overline{S}_n \geq x) < +\infty$ for any $n \geq 1$; thus we can apply the subadditive lemma 1.1 to get the existence of the limit defining $J(x)$. Next, for $x \in \mathbb{R}$ and $n \geq 1$,

$$J(x) \leq -\frac{1}{n} \ln P(\overline{S}_n \geq x) \leq -\ln P(X_1 \geq x).$$

Sending $x$ to $-\infty$, we see that $\lim_{x \to -\infty} J(x) = 0$.

We next prove that $J$ is convex. Let $\lambda \in [0,1]$ and let $x, y \in \mathbb{R}$ such that $x < y$. As noted above, we have either $P(\overline{S}_n \geq y) = 0$ for all $n \geq 1$ or $P(\overline{S}_n \geq y) > 0$ for all $n \geq 1$. If the first case occurs, we have $J(y) = +\infty$ and certainly $J\big(\lambda x + (1-\lambda)y\big) \leq \lambda J(x) + (1-\lambda)J(y)$. Suppose that the second case occurs. Let $m \leq n$ be two integers and let $n = mq + r$ be the Euclidean division of $n$ by $m$. We set $s = \lfloor \lambda q \rfloor$, the largest integer less than or equal to $\lambda q$. On the event

$$\bigcap_{1 \leq i \leq s} \Big\{ \sum_{j=(i-1)m+1}^{im} X_j \geq mx \Big\} \cap \bigcap_{s < i \leq q} \Big\{ \sum_{j=(i-1)m+1}^{im} X_j \geq my \Big\} \cap \Big\{ \sum_{j=qm+1}^{n} X_j \geq ry \Big\}$$

the following inequality holds (recall that $x < y$):

$$\sum_{j=1}^{n} X_j \geq sxm + (q-s)ym + ry \geq \big(\lambda x + (1-\lambda)y\big)n.$$

Therefore

$$P\big(\overline{S}_n \geq \lambda x + (1-\lambda)y\big) \geq P(\overline{S}_m \geq x)^s \, P(\overline{S}_m \geq y)^{q-s} \, P(\overline{S}_r \geq y).$$

Notice that $P(\overline{S}_r \geq y) > 0$ for $r \in \{1 \cdots m\}$. Taking logarithms and sending successively $n$ and $m$ to $\infty$, we get

$$-J\big(\lambda x + (1-\lambda)y\big) \geq -\lambda J(x) - (1-\lambda)J(y).$$

We prove now that $J$ is lower semicontinuous. Since it is convex, it is continuous on the interior of its domain, and the only delicate point is the boundary point

$$c = \sup \{ x \in \mathbb{R} : J(x) < +\infty \}.$$

If $c = +\infty$, we have nothing more to prove. Let us suppose that $c < +\infty$. For any $\varepsilon > 0$, we have

$$+\infty = J(c+\varepsilon) \leq -\ln P(X_1 \geq c+\varepsilon),$$

so that $P(X_1 > c) = 0$ and the support of the common law of the sequence $(X_n)_{n \in \mathbb{N}}$ is included in $]-\infty, c]$. Therefore

$$\forall n \geq 1 \qquad P(\overline{S}_n \geq c) = P(X_1 = c)^n$$

and $J(c) = -\ln P(X_1 = c)$. For $\delta > 0$, let

$$N = \big|\{ i : 1 \le i \le n,\, X_i < c - \delta \}\big| .$$

With probability 1, we have

$$n\overline{S}_n \le (n-N)c + N(c-\delta) = nc - N\delta ,$$

thus for $\varepsilon \in ]0, \delta/2[$ sufficiently small,

$$\begin{aligned} P(\overline{S}_n \ge c-\varepsilon) \le P(N \le \varepsilon n/\delta) &\le \sum_{i=1}^{\lfloor \varepsilon n/\delta \rfloor} \binom{n}{i} P(X_1 \ge c-\delta)^{n-i} \\ &\le \Big\lfloor \frac{\varepsilon n}{\delta} \Big\rfloor \binom{n}{\lfloor \varepsilon n/\delta \rfloor} P(X_1 \ge c-\delta)^{n - \lfloor \varepsilon n/\delta \rfloor} . \end{aligned}$$

We need now the following standard expansion of the binomial coefficient.

**Lemma 1.4.** *For any $n \ge 1$, any $k \in \{0 \cdots n\}$, we have*

$$\Big| \frac{1}{n} \ln \frac{n!}{k!(n-k)!} + \frac{k}{n} \ln \frac{k}{n} + \frac{n-k}{n} \ln \frac{n-k}{n} \Big| \le \frac{2}{n} (\ln n + 2) .$$

*Proof.* Setting, for $n \in \mathbb{N}$, $f(n) = \ln n! - n \ln n + n$, we have

$$\begin{aligned} \ln \frac{n!}{k!(n-k)!} &= \ln n! - \ln k! - \ln(n-k)! \\ &= -k \ln \frac{k}{n} - (n-k) \ln \frac{n-k}{n} + f(n) - f(k) - f(n-k) . \end{aligned}$$

Comparing the discrete sum $\ln n! = \sum_{1 \le k \le n} \ln k$ to the integral $\int_1^n \ln x \, dx$, we see that $1 \le f(n) \le \ln n + 2$ for all $n \ge 1$, whence

$$\big| f(n) - f(k) - f(n-k) \big| \le 2(\ln n + 2) ,$$

and we have the desired inequalities. □

Taking logarithms, sending $n$ to $+\infty$, we get with the help of lemma 1.4,

$$J(c-\varepsilon) \ge \frac{\varepsilon}{\delta} \ln \frac{\varepsilon}{\delta} + \Big(1 - \frac{\varepsilon}{\delta}\Big) \ln \Big(1 - \frac{\varepsilon}{\delta}\Big) - \Big(1 - \frac{\varepsilon}{\delta}\Big) \ln P(X_1 \ge c-\delta) .$$

Sending successively $\varepsilon$ and then $\delta$ to 0, we obtain that

$$\lim_{\substack{\varepsilon \to 0 \\ \varepsilon > 0}} J(c-\varepsilon) \ge -\ln P(X_1 = c) = J(c)$$

so that $J$, being non–decreasing, is indeed left continuous at $c$. Similarly, the map $K$ is convex lower semicontinuous non–increasing. We have

$$\forall x \in \mathbb{R} \qquad P(\overline{S}_n \leq x) + P(\overline{S}_n \geq x) \geq 1\,.$$

Applying lemma 6.7, we see that $J(x) = 0$ or $K(x) = 0$ for any $x \in \mathbb{R}$. Let us define

$$a = \inf\{\, x \in \mathbb{R} : K(x) = 0\,\}\,, \quad b = \sup\{\, x \in \mathbb{R} : J(x) = 0\,\}\,.$$

The numbers $a, b$ can be equal to $-\infty$ or $+\infty$. The previous remark implies that $a \leq b$. Certainly $J(x) + K(x) = 0$ for $x \in [a, b]$.

We next prove the large deviation lower bound, i.e., for $x \in \mathbb{R}$,

$$\forall \varepsilon > 0 \qquad \liminf_{n\to\infty} \frac{1}{n} \ln P\big(\overline{S}_n \in ]x-\varepsilon, x+\varepsilon[\big) \geq -J(x) - K(x)\,.$$

Let $x \in \mathbb{R}$. If $J(x) + K(x) = +\infty$, the large deviation lower bound holds trivially. Suppose now that $J(x) + K(x) < \infty$. If $x \geq b$, then $K(x) = 0$ and we write, for $\varepsilon > 0$,

$$P(\overline{S}_n \in ]x-\varepsilon, x+\varepsilon[) \geq P(\overline{S}_n \in [x, x+\varepsilon[) = P(\overline{S}_n \geq x) - P(\overline{S}_n \geq x+\varepsilon)\,.$$

Yet $J$ is convex, it vanishes at the left of $b$ and is positive at $x+\varepsilon$. Therefore it is strictly increasing on $[x, x+\varepsilon]$ and $J(x+\varepsilon) > J(x)$. In case $b = -\infty$, we reach the same conclusion using the fact that $\lim_{-\infty} J = 0$. In particular, $P(\overline{S}_n \geq x+\varepsilon)$ is negligible compared to $P(\overline{S}_n \geq x)$, so that taking logarithms and sending $n$ to $\infty$, we get the desired inequality. The case where $x \leq a$ is similar. Let us finally consider the case where $x \in ]a, b[$. Let $\lambda \in [0, 1]$ be such that $x = \lambda a + (1-\lambda) b$. Let also $m \leq n$ be two integers and let $n = mq + r$ be the Euclidean division of $n$ by $m$. We set $s = \lfloor \lambda q \rfloor$. Let $M > 0$ be large enough so that $P(|X_1| \leq M) > 0$. On the event

$$\bigcap_{1\leq i\leq s} \Big\{ \frac{1}{m} \sum_{j=(i-1)m+1}^{im} X_j \in ]a-\varepsilon, a+\varepsilon[ \Big\} \cap \bigcap_{s<i\leq q} \Big\{ \frac{1}{m} \sum_{j=(i-1)m+1}^{im} X_j \in ]b-\varepsilon, b+\varepsilon[ \Big\}$$
$$\cap \bigcap_{qm<j\leq n} \{\, |X_j| \leq M \,\}$$

the following inequalities hold:

$$msa + m(q-s)b - n\varepsilon - mM \leq \sum_{j=1}^{n} X_j \leq msa + m(q-s)b + n\varepsilon + mM\,.$$

For $n$ sufficiently large, we have $mM + |msa + m(q-s)b - nx| < n\varepsilon$ and

$$P\big(\overline{S}_n \in ]x-2\varepsilon, x+2\varepsilon[\big) \geq$$
$$P\big(\overline{S}_m \in ]a-\varepsilon, a+\varepsilon[\big)^s \, P\big(\overline{S}_m \in ]b-\varepsilon, b+\varepsilon[\big)^{q-s} \, P(|X_1| \leq M)^m\,,$$

whence, taking logarithms and sending $n$ to $\infty$,

$$\liminf_{n\to\infty} \frac{1}{n} \ln P\big(\overline{S}_n \in ]x-2\varepsilon, x+2\varepsilon[\big) \geq$$
$$\frac{\lambda}{m} \ln P\big(\overline{S}_m \in ]a-\varepsilon, a+\varepsilon[\big) + \frac{1-\lambda}{m} \ln P\big(\overline{S}_m \in ]b-\varepsilon, b+\varepsilon[\big)\,.$$

We send next $m$ to $\infty$ and we use the large deviation lower bound for neighborhoods of $a$ and $b$. Taking into account that $K(a) = J(b) = 0$, we get

$$\liminf_{n\to\infty} \frac{1}{n} \ln P\big(\overline{S}_n \in ]x-2\varepsilon, x+2\varepsilon[\big) \geq -\lambda K(a) - (1-\lambda)J(b) = 0\,.$$

We next prove the large deviation upper bound. Let $F$ be a closed subset of $\mathbb{R}$. If $F \cap [a,b] \neq \emptyset$, then $\inf_F J + K = 0$ and the large deviation upper bound holds trivially. Suppose now that $F \cap [a,b] = \emptyset$. Let us define

$$c = \sup F \cap ]-\infty, a]\,, \qquad d = \inf F \cap [b, +\infty[\,,$$

where we use the following convention:
- If $a = -\infty$ then $c = -\infty$ and $K(-\infty) = +\infty$.
- If $b = +\infty$ then $d = +\infty$ and $J(+\infty) = +\infty$.

We have
$$P(\overline{S}_n \in F) \leq P(\overline{S}_n \leq c) + P(\overline{S}_n \geq d)\,.$$

Taking logarithms and sending $n$ to $\infty$, we get with the help of lemma 6.7

$$\limsup_{n\to\infty} \frac{1}{n} \ln P\big(\overline{S}_n \in F\big) \leq -\min(K(c), J(d))$$
$$= -\min\big(K(c)+J(c), K(d)+J(d)\big)\,.$$

The last equality holds because $J(c) = K(d) = 0$. The right–hand quantity is equal to the minimum of $J+K$ over $F$, so that we have the large deviation upper bound. Let us prove that $I = J+K$. First, for $x \in \mathbb{R}$,

$$\forall y < x \qquad I(x) \geq \lim_{n\to\infty} -\frac{1}{n} \ln P(\overline{S}_n \geq y) = J(y)\,.$$

Since $J$ is lower semicontinuous, letting $y$ go to $x$ from below, we obtain that $I(x) \geq J(x)$. We prove similarly that $I(x) \geq K(x)$ for $x \in \mathbb{R}$. It follows that $I \geq \max(J,K) = J+K$. Let now $x > b$. Since $J$ is increasing on $[b, +\infty[$, then for $y, z$ such that $b < y < x < z$, we have $J((x+y)/2) < J(z)$. Moreover

$$P(y < \overline{S}_n < z) \geq P\Big(\overline{S}_n \geq \frac{x+y}{2}\Big) - P(\overline{S}_n \geq z)\,,$$

whence, taking logarithms and sending $n$ to $\infty$, we obtain

$$\lim_{n\to\infty} -\frac{1}{n} \ln P(y < \overline{S}_n < z) \leq J\Big(\frac{x+y}{2}\Big) \leq J(x)\,.$$

Taking the supremum over $y, z$ such that $y < x < z$, we conclude that $I(x) \leq J(x)$. Similarly, we prove that $I(x) \leq K(x)$ for $x < a$. Finally, the large deviation lower bound implies that $I(x) = 0$ for $x \in [a,b]$. Therefore $I = J+K$.
□

A specific feature of this large deviation principle is that the rate function is given by an operational formula. We first define the Log–Laplace of the law of $X_1$ as the function $\Lambda : \mathbb{R} \to \mathbb{R} \cup \{+\infty\}$ given by

$$\forall \lambda \in \mathbb{R} \qquad \Lambda(\lambda) = E\big(\exp(\lambda X_1)\big) = \ln \int \exp(\lambda X_1)\, dP$$

and second its Fenchel–Legendre transform $\Lambda^* : \mathbb{R} \to [0, +\infty]$ by

$$\forall x \in \mathbb{R} \qquad \Lambda^*(x) = \sup_{\lambda \in \mathbb{R}} \big(\lambda x - \Lambda(\lambda)\big)\,.$$

Both functions $\Lambda$ and $\Lambda^*$ are convex lower semicontinuous. If $\Lambda$ is finite everywhere, then $\Lambda^*$ vanishes only at $E(X_1)$.

**Theorem 1.5.** *The rate function $I$ governing the large deviations of $(\overline{S}_n)_{n\geq 1}$ is equal to $\Lambda^*$.*

*Proof.* The proof in the general case requires a delicate approximation argument [40]. Therefore we will only consider the case where the support of $\mu$ is bounded. In this case $I$ has compact level sets and a direct application of Varadhan's lemma (see theorem 6.9) yields

$$\forall \lambda \in \mathbb{R} \qquad \lim_{n \to \infty} \frac{1}{n} \ln E\big(\exp\big(n\lambda \overline{S}_n\big)\big) = \sup_{\alpha \in \mathbb{R}} \big(\lambda\alpha - I(\alpha)\big)\,.$$

The variables $X_n$, $n \geq 1$, being independent, the quantity appearing inside the limit is equal to $\Lambda(\lambda)$. Since $I$ is convex lower semicontinuous, it is equal to the Fenchel–Legendre transform of its Fenchel–Legendre transform $\Lambda$. □

# Part II

# Presentation of the models

# 2

# Ising model

Let us try to set up a simple model of our experiment with water and oil (see section 1.1). A convenient choice is a lattice model: each site of the lattice is occupied either by a water particle or by an oil particle, which we indicate respectively by $+$ or $-$. The interaction between different particles is repulsive and occurs when the substances are in immediate contact. Hence a repulsive nearest neighbour interaction is a sensible choice. Since we focus only on the repulsive interaction between different molecules, we can assume that the two substances are symmetric and that their self–interactions are of equal magnitude, or equivalently, equal to zero. We do not assume that the self–interactions between two particles of the same type are negligible compared to the repulsive effect; rather, we say that because of the symmetry, the global effect of the self–interactions cancels out. Thus the total energy of a configuration should be simply the sum of all nearest neighbour pairs with different signs. We end up exactly with the Hamiltonian of the famous Ising model (to be defined precisely in the next section). In our experiment the density of oil is fixed, therefore we have a constraint on the possible configurations: the proportion of pluses and minuses has to be fixed. This situation amounts to consider the Ising model with plus boundary conditions (guaranteeing the water dominance) conditioned on the event that the average magnetization is equal to a fixed value smaller than the spontaneous magnetization at the given temperature.

## 2.1 Construction of the model

For reasons of technical simplicity, it is easier to build our model on a lattice. We will work with the lattice $\mathbb{Z}^d$; each site of the lattice is occupied by one of the two types of particles, which we denote by $-$ and $+$. Let $\Lambda \subset \mathbb{Z}^d$ be a cubic box. A configuration in $\Lambda$ is a map $\sigma : \Lambda \to \{-,+\}$ and for $x \in \Lambda$, we denote by $\sigma(x)$ the type of the particle present at $x$. The energy or Hamiltonian $H_\Lambda(\sigma)$ of the configuration $\sigma$ in $\Lambda$ is, up to a constant, twice the number of interfaces between the minuses and the pluses, that is,

$$H_\Lambda(\sigma) = -\frac{1}{2} \sum_{x,y\in\Lambda,|x-y|=1} \sigma(x)\sigma(y) = - \sum_{\{x,y\}\in\Lambda^2,|x-y|=1} \sigma(x)\sigma(y)\,.$$

We use the usual rules to multiply signs: $++ = -- = +$, $-+ = +- = -$. The first sum is above ordered pairs (whence the factor 1/2) while the second is above unordered pairs. We need also a mechanism to ensure the dominance of one type of particles. This is achieved through boundary conditions. We consider only two types of boundary conditions, by putting either a layer of pluses or of minuses around the box $\Lambda$. The energy or Hamiltonian $H^*_\Lambda(\sigma)$ with boundary conditions $*$ (where $*$ stands for $-$ or $+$) is defined as above for the configurations $\sigma$ such that $\sigma(x) = *$ for all $x \in \partial^{in}\Lambda$ (where $\partial^{in}\Lambda$ denotes the sites in $\Lambda$ which are at a distance less than or equal to 1 from the complement of $\Lambda$) and $H^*_\Lambda(\sigma) = +\infty$ otherwise. Next we add some randomness in the model. Let $T > 0$ be the temperature. We build a probability law on the space $\{-,+\}^\Lambda$ of the configurations. This space is huge but finite, hence to define the law we need to specify the individual probability of each possible configuration. The natural way to do this is to use the Boltzmann factor. So, the Gibbs measure $\mu^*_{\Lambda,T}$ in $\Lambda$ at temperature $T$ with boundary conditions $*$ is given by

$$\forall \sigma \in \{-,+\}^\Lambda \qquad \mu^*_{\Lambda,T}(\sigma) = \frac{1}{Z^*_{\Lambda,T}} \exp -\frac{H^*_\Lambda(\sigma)}{T}$$

where the normalizing factor $Z^*_{\Lambda,T}$, called the partition function, is equal to

$$Z^*_{\Lambda,T} = \sum_{\sigma\in\{-,+\}^\Lambda} \exp -\frac{H^*_\Lambda(\sigma)}{T}\,.$$

Whenever the superscript $*$ is absent, the boundary conditions are not specified and we have the Ising Gibbs measure $\mu_{\Lambda,T}$ with free boundary conditions, associated to the Hamiltonian $H_\Lambda$. Let us take a closer look at this formula. The elements $\Lambda$, $T$, $* \in \{-,+\}$ being fixed, the most likely configurations are those having a small energy, i.e., those for which the contacts between the minuses and the pluses are reduced. Thus we have built a complex probability law with strong spatial correlations. We shall next play a bit with the elements controlling the Gibbs measures $\mu^\pm_{\Lambda,T}$ in order to get some feeling for their influence.

## 2.2 First asymptotics

Imagine that we fix the box $\Lambda$ and that we set the boundary conditions to $+$. If we send $T$ to 0, then the measure $\mu^+_{\Lambda,T}$ concentrates on the configuration which realizes the global minimum of the Hamiltonian $H^+_\Lambda$, in this case the configuration where all the sites are pluses. On the contrary, if we send $T$ to $\infty$, the value of the Hamiltonian becomes irrelevant and $\mu^+_{\Lambda,T}$ converges towards the Bernoulli product law where all the sites are independent. The case of $\mu^-_{\Lambda,T}$ being symmetric, we see that

$$\begin{array}{ccccc} \text{dirac mass at "all pluses"} & \longleftarrow & \mu^+_{\Lambda,T} & \longrightarrow & \text{i.i.d. Bernoulli} \\ & T\downarrow 0 & & T\uparrow\infty & \\ \text{dirac mass at "all minuses"} & \longleftarrow & \mu^-_{\Lambda,T} & \longrightarrow & \text{i.i.d. Bernoulli} \end{array}$$

Something remarkable has already happened: as $T\uparrow\infty$, the boundary conditions are forgotten, while as $T\downarrow 0$, they completely determine the limit. However, we wish to work at a fixed positive temperature $T$. In order to observe a sharp mathematical phenomenon, we consider another kind of limit, namely the thermodynamic limit where the number of particles goes to infinity. This is achieved by letting the box $\Lambda$ grow and invade the whole lattice $\mathbb{Z}^d$. As $\Lambda$ increases to $\mathbb{Z}^d$, the expectation $\mu^+_{\Lambda,T}(\sigma(0))$ decreases and converges towards a limiting quantity $m^*(T)$:

$$\lim_{\Lambda\uparrow\mathbb{Z}^d}\mu^+_{\Lambda,T}(\sigma(0)) \;=\; m^*(T) \;=\; -\lim_{\Lambda\uparrow\mathbb{Z}^d}\mu^-_{\Lambda,T}(\sigma(0))\,.$$

Here is a heuristic explanation for this monotone convergence. Let us consider a huge box $\Lambda$ and the site at the center of the box $\Lambda$. With free boundary conditions, the law of $\sigma(0)$ under $\mu_{\Lambda,T}$ is symmetric, hence it is the one of a fair coin, i.e.,

$$\mu_{\Lambda,T}(\sigma(0)=+) \;=\; 1/2 \;=\; \mu_{\Lambda,T}(\sigma(0)=-)\,.$$

If we put $+$ boundary conditions, these boundary conditions start to influence positively the sites at distance 1 from the boundary of the box, which themselves influence the sites at distance 2 from the boundary. This effect propagates and reaches the origin, so that the law of $\sigma(0)$ under $\mu^+_{\Lambda,T}$ is slightly biased towards $+$:

$$\mu^+_{\Lambda,T}(\sigma(0)=+) \;>\; 1/2 \;>\; \mu^+_{\Lambda,T}(\sigma(0)=-)\,.$$

The larger the box $\Lambda$ is, the smaller is the resulting effect at the origin, hence the influence of the boundary conditions decreases as the box increases and the following monotone limit exists:

$$m^*(T) \;=\; \lim_{\Lambda\uparrow\mathbb{Z}^d}\mu^+_{\Lambda,T}(\sigma(0))\,.$$

The fundamental and basic question is whether something of the influence of the boundary conditions still remains after we have sent them to infinity. Equivalently, is $m^*(T)$ equal to 0?

The quantity $m^*(T)$ is called the spontaneous magnetization at temperature $T$. This terminology stems from the fact that the Ising model was originally introduced as a model of ferromagnetism: under some adequate conditions, a magnet submitted to the influence of a magnetic field will remember the sign of the field even after it has disappeared (see [66] and the references therein for a serious physical introduction to the Ising model).

## 2.3 Phase transition

We say that there is a phase transition at temperature $T$ if $m^*(T) > 0$. The first fundamental result concerning the phase transition in the Ising model is the following.

**Theorem 2.1.** *In any dimension $d \geq 2$, there exists a positive and finite critical temperature $T_c(d)$ such that the Ising model exhibits a phase transition for $T < T_c(d)$ and it does not for $T > T_c(d)$.*

The proof of this theorem will be completed in section 4.3. It is also possible to take the thermodynamic limit of the finite volume Gibbs measure $\mu^+_{\Lambda,T}$, and not only of the expected value $\mu^+_{\Lambda,T}(\sigma(0))$. As $\Lambda$ increases to $\mathbb{Z}^d$, the measure $\mu^+_{\Lambda,T}$ decreases stochastically and converges weakly towards the infinite volume Gibbs measure $\mu^+_T$, which is a probability measure on the space of infinite volume configurations $\{-,+\}^{\mathbb{Z}^d}$. Similarly, $\mu^-_{\Lambda,T}$ increases weakly towards a measure $\mu^-_T$:

$$\lim_{\Lambda\uparrow\mathbb{Z}^d} \mu^-_{\Lambda,T} = \mu^-_T\,, \qquad \lim_{\Lambda\uparrow\mathbb{Z}^d} \mu^+_{\Lambda,T} = \mu^+_T\,.$$

The spontaneous magnetization $m^*(T)$ is equal to the expected value of $\sigma(0)$ under $\mu^+_T$ and there is a phase transition at temperature $T$ if and only if $\mu^-_T$ and $\mu^+_T$ are distinct. In other words, we have

$$\forall T < T_c(d) \qquad m^*(T) > 0\,, \quad \mu^-_T \neq \mu^+_T$$

whereas the infinite volume Gibbs measure is unique for $T > T_c(d)$ and

$$\forall T > T_c(d) \qquad m^*(T) = 0\,, \quad \mu^-_T = \mu^+_T\,.$$

## 2.4 Proofs of the heuristics

Let us try to prove rigorously the heuristics of section 2.2. We need two basic tools: the Markov property and the FKG inequality.

### 2.4.1 General boundary conditions.

Let $A$ be a subset of $\mathbb{Z}^d$. Its inner vertex boundary $\partial^{in} A$ consists of the points of $A$ having a neighbour outside $A$:

$$\partial^{in} A = \big\{ x \in A : \exists y \in \mathbb{Z}^d \setminus A \quad |x-y| = 1 \big\} .$$

Let $\eta$ be a spin configuration in $A$. The Hamiltonian in $A$ with boundary conditions $\eta$ is given by

$$\forall \sigma \in \{-,+\}^A \qquad H^\eta_A(\sigma) = \begin{cases} H_A(\sigma) & \text{if } \sigma(x) = \eta(x) \text{ for } x \in \partial^{in} A \\ +\infty & \text{otherwise} \end{cases}$$

The Ising Gibbs measure in $A$ at temperature $T$ with boundary conditions $\eta$ is then

$$\forall \sigma \in \{-,+\}^A \qquad \mu^\eta_{A,T}(\sigma) = \frac{1}{Z^\eta_{A,T}} \exp -\frac{H^\eta_A(\sigma)}{T} ,$$

where the normalizing factor $Z^\eta_{A,T}$, called the partition function, is equal to

$$Z^\eta_{A,T} = \sum_{\sigma \in \{-,+\}^A} \exp -\frac{H^\eta_A(\sigma)}{T} .$$

The range of the interactions being equal to 1, the Ising Gibbs measure enjoys the following spatial Markov property.

### 2.4.2 Markov property.

For any finite subsets $A, B$ of $\mathbb{Z}^d$ such that $A \subset B$, any configurations $\eta, \xi$ in $B$, any function $f$ depending only on the sites in $A$,

$$\mu^\xi_{B,T}\Big( f(\sigma) \mid \forall x \in (B \setminus A) \cup \partial^{in} A \quad \sigma(x) = \eta(x) \Big) =$$
$$\mu^\xi_{B,T}\Big( f(\sigma) \mid \forall x \in \partial^{in} A \quad \sigma(x) = \eta(x) \Big) = \mu^\eta_{A,T}\big(f(\sigma)\big) .$$

*Proof.* Let $\sigma$ be a configuration in $A$ such that $\sigma(x) = \eta(x)$ for $x \in \partial^{in} A$ and let $\varrho$ be a configuration in $B \setminus A$. We denote by $\sigma\varrho$ the configuration in $B$ which is equal to $\sigma$ on $A$ and to $\varrho$ on $B \setminus A$. We decompose $H^\xi_B(\sigma\varrho)$ as

$$H^\xi_B(\sigma\varrho) = H^\eta_A(\sigma) - \frac{1}{2} \sum_{\substack{x,y \in B\setminus A \\ |x-y|=1}} \varrho(x)\varrho(y) - \sum_{\substack{x \in A, y \in B\setminus A \\ |x-y|=1}} \eta(x)\varrho(y) .$$

To compute the first conditional expectation, we take $\varrho = \eta$ in the above formula and we sum over $\sigma$ the factor $f(\sigma)\exp(-H^\xi_B(\sigma\eta)/T)$. To compute the second conditional expectation, we sum over $\sigma$ and $\varrho$ the factor $f(\sigma)\exp(-H^\xi_B(\sigma\varrho)/T)$. Thanks to the above additive decomposition of

$H^{\xi}_{B}(\sigma\varrho)$, we can factorize the sums over $\sigma$ and over $\varrho$. In both cases, we obtain a result of the form

$$\text{constant} \times \sum_{\sigma\in\{-,+\}^{\Lambda}} f(\sigma)\exp -\frac{H^{\eta}_{\Lambda}(\sigma)}{T},$$

where the constant does not depend on $f$. Taking $f=1$ in this formula, we see that this constant has to be equal to $1/Z^{\eta}_{\Lambda,T}$, so that the desired conclusion follows. □

### 2.4.3 FKG inequality.

Let $\Lambda$ be a box included in $\mathbb{Z}^d$. There is a natural partial order on the configurations in $\Lambda$. For any $\sigma,\eta$ in $\{-,+\}^{\Lambda}$, we say that $\sigma$ is less than or equal to $\eta$ if and only if each site which is set to $+$ in $\sigma$ is also set to $+$ in $\eta$. Let $f,g:\{-,+\}^{\Lambda}\to\mathbb{R}$ be two non–decreasing functions. For any temperature $T>0$ and any boundary conditions $\xi$,

$$\mu^{\xi}_{\Lambda,T}(fg)\,\geq\,\mu^{\xi}_{\Lambda,T}(f)\,\mu^{\xi}_{\Lambda,T}(g)\,.$$

*Remark 2.2.* The FKG inequality holds in any finite subset of $\mathbb{Z}^d$. We present here a proof in the case of a box, but the argument can be easily adapted to the case of an arbitrary set.

*Proof.* We fix the temperature $T$ throughout the proof and we remove it from the notation. Suppose first that $f,g$ depend on a single site $x$. The monotonicity of $f$ and $g$ implies that for any configurations $\sigma,\eta$ in $\Lambda$,

$$\big(f(\sigma)-f(\eta)\big)\big(g(\sigma)-g(\eta)\big)\,\geq\,0\,.$$

By developing this identity and averaging with respect to the product measure $\mu^{\xi}_{\Lambda,T}\otimes\mu^{\xi}_{\Lambda,T}$, we obtain the required inequality. Let $c$ be the center of the box $\Lambda$ and let $(x_n)_{n\in\mathbb{N}}$ be an ordering of the points of $\mathbb{Z}^d$ such that

$$\forall n\in\mathbb{N}\qquad |x_n-c|_\infty\,\leq\,|x_{n+1}-c|_\infty$$

where $|\cdot|_\infty$ is the usual supremum norm. We set

$$\forall n\in\mathbb{N}\qquad A_n\,=\,\{\,x_0,\dots,x_n\,\}\,.$$

The structure of the sequence $(x_n)_{n\in\mathbb{N}}$ implies that $\Lambda=A_n$ for some $n\in\mathbb{N}$ and moreover for any $n\in\mathbb{N}$, we have

- either $\partial^{in}A_{n+1}=\partial^{in}A_n\cup\{x_{n+1}\}$
- or $\exists\, y_{n+1}\in\partial^{in}A_n\qquad \partial^{in}A_{n+1}=\partial^{in}A_n\cup\{x_{n+1}\}\setminus\{y_{n+1}\}$.

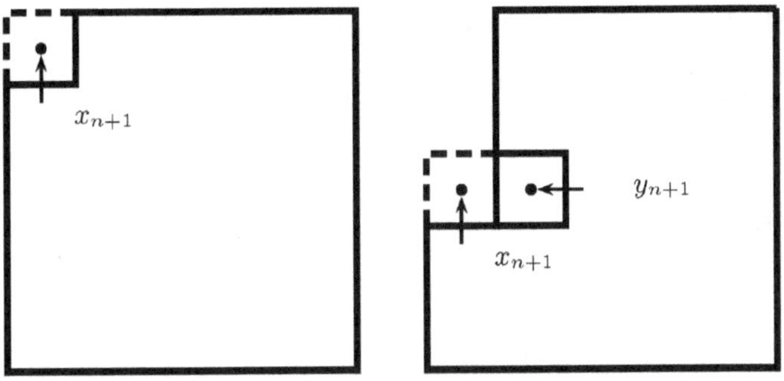


We proceed by induction on $n$ to show that the FKG inequality holds in $A_n$: for any non–decreasing functions $f, g : \{-,+\}^{A_n} \to \mathbb{R}$,

$$\forall \xi \in \{-,+\}^{A_n} \qquad \mu^{\xi}_{A_n}(fg) \geq \mu^{\xi}_{A_n}(f)\,\mu^{\xi}_{A_n}(g)\,.$$

The case $n = 0$ has already been examined. Suppose that the result is true until some rank $n \geq 0$ and let $f, g$ be two non–decreasing functions depending only on the sites in $A_{n+1}$. Let also $\xi$ be a configuration in $A_{n+1}$. Suppose first that $\partial^{in} A_{n+1} = \partial^{in} A_n \cup \{x_{n+1}\}$. Let $\widehat{f}, \widehat{g}$ be the functions on $A_n$ defined for $\sigma \in \{-,+\}^{A_n}$ by

$$\begin{aligned}\widehat{f}\big(\sigma(x_0), \dots, \sigma(x_n)\big) &= f\big(\sigma(x_0), \dots, \sigma(x_n), \xi(x_{n+1})\big)\,,\\ \widehat{g}\big(\sigma(x_0), \dots, \sigma(x_n)\big) &= g\big(\sigma(x_0), \dots, \sigma(x_n), \xi(x_{n+1})\big)\,.\end{aligned}$$

Using the FKG inequality in $A_n$, we have

$$\begin{aligned}\mu^{\xi}_{A_{n+1}}(fg) \;=\; \mu^{\xi}_{A_{n+1}}(\widehat{f}\widehat{g}) \;&=\; \mu^{\xi}_{A_n}(\widehat{f}\widehat{g})\\ &\geq\; \mu^{\xi}_{A_n}(\widehat{f})\,\mu^{\xi}_{A_n}(\widehat{g}) \;=\; \mu^{\xi}_{A_{n+1}}(f)\,\mu^{\xi}_{A_{n+1}}(g)\,.\end{aligned}$$

Suppose next that: $\exists\, y_{n+1} \in \partial^{in} A_n \qquad \partial^{in} A_{n+1} = \partial^{in} A_n \cup \{x_{n+1}\} \setminus \{y_{n+1}\}$. Let $\varepsilon \in \{-,+\}$. If $\sigma$ is a configuration in $A_n \setminus \{y_{n+1}\}$, we denote by $\sigma^{\varepsilon}$ the configuration in $A_{n+1}$ such that

$$\forall x \in A_{n+1} \qquad \sigma^{\varepsilon}(x) = \begin{cases} \sigma(x) & \text{if } x \in A_n \setminus \{y_{n+1}\} \\ \varepsilon & \text{if } x = y_{n+1} \\ \xi(x) & \text{if } x = x_{n+1} \end{cases}$$

Let $\widehat{f}, \widehat{g}$ be the functions on $A_n \setminus \{y_{n+1}\}$ defined by

$$\forall \sigma \in \{-,+\}^{A_n \setminus \{y_{n+1}\}} \qquad \widehat{f}(\sigma) = f(\sigma^{\varepsilon})\,, \quad \widehat{g}(\sigma) = g(\sigma^{\varepsilon})\,.$$

Then

$$\mu^{\xi}_{A_{n+1}}(fg \,|\, \sigma(y_{n+1}) = \varepsilon) \;=\; \mu^{\xi}_{A_{n+1}}(\widehat{f}\widehat{g} \,|\, \sigma(y_{n+1}) = \varepsilon)\,.$$

Yet the functions $\widehat{f}, \widehat{g}$ depend only on $A_n$, and $\partial^{in} A_n \subset \{y_{n+1}\} \cup \partial^{in} A_{n+1}$.

By the Markov property and the FKG inequality in $A_n$, we have

$$\begin{aligned}\mu^{\xi}_{A_{n+1}}(\widehat{f}\widehat{g}\,|\,\sigma(y_{n+1})=\varepsilon) \;=\; \mu^{\xi^{\varepsilon}}_{A_n}(\widehat{f}\widehat{g}) \geq \mu^{\xi^{\varepsilon}}_{A_n}(\widehat{f})\,\mu^{\xi^{\varepsilon}}_{A_n}(\widehat{g}) \\ = \mu^{\xi}_{A_{n+1}}(\widehat{f}\,|\,\sigma(y_{n+1})=\varepsilon)\,\mu^{\xi}_{A_{n+1}}(\widehat{g}\,|\,\sigma(y_{n+1})=\varepsilon)\,.\end{aligned}$$

The functions

$$\sigma \mapsto \mu^{\xi}_{A_{n+1}}(\widehat{f}\,|\,\sigma(y_{n+1}))\,, \qquad \sigma \mapsto \mu^{\xi}_{A_{n+1}}(\widehat{g}\,|\,\sigma(y_{n+1}))$$

depend only on $\sigma(y_{n+1})$ and they are non–decreasing (they are the conditional expectations of non–decreasing functions). We take the expectation in the above inequality with respect to $\mu^{\xi}_{A_{n+1}}$ and we apply the FKG inequality for functions depending only on one site. We get

$$\begin{aligned}\mu^{\xi}_{A_{n+1}}(fg) \;&=\; \mu^{\xi}_{A_{n+1}}\Big(\mu^{\xi}_{A_{n+1}}(fg\,|\,\sigma(y_{n+1}))\Big) \\ &\geq \mu^{\xi}_{A_{n+1}}\Big(\mu^{\xi}_{A_{n+1}}(f\,|\,\sigma(y_{n+1}))\,\mu^{\xi}_{A_{n+1}}(g\,|\,\sigma(y_{n+1}))\Big) \\ &\geq \mu^{\xi}_{A_{n+1}}\Big(\mu^{\xi}_{A_{n+1}}(f\,|\,\sigma(y_{n+1}))\Big)\,\mu^{\xi}_{A_{n+1}}\Big(\mu^{\xi}_{A_{n+1}}(g\,|\,\sigma(y_{n+1}))\Big) \\ &= \mu^{\xi}_{A_{n+1}}(f)\,\mu^{\xi}_{A_{n+1}}(g)\,,\end{aligned}$$

thereby proving the FKG inequality in $A_{n+1}$. $\square$

We can now prove rigorously the heuristics of section 2.2. Let $\Lambda, \Gamma$ be two boxes containing 0 and such that $\Lambda \subset \Gamma$. By the FKG inequality and the Markov property, we have

$$\mu^{+}_{\Gamma,T}(\sigma(0)) \;\leq\; \mu^{+}_{\Gamma,T}(\sigma(0)\,|\,\sigma(x)=+ \text{ for } x\in(\Gamma\setminus\Lambda)\cup\partial^{in}\Lambda) \;=\; \mu^{+}_{\Lambda,T}(\sigma(0))\,.$$

Therefore the expectation $\mu^{+}_{\Lambda,T}(\sigma(0))$ decreases and converges towards a limiting quantity $m^*(T)$ as $\Lambda$ increases to $\mathbb{Z}^d$.

subcritical Ising model: $T = 2.2$, 3 hours

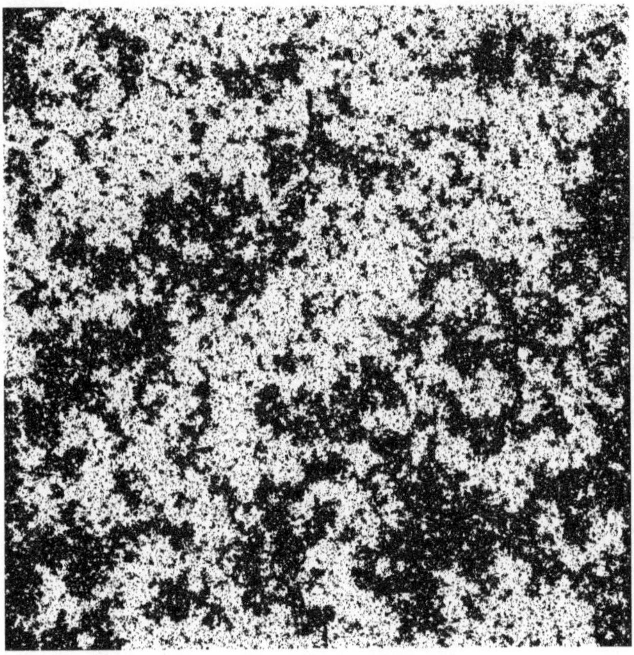

supercritical Ising model: $T = 2.3$, 5 days

Simulations on a $1024 \times 1024$ grid with a 1 Ghz PC

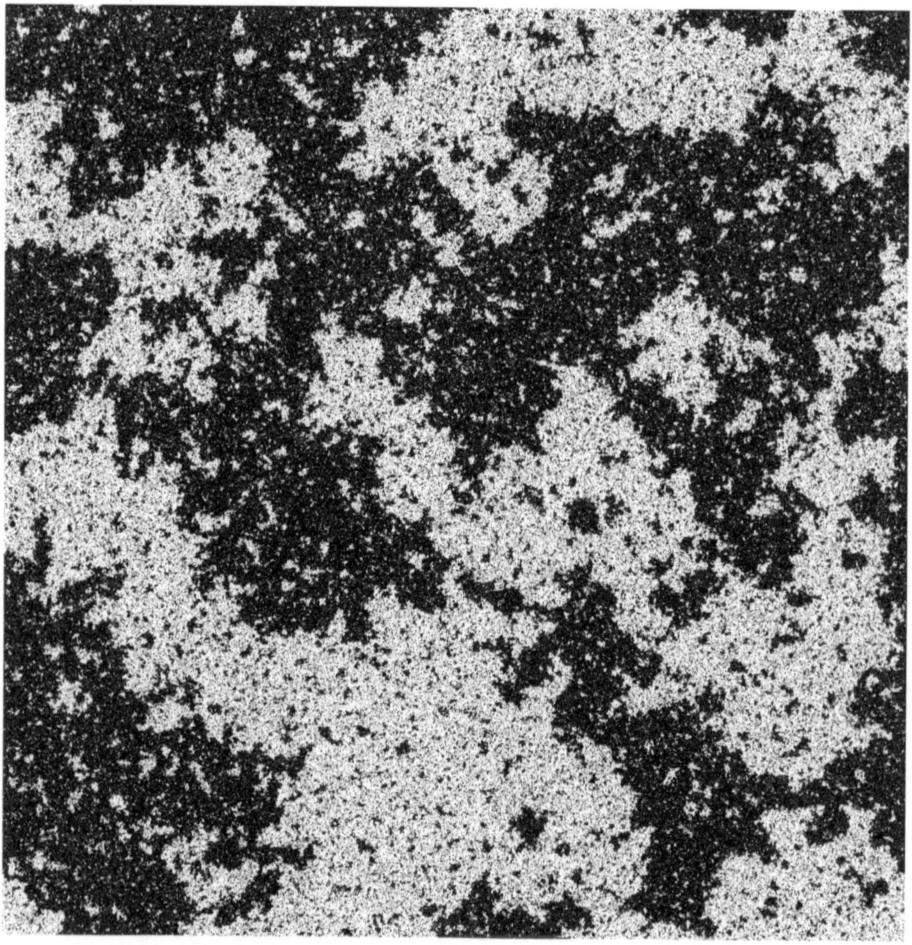

critical Ising model: $T = 2.269$

32 days of simulation on a $4096 \times 4096$ grid with a 1 Ghz PC

# 3

# Bernoulli percolation

We present in this chapter another beautiful and fundamental model, namely the percolation model. This model was invented by Broadbent and Hammersley [32] to model the flow of a liquid in a random medium. The standard mathematical reference for percolation is Grimmett's book [78]. It turns out that the questions of phase coexistence which are the theme of these lecture notes were first solved in the percolation case (at least in dimension 3).

We turn $\mathbb{Z}^d$ into a graph with vertex set $\mathbb{Z}^d$ and edge set

$$\mathbb{E}^d = \{\, \{x,y\} : x \in \mathbb{Z}^d,\, y \in \mathbb{Z}^d,\, |x-y| = 1 \,\}$$

where $|\cdot|$ is the usual Euclidean norm. This graph is called the $d$–dimensional cubic lattice. A path $\gamma$ in $(\mathbb{Z}^d, \mathbb{E}^d)$ is an alternating sequence

$$x_0, e_0, x_1, e_1, \cdots, e_{n-1}, x_n, \cdots$$

of distinct vertices $x_i$ and edges $e_i$, where $e_i$ is the edge between $x_i$ and $x_{i+1}$. The path is said to connect every pair of its vertices. If the path terminates at some vertex $x_n$ it is said to have length $n$, otherwise it is infinite.

## 3.1 The probability space

The nearest neighbour Bernoulli bond percolation model on the cubic lattice at density $p$ is defined by independently choosing each edge of $\mathbb{E}^d$ to be open with probability $p$ or closed with probability $1-p$. We denote by $P$ the corresponding product probability measure on the configuration space $\Omega$ consisting of all the functions from $\mathbb{E}^d$ to $\{0,1\}$ (1 stands for open and 0 for closed). We equip $\Omega$ with the product $\sigma$–field $\mathcal{P}(\{0,1\})^{\otimes \mathbb{E}^d}$ (where $\mathcal{P}(\{0,1\})$ is the collection of the subsets of $\{0,1\}$). Let $\omega$ be a configuration. We consider the random graph having as vertices the points of $\mathbb{Z}^d$ and edge set the open edges of $\omega$ only; equivalently, we remove the closed edges. The open clusters in $\omega$ are the connected components of this random graph. We denote by $C(0)$

the open cluster containing the origin. Two subsets $A, B$ of $\mathbb{Z}^d$ are connected in the configuration $\omega$ if there is a path of open edges in $\omega$ connecting a site of $A$ to a site of $B$; we denote this event by $\{ A \longleftrightarrow B \}$.

## 3.2 Order on $\Omega$

There is a natural order on $\Omega$ defined by the relation: $\omega_1 \leq \omega_2$ if and only if all the edges open in $\omega_1$ are open in $\omega_2$. An event is said to be increasing (respectively decreasing) if its characteristic function is non–decreasing (respectively non–increasing) with respect to this partial order. Suppose the events $A, B$ are both increasing or both decreasing. The Harris–FKG inequality [78] says that

$$P(A \cap B) \geq P(A)P(B) .$$

Bernoulli percolation can be easily compared with more complicated models with the help of stochastic domination. Let $\mu, \nu$ be two probability measures on $\Omega$. We say that $\mu$ stochastically dominates $\nu$ if $\mu(f) \geq \nu(f)$ for any bounded measurable non–decreasing function $f : \Omega \to \mathbb{R}$. Let $\delta > 0$ and let $(X(x), x \in \mathbb{Z}^d)$ be a process such that

$$\forall x \in \mathbb{Z}^d \qquad P\big(X(x) = 1 \,\big|\, X(y),\ y \neq x\big) \geq \delta .$$

Then the law of $(X(x), x \in \mathbb{Z}^d)$ stochastically dominates the Bernoulli product measure with parameter $\delta$ (see [78], p. 179).

## 3.3 Phase transition

The box of diameter $n \in \mathbb{N}$ centered at the origin is the set

$$\Lambda(n) = \big\{ (x_1, \cdots, x_d) \in \mathbb{R}^d : \forall i \in \{1 \cdots d\} \ -n/2 < x_i \leq n/2 \big\} .$$

For $\Lambda$ a box, we denote by $\partial^{in} \Lambda$ the sites of $\Lambda$ which have a nearest neighbour in $\mathbb{Z}^d \setminus \Lambda$. Since the events $\{ 0 \longleftrightarrow \partial^{in} \Lambda(n) \}$, $n \geq 1$, are decreasing with $n$, then

$$P\Big( \bigcap_{n \geq 1} \{ 0 \longleftrightarrow \partial^{in} \Lambda(n) \} \Big) = \lim_{n \to \infty} P\big(0 \longleftrightarrow \partial^{in} \Lambda(n)\big) .$$

The limiting event $\bigcap_{n \geq 1} \{ 0 \longleftrightarrow \partial^{in} \Lambda(n) \}$ is the event that there exists an infinite open path starting at 0. We denote it by $\{ 0 \longleftrightarrow \infty \}$ and we set

$$\forall p \in [0,1] \qquad \theta(p) = P(0 \longleftrightarrow \infty) .$$

The map $\theta : [0,1] \to [0,1]$ is non–decreasing and $\theta(0) = 0$, $\theta(1) = 1$. It is known that the Bernoulli bond percolation model exhibits a phase transition in dimensions $d \geq 2$.

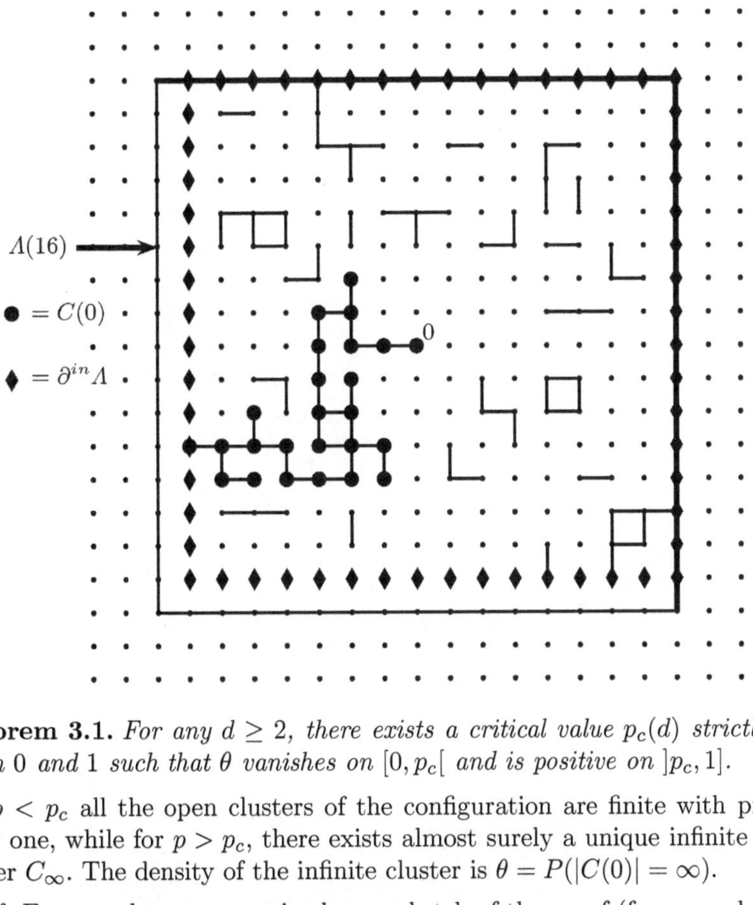


**Theorem 3.1.** *For any $d \geq 2$, there exists a critical value $p_c(d)$ strictly between* 0 *and* 1 *such that $\theta$ vanishes on $[0, p_c[$ and is positive on $]p_c, 1]$.*

For $p < p_c$ all the open clusters of the configuration are finite with probability one, while for $p > p_c$, there exists almost surely a unique infinite open cluster $C_\infty$. The density of the infinite cluster is $\theta = P(|C(0)| = \infty)$.

*Proof.* For completeness, we give here a sketch of the proof (for more details, see [78]). For any $n \geq 1$,

$$\theta(p) \leq P\big(\text{there exists an open path of length } n \text{ starting at } 0\big)\,.$$

The number of such paths is less than $2d(2d-1)^{n-1}$ and the probability for a fixed path to be open is $p^n$, thus $\theta(p) \leq 2d(2d-1)^{n-1}p^n$. Sending $n$ to $\infty$, we see that $\theta(p) = 0$ for $p < 1/(2d-1)$. Next, if $\{\, 0 \longleftrightarrow \infty \,\}$ does not occur, then there exists a set of closed edges surrounding 0 which separates 0 from $\infty$. This set can be chosen to be connected for the following relation: two edges are in relation if the unit squares orthogonal to them centered in the middle of their segments share a common boundary. This is a geometrical intuitive fact (but not that simple to prove!). There exists a constant $c$ depending on the dimension only such that the number of connected sets of edges of cardinality $n$ surrounding 0 is less than $nc^n$. Therefore $1 - \theta(p) \leq \sum_{n \geq 1} nc^n(1-p)^n$. It follows that $\lim_{p \to 1} \theta(p) = 1$, therefore $\theta$ is positive for $p$ close to 1. □

In the following pictures of bond percolation configurations, each gray level corresponds to the vertices of one cluster.

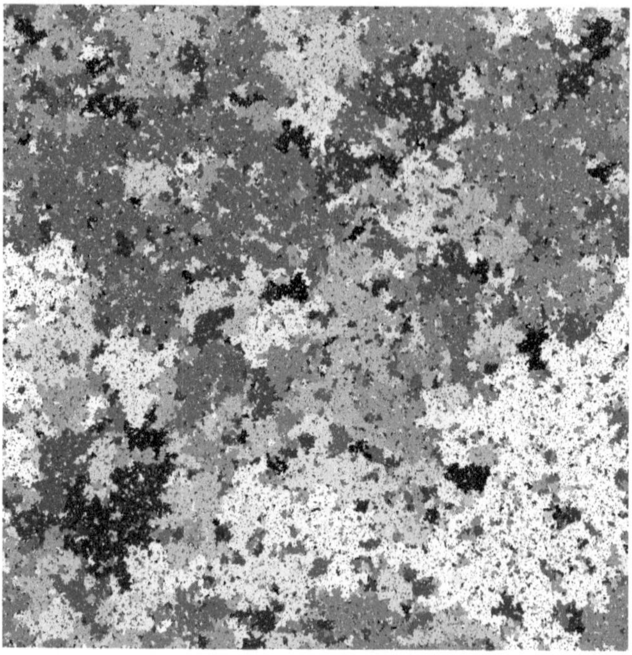

subcritical bond percolation: $p = 0.49$, $1024 \times 1024$ grid

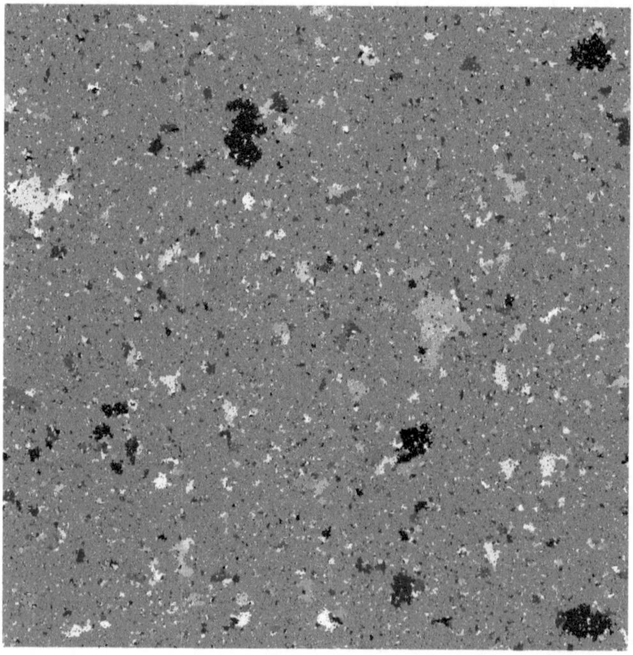

supercritical bond percolation: $p = 0.51$, $1024 \times 1024$ grid

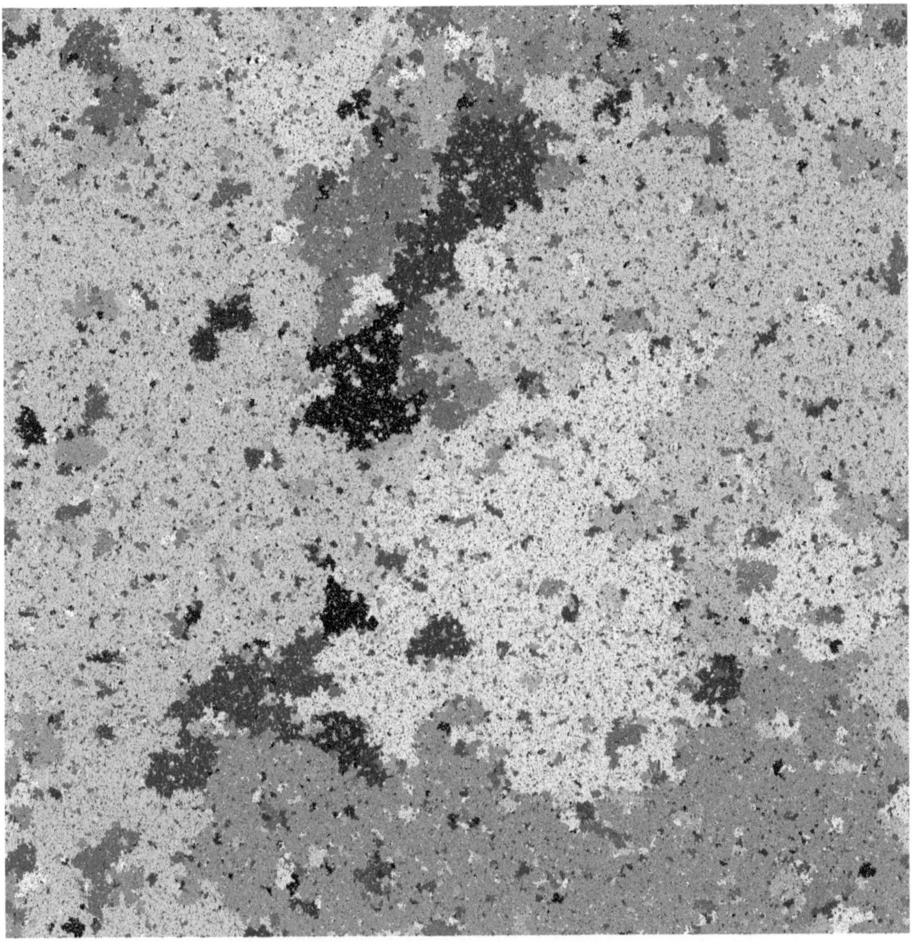

critical bond percolation: $p = 0.5$, $4096 \times 4096$ grid

# 4

# FK or random cluster model

The random cluster model, also called the FK percolation model, is a dependent percolation process. It was invented by Fortuin and Kasteleyn [74]. For a beautiful survey, we refer the reader to [80]. There exists a general mechanism to transfer percolation results to the Ising model. One should first try to prove the results in the FK percolation model and then transport them through a simple colouring operation.

## 4.1 Finite volume FK measures

We first define finite volume FK measures. Let $\Lambda$ be a box included in $\mathbb{Z}^d$. We will build probability measures on the space of the percolation configurations in $\Lambda$, that is $\Omega_\Lambda = \{0,1\}^{\mathbb{E}^d(\Lambda)}$, where $\mathbb{E}^d(\Lambda)$ is the set of the edges of $\mathbb{E}^d$ which are included in $\Lambda$ (an edge is included in $\Lambda$ if both its endpoints belong to $\Lambda$). Let $\omega \in \Omega_\Lambda$ be such a percolation configuration. We define $cl(\omega)$ as the number of clusters of the configuration $\omega$. For fixed $p \in [0,1]$ and $q \geq 1$, the FK measure in $\Lambda$ with parameters $p, q$ and free boundary conditions is the probability measure $\Phi_\Lambda^{f,p,q}$ on $\Omega_\Lambda$ defined by

$$\forall \omega \in \Omega_\Lambda \qquad \Phi_\Lambda^{f,p,q}(\omega) = \frac{1}{Z_\Lambda^{f,p,q}} \Big( \prod_{e \in \mathbb{E}^d(\Lambda)} p^{\omega(e)}(1-p)^{1-\omega(e)} \Big) q^{cl(\omega)}$$

where $Z_\Lambda^{f,p,q}$ is the appropriate normalization factor, so that $\Phi_\Lambda^{f,p,q}(\Omega_\Lambda) = 1$. Here the superscript f refers to the "free boundary conditions", as opposed to the "wired boundary conditions". The boundary conditions influence the way we count the clusters in $\Lambda$. In the wired case, denoted by the superscript w, all the clusters intersecting the boundary $\partial^{in}\Lambda$ count only for one cluster, as if all the vertices of $\partial^{in}\Lambda$ were wired together ($\partial^{in}\Lambda$ denotes the sites of $\Lambda$ which have a neighbour in $\mathbb{Z}^d \setminus \Lambda$). This way we define a different counting function $cl^w(\omega)$. The associated FK measure is denoted by $\Phi_\Lambda^{w,p,q}$ and is called the FK measure in $\Lambda$ with parameters $p, q$ and wired boundary conditions. It is defined with the same formula as for $\Phi_\Lambda^{f,p,q}$, where the counting function is now $cl^w(\omega)$ instead of $cl(\omega)$. A direct computation shows that for any box $\Lambda$, any configuration $\eta$ in $\Lambda$ and any edge $e \in \mathbb{E}^d(\Lambda)$,

$$\Phi_\Lambda^{f,p,q}\Big(\omega(e)=1 \,\Big|\, \omega=\eta \text{ on } \mathbb{E}^d(\Lambda)\setminus\{e\}\Big) = \begin{cases} p & \text{if the endpoints of } e \text{ are connected in } \mathbb{E}^d(\Lambda)\setminus\{e\} \text{ by a path of edges open in } \eta \\[2ex] \dfrac{p}{p+q(1-p)} & \text{otherwise} \end{cases}$$

With wired boundary conditions, the above conditional probability is also equal to $p$ if both endpoints of $e$ are connected to $\partial^{in}\Lambda$ in $\mathbb{E}^d(\Lambda)\setminus\{e\}$ by paths of edges open in $\eta$. The above conditional probability is non–decreasing with respect to the configuration $\eta$. This implies that the finite volume FK measures satisfy the FKG inequality (see section 3.2).

Let $\mu$, $\nu$ be two measures on $\Omega_\Lambda$. We say that $\mu$ is stochastically dominated by $\nu$, which we denote by $\mu \preceq \nu$, if $\mu(f) \leq \nu(f)$ for any bounded non–decreasing measurable function $f : \Omega \to \mathbb{R}$. The expression of the above conditional probability, together with a classical result of Holley [86], yields the following stochastic comparisons. For any $q \geq 1$, any $p \in [0,1]$,

$$P_{\frac{p}{p+q(1-p)}} \preceq \Phi_\Lambda^{f,p,q} \preceq \Phi_\Lambda^{w,p,q} \preceq P_p\,,$$

where $P_p$ is the law of the Bernoulli percolation model with parameter $p$. Moreover, for $* = f$ or $* = w$, $q \geq 1$ and $p_1 \leq p_2$, we have $\Phi_\Lambda^{*,p_1,q} \preceq \Phi_\Lambda^{*,p_2,q}$.

## 4.2 Phase transition

We wish to take the thermodynamic limit where the box $\Lambda$ grows and invades the whole lattice $\mathbb{Z}^d$. The finite volume FK measures $\Phi_\Lambda^{f,p,q}$ and $\Phi_\Lambda^{w,p,q}$ satisfy the FKG inequality whenever $p \in ]0,1[$, $q \geq 1$ (see sections 3.2 and 4.1). As a consequence, when $\Lambda$ increases to $\mathbb{Z}^d$, the probability $\Phi_\Lambda^{w,p,q}(0 \longleftrightarrow \partial^{in}\Lambda)$ decreases and converges towards a limiting value $\theta^w(p,q)$:

$$\theta^w(p,q) = \lim_{\Lambda\uparrow\mathbb{Z}^d} \Phi_\Lambda^{w,p,q}(0 \longleftrightarrow \partial^{in}\Lambda)\,.$$

We define the critical point by

$$p_c(q) = \sup\{\, p : \theta^w(p,q) = 0\,\}\,.$$

Theorem 3.1 and the stochastic comparison between FK and Bernoulli percolation imply that the FK model exhibits also a phase transition phenomenon.

**Proposition 4.1.** *For $d \geq 2$ and $q \geq 1$, the critical point $p_c(q)$ is strictly between 0 and 1.*

It is also possible to consider the thermodynamic limits of the finite volume FK measures. When $\Lambda$ increases to $\mathbb{Z}^d$, the measure $\Phi_\Lambda^{w,p,q}$ decreases and converges weakly towards the infinite volume FK measure $\Phi_\infty^{w,p,q}$, which is a probability measure on the space of infinite volume configurations $\Omega = \{0,1\}^{\mathbb{E}^d}$. Similarly, $\Phi_\Lambda^{f,p,q}$ increases weakly towards $\Phi_\infty^{f,p,q}$:

$$\lim_{\Lambda\uparrow\mathbb{Z}^d} \Phi_\Lambda^{f,p,q} = \Phi_\infty^{f,p,q}, \qquad \lim_{\Lambda\uparrow\mathbb{Z}^d} \Phi_\Lambda^{w,p,q} = \Phi_\infty^{w,p,q}.$$

The infinite volume FK measures $\Phi_\infty^{f,p,q}$, $\Phi_\infty^{w,p,q}$ are translation–invariant and ergodic, i.e., any translation–invariant event has probability 0 or 1. More generally, the infinite volume FK measures corresponding to the parameters $p, q$ are all the weak limit points of finite volume FK measures subject to general arbitrary boundary conditions (see section 4.4 for the definition).

**Theorem 4.2.** *Let $q \geq 1$. There exists a subset $\mathcal{U}(q)$ of $[0,1]$, which is at most countable, whose closure does not contain 0 nor 1, such that $\Phi_\infty^{f,p,q} = \Phi_\infty^{w,p,q}$ for $p \notin \mathcal{U}(q)$.*

See [79] for a proof of this theorem. The set $\mathcal{U}(q)$ coincides with the set of the discontinuity points of the map $p \mapsto \theta^w(p,q)$. In the region $[0,1]\setminus\mathcal{U}(q)$, there exists a unique infinite volume FK measure which we denote by $\Phi_\infty^{p,q}$. The percolation probability $\theta^w(p,q)$ can then be rewritten as

$$\theta^w(p,q) = \theta(p,q) = \Phi_\infty^{p,q}(0 \longleftrightarrow \infty)$$

and we can drop the superscript $w$ in the notation $\theta^w(p,q)$. A challenging conjecture is that $\mathcal{U}(q)$ is either empty or reduces to the critical point. The case $q = 1$ is obvious. This result is known in two dimensions for $q \geq 1$ [79]. For $q = 2$, the conjecture has been proved in dimensions $d \geq 3$ by Bodineau with the help of results specific to the Ising model [26, 27].

## 4.3 FK Ising coupling

A simple colouring operation enables us to go from FK percolation to the Ising model. We describe next this fundamental coupling between the FK measures and the Ising Gibbs measures (see [65, 74, 107] for more details). Let $\Lambda \subset \mathbb{Z}^d$ be a box. We recall that $\partial^{in}\Lambda$ denotes the sites of $\Lambda$ which have a neighbour in $\mathbb{Z}^d \setminus \Lambda$. An edge–spin configuration in $\Lambda$ is an element $(\omega, \sigma)$ of $\{0,1\}^{\mathbb{E}^d(\Lambda)} \times \{-,+\}^\Lambda$. Let $p$ belong to $[0,1]$. Let $\mathbb{P}_\Lambda$ be the probability measure on the space of edge–spin configurations in $\Lambda$ obtained through the following procedure:

- the edges in $\Lambda$ are opened with probability $p$ and closed with probability $1-p$;

• the spin value of the sites in $\Lambda$ is drawn randomly with the uniform distribution on $\{-,+\}$;
• the previous operations are performed independently;
• finally the measure is conditioned on the event that there is no open edge in $\Lambda$ between two sites with different spin values.

The support of $\mathbb{P}_\Lambda$ consists of the edge–spin configurations $(\omega,\sigma)$ in $\Lambda$ such that all the sites belonging to a given $\omega$–cluster $C$ have the same spin value, which we denote by $\sigma(C)$. To sum up this construction with a formula, we have

$$\forall(\omega,\sigma)\in\{0,1\}^{\mathbb{E}^d(\Lambda)}\times\{-,+\}^\Lambda$$
$$\mathbb{P}_\Lambda(\omega,\sigma)=\frac{1}{Z}\prod_{e\in\mathbb{E}^d(\Lambda)}p^{\omega(e)}(1-p)^{1-\omega(e)}1_{\{(\sigma(x)-\sigma(y))\omega(e)=0\}}$$

where $Z$ is an appropriate normalization factor. Let us compute the marginals of $\mathbb{P}_\Lambda$. For any $\omega\in\{0,1\}^{\mathbb{E}^d(\Lambda)}$,

$$\sum_{\sigma\in\{-,+\}^\Lambda}\mathbb{P}_\Lambda(\omega,\sigma)=\sum_{\sigma\in\{-,+\}^\Lambda}\frac{1}{Z}\prod_{e\in\mathbb{E}^d(\Lambda)}p^{\omega(e)}(1-p)^{1-\omega(e)}$$

where the sum runs over the spin configurations $\sigma$ which are compatible with $\omega$, meaning that $\sigma(x)=\sigma(y)$ whenever $x,y$ are the endpoints of an edge which is open in $\omega$. Equivalently, all the vertices belonging to an open cluster of $\omega$ must have the same spin. There are $2^{cl(\omega)}$ such spin configurations, so that

$$\sum_{\sigma\in\{-,+\}^\Lambda}\mathbb{P}_\Lambda(\omega,\sigma)=\frac{1}{Z}2^{cl(\omega)}\prod_{e\in\mathbb{E}^d(\Lambda)}p^{\omega(e)}(1-p)^{1-\omega(e)}=\Phi_\Lambda^{f,p,2}(\omega)\,.$$

Next, for any $\sigma\in\{-,+\}^\Lambda$,

$$\sum_{\omega\in\{0,1\}^{\mathbb{E}^d(\Lambda)}}\mathbb{P}_\Lambda(\omega,\sigma)=\sum_{\omega\in\{0,1\}^{\mathbb{E}^d(\Lambda)}}\frac{1}{Z}\prod_{e\in\mathbb{E}^d(\Lambda)}p^{\omega(e)}(1-p)^{1-\omega(e)}$$

where the sum runs over the percolation configurations $\omega$ which are compatible with $\sigma$, meaning that $\omega(e)=0$ whenever the endpoints of the edge $e$ have different spins. By factorizing, we see that the above sum is equal to $Z^{-1}(1-p)^N$, where $N$ is the number of such edges. Up to an additive constant, this number $N$ is equal to half of the Ising Hamiltonian $H_\Lambda(\sigma)$, hence

$$\sum_{\omega\in\{0,1\}^{\mathbb{E}^d(\Lambda)}}\mathbb{P}_\Lambda(\omega,\sigma)=\frac{1}{Z_\Lambda}(1-p)^{\frac{1}{2}H_\Lambda(\sigma)}=\mu_{\Lambda,-2/\ln(1-p)}(\sigma)\,.$$

Thus the first marginal of $\mathbb{P}_\Lambda$ on $\{0,1\}^{\mathbb{E}^d(\Lambda)}$ is the FK measure $\Phi_\Lambda^{f,p,2}$ with free boundary conditions, while its second marginal on $\{-,+\}^\Lambda$ is the Ising Gibbs measure $\mu_{\Lambda,T}$ with free boundary conditions, where

$$\frac{1}{T} = -\frac{1}{2}\ln(1-p)\,.$$

It is also possible to build a coupling involving different boundary conditions. For instance, let us add the following rule in the definition of the probability measure on the space of edge–spin configurations in $\Lambda$:

• the spin value of the sites in $\partial^{in}\Lambda$ is set to $+$.

Let us denote by $\mathbb{P}_\Lambda^+$ the resulting measure. The first marginal of $\mathbb{P}_\Lambda^+$ on $\{0,1\}^{\mathbb{E}^d(\Lambda)}$ is the FK measure $\Phi_\Lambda^{w,p,2}$ with wired boundary conditions, while its second marginal on $\{-,+\}^\Lambda$ is the Ising Gibbs measure $\mu_{\Lambda,T}^+$ with $+$ boundary conditions, where $T = -2/\ln(1-p)$. With the help of this coupling, we can generate an Ising configuration from an FK configuration, and vice–versa.

### 4.3.1 From FK to Ising.

To draw a spin configuration in $\Lambda$ according to $\mu_{\Lambda,T}^+$ we can proceed as follows. First we draw an edge configuration in $\Lambda$ according to $\Phi_\Lambda^{w,p,2}$. Second we colour each open cluster independently, with $+$ for the clusters intersecting $\partial^{in}\Lambda$ and with the uniform distribution on $\{-,+\}$ for the other clusters.

### 4.3.2 From Ising to FK.

To draw an edge configuration in $\Lambda$ according to $\Phi_\Lambda^{w,p,2}$ we can proceed as follows. First we draw a spin configuration in $\Lambda$ according to $\mu_{\Lambda,T}^+$. Second we declare closed the edges whose endpoints have different spins. The edges whose endpoints have the same spin are open independently with probability $p$ and closed with probability $1-p$.

This coupling provides also a correspondence between the fundamental quantities describing the two models.

### 4.3.3 Phase transition.

For any box $\Lambda$, $p \in [0,1]$, setting $T = -2/\ln(1-p)$, we have

$$\mu_{\Lambda,T}^+(\sigma(0)) = \sum_{\omega\in\{0,1\}^{\mathbb{E}^d(\Lambda)}} \sum_{\sigma\in\{-,+\}^\Lambda} \sigma(0)\,\mathbb{P}_\Lambda^+(\omega,\sigma)\,.$$

If $\omega$ is a percolation configuration such that 0 is not connected to $\partial^{in}\Lambda$, then the cluster containing 0 is coloured $-$ or $+$ with probability $1/2$, and the sum $\sum_{\sigma\in\{-,+\}^\Lambda}\sigma(0)\,\mathbb{P}_\Lambda^+(\omega,\sigma)$ vanishes. Hence only the percolation configurations such that $0 \longleftrightarrow \partial^{in}\Lambda$ contribute to the sum. For such configurations, the spin at the origin is set to $+$ by the coupling rule. Thus

$$\mu_{\Lambda,T}^+(\sigma(0)) = \Phi_\Lambda^{w,p,2}(0 \longleftrightarrow \partial^{in}\Lambda)\,.$$

Letting $\Lambda$ grow to $\mathbb{Z}^d$, we find also that $m^*(T) = \theta^w(p,2)$. Theorem 2.1 follows from this identity and proposition 4.1.

### 4.3.4 Correlation and connection.

For any box $\Lambda$, $p \in [0,1]$, $x,y \in \Lambda$, setting $T = -2/\ln(1-p)$, we have

$$\mu^+_{\Lambda,T}(\sigma(x)\sigma(y)) = \sum_{\omega\in\{0,1\}^{\mathbb{E}^d(\Lambda)}} \sum_{\sigma\in\{-,+\}^\Lambda} \sigma(x)\sigma(y)\,\mathbb{P}^+_\Lambda(\omega,\sigma)\,.$$

If $\omega$ is a percolation configuration such that $x,y$ are not connected, then the clusters containing $x$ and $y$ are coloured independently $-$ or $+$ with probability 1/2, and the sum $\sum_{\sigma\in\{-,+\}^\Lambda} \sigma(x)\sigma(y)\,\mathbb{P}^+_\Lambda(\omega,\sigma)$ vanishes. Hence only the percolation configurations such that $x \longleftrightarrow y$ contribute to the sum. For such configurations, the coupling rule ensures that the spins at $x$ and $y$ are equal. Thus

$$\mu^+_{\Lambda,T}(\sigma(x)\sigma(y)) = \Phi^{w,p,2}_\Lambda(x \longleftrightarrow y)\,.$$

Letting $\Lambda$ grow to $\mathbb{Z}^d$, we find also that $\mu^+_T(\sigma(x)\sigma(y)) = \Phi^{w,p,2}_\infty(x \longleftrightarrow y)$. In conclusion, the two points correlation function of the Ising model is equal to the two points connectivity function of the FK percolation model.

Ising
$T = 2.2$

FK
$p = 0.597$
$q = 2$

Ising FK coupling: 3 hours, $1024 \times 1024$ grid

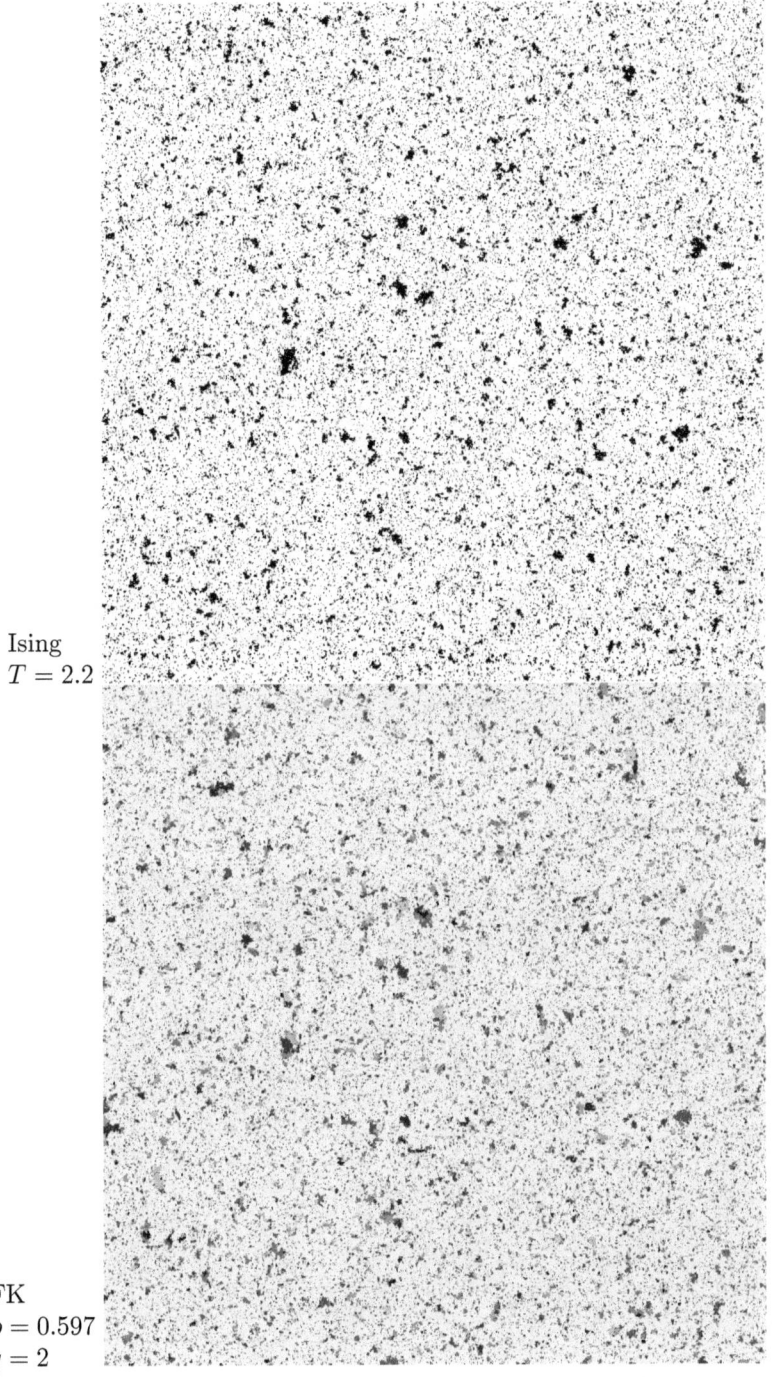

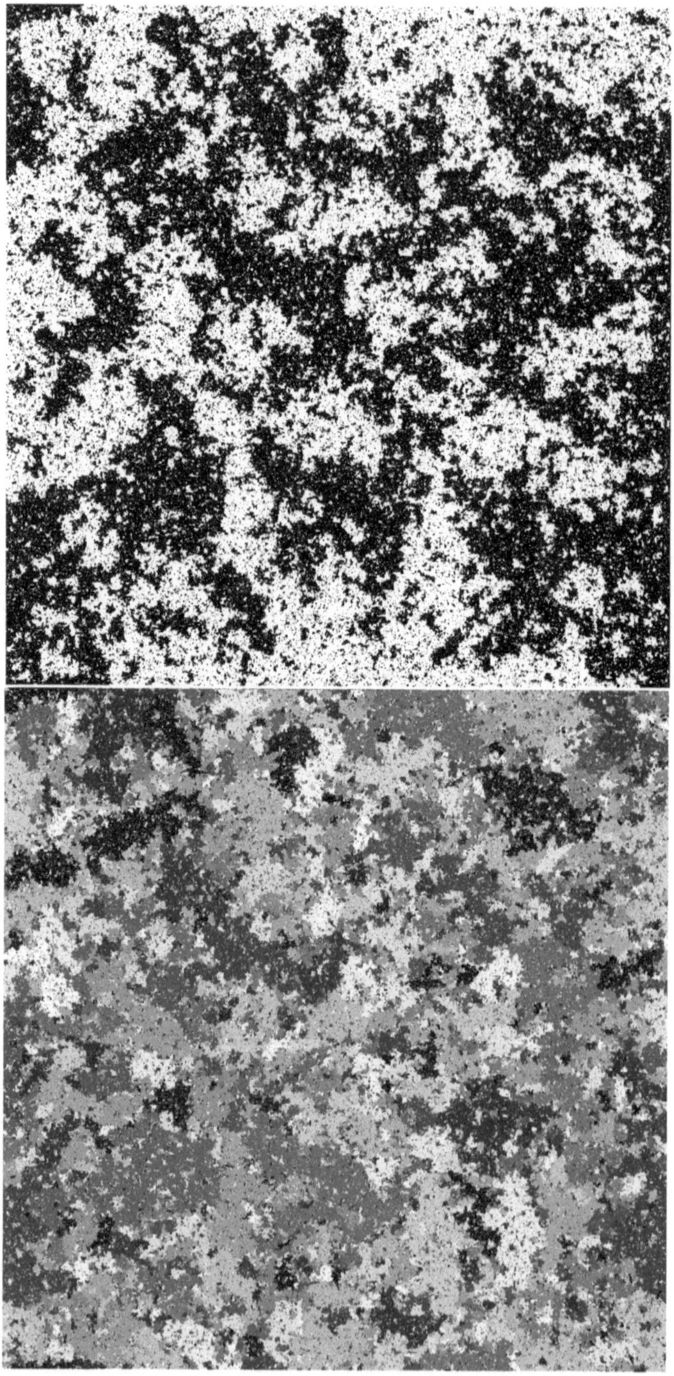

Ising
$T = 2.3$

FK
$p = 0.581$
$q = 2$

Ising FK coupling: 5 days, $1024 \times 1024$ grid

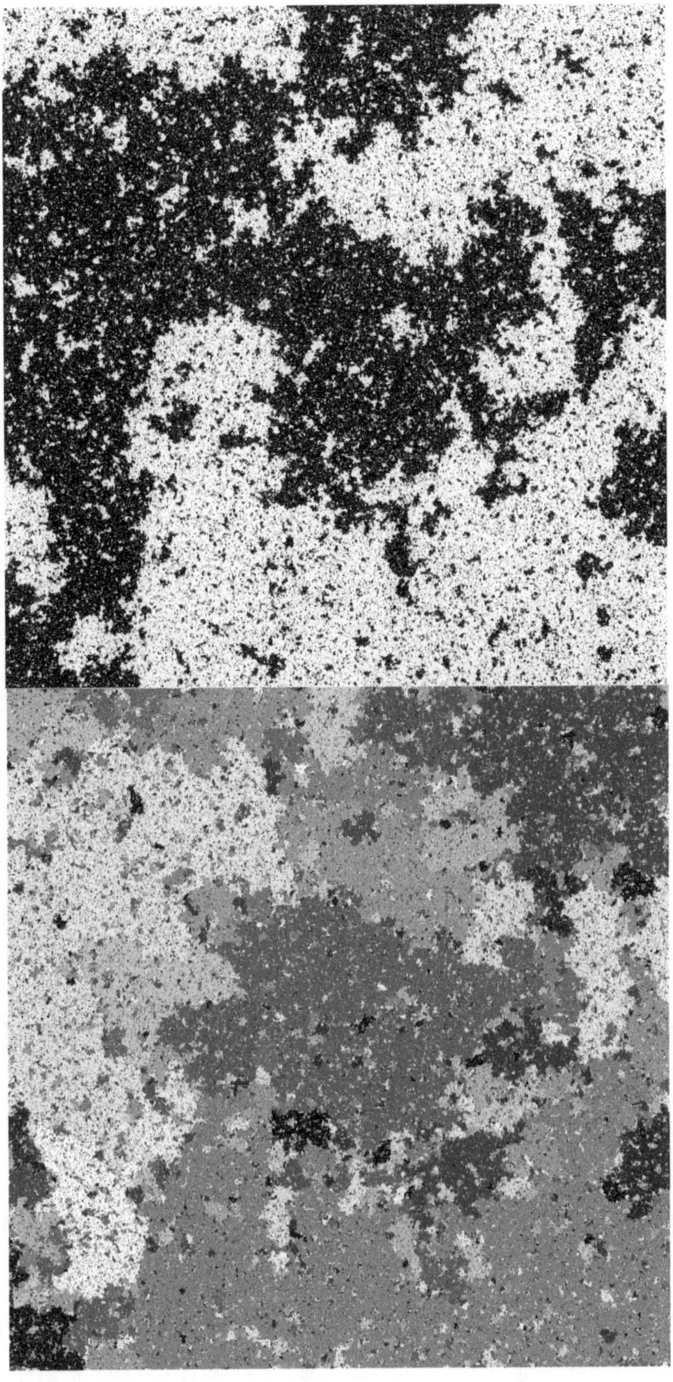


Ising FK coupling: 5 days, $2048 \times 2048$ grid

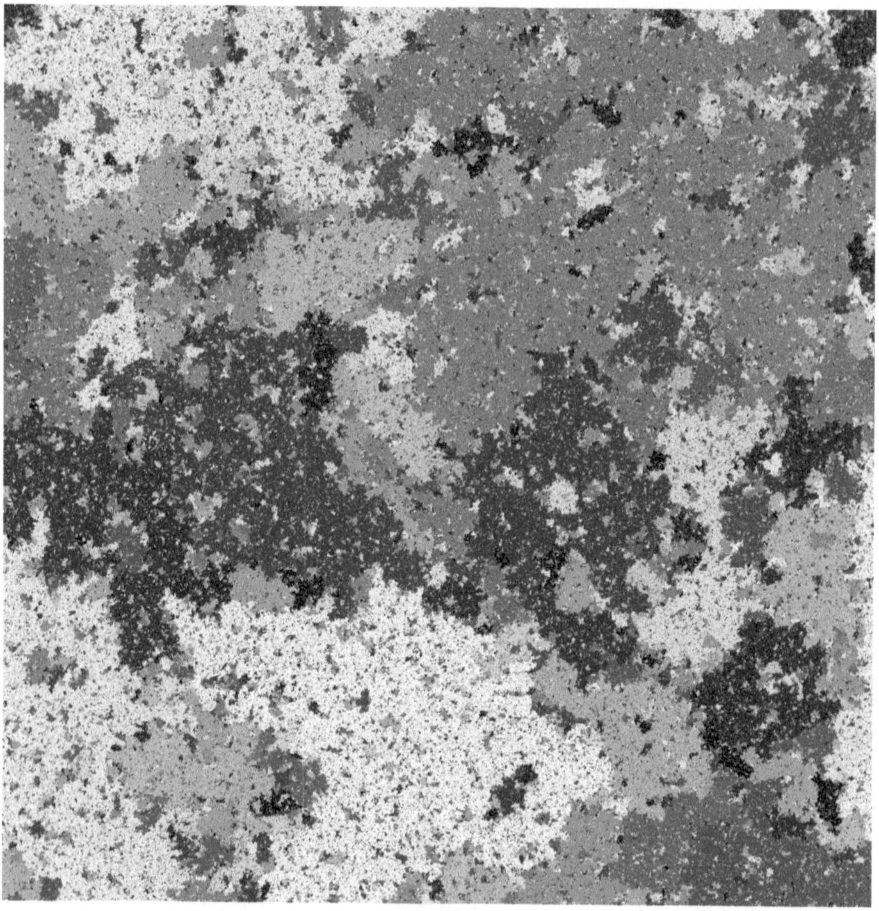

critical FK model: $p = 0.586$, $q = 2$

32 days of simulation on a $4096 \times 4096$ grid with a 1 Ghz PC

## 4.4 Boundary conditions

We introduce partially wired boundary conditions as in [112]. Let $V$ be a finite subset of $\mathbb{Z}^d$ and let $\mathbb{E}^d(V)$ be the set of edges whose both endpoints are included in $V$. Consider a partition $\pi$ of the set $\partial^{in}V$, say $\pi = \{B_1, ..., B_n\}$ (the sets $B_i$ are disjoint non–empty subsets of $\partial^{in}V$ with $B_1 \cup \cdots \cup B_n = \partial^{in}V$). We say that $x, y \in \partial^{in}V$ are $\pi$–wired if $x, y \in B_i$ for some $i \in \{1, ..., n\}$. Fix a configuration $\eta \in \Omega_V = \{0,1\}^{\mathbb{E}^d(V)}$. We want to count the $\eta$–clusters in $V$ in such a way that $\pi$–wired sites are considered to be connected. This can be done in the following formal way. We introduce an equivalence relation on $V$: two sites $x, y$ in $V$ are said to be $\pi\cdot\eta$–wired if they are both joined by $\eta$–open paths to sites $x', y' \in \partial^{in}V$ which are themselves $\pi$–wired, or if they are connected together by an $\eta$–open path. The equivalence classes of this relation are called $\pi\cdot\eta$–clusters, or $\eta$–clusters in $V$ with respect to the boundary condition $\pi$. The number of $\pi\cdot\eta$–clusters is denoted by $cl^\pi(\eta)$.

For fixed $p \in [0,1]$ and $q \geq 1$, the FK measure with parameters $(p,q)$ and boundary conditions $\pi$ is the probability measure $\Phi_V^{\pi,p,q}$ on $\Omega_V$ defined by the formula

$$\forall \eta \in \Omega_V \qquad \Phi_V^{\pi,p,q}(\eta) = \frac{1}{Z_V^{\pi,p,q}} \Big( \prod_{e \in \mathbb{E}^d(V)} p^{\eta(e)}(1-p)^{1-\eta(e)} \Big) \, q^{cl^\pi(\eta)}$$

where $Z_V^{\pi,p,q}$ is the appropriate normalization factor, so that $\Phi_V^{\pi,p,q}(\Omega_V) = 1$. There are two extremal boundary conditions: the free boundary condition corresponds to the partition $f$ defined to have exactly $|\partial^{in}V|$ classes, and the wired boundary condition corresponds to the partition $w$ with only one class. The set of all the FK measures corresponding to different boundary conditions will be denoted by $\mathcal{FK}(p,q,V)$, and we define $c\mathcal{FK}(p,q,V)$ to be its convex hull.

We will list some useful properties of FK measures.

### 4.4.1 FKG inequality.

We use here some definitions introduced in section 3.2. A property of crucial importance is that for $q \geq 1$, every $\Phi \in \mathcal{FK}(p,q,V)$ is strong FKG. This means that for any collection $(e_i, i \in I)$ of edges included in $V$, any non–decreasing functions $f, g$, we have

$$\Phi\big(fg \,|\, \omega(e_i),\, i \in I\big) \geq \Phi\big(f \,|\, \omega(e_i),\, i \in I\big)\, \Phi\big(g \,|\, \omega(e_i),\, i \in I\big)\,.$$

In some cases it is possible to compare FK measures with different boundary conditions. There is a partial order on the set of partitions of $\partial^{in}V$. We say that $\pi$ dominates $\pi'$ if $x$, $y$ are $\pi$–wired whenever they are $\pi'$–wired. We then have $\Phi_V^{\pi',p,q} \preceq \Phi_V^{\pi,p,q}$. This implies immediately that

$$\forall \Phi \in \mathcal{FK}(p,q,V) \qquad \Phi_V^{f,p,q} \preceq \Phi \preceq \Phi_V^{w,p,q}\,.$$

Next we discuss properties of conditional FK measures.

### 4.4.2 Markov property.

Let $U \subset V$ be two finite subsets of $\mathbb{Z}^d$ and let $\pi$ be a partition of $\partial^{in} V$. Let also $\omega \in \Omega_V$. We define a new partition of $\partial^{in} U$, denoted by $\pi \cdot W_V^U(\omega)$, by considering $x, y \in \partial^{in} U$ to be $\pi \cdot W_V^U(\omega)$–wired if they are joined by $\omega$–open paths in $\mathbb{E}^d(V) \setminus \mathbb{E}^d(U)$ to sites $x'$, $y'$ which are themselves $\pi$–wired, or if they are joined together by an $\omega$–open path in $\mathbb{E}^d(V) \setminus \mathbb{E}^d(U)$. For any function $f$ depending only on the edges $\mathbb{E}^d(U)$ and any configuration $\eta \in \Omega_V$, we have

$$\Phi_V^{\pi,p,q}\big(f(\omega) \mid \forall e \in \mathbb{E}^d(V) \setminus \mathbb{E}^d(U) \quad \omega(e) = \eta(e)\big) = \Phi_U^{\pi \cdot W_V^U(\eta),p,q}(f)\,.$$

*Proof.* Let $\omega$ be a percolation configuration in $V$. We denote by $cl_U^\pi(\omega)$ the number of the $\pi \cdot \omega$–clusters which intersect $U$ and by $cl_{V \setminus U}^\pi(\omega)$ the number of the $\pi \cdot \omega$–clusters which don't intersect $U$. Then

$$cl^\pi(\omega) = cl_U^\pi(\omega) + cl_{V \setminus U}^\pi(\omega)\,.$$

Suppose that $\omega(e) = \eta(e)$ for $e \in \mathbb{E}^d(V) \setminus \mathbb{E}^d(U)$. Then $cl_U^\pi(\omega) = cl^{\pi \cdot W_V^U(\eta)}(\omega)$ and

$$\Phi_V^{\pi,p,q}(\omega) = \frac{1}{Z_V^{\pi,p,q}} \Big( \prod_{e \in \mathbb{E}^d(V) \setminus \mathbb{E}^d(U)} p^{\eta(e)}(1-p)^{1-\eta(e)} \Big)\, q^{cl_{V \setminus U}^\pi(\eta)}$$
$$\times \Big( \prod_{e \in \mathbb{E}^d(U)} p^{\omega(e)}(1-p)^{1-\omega(e)} \Big)\, q^{cl_U^\pi(\omega)} = c \times \Phi_U^{\pi \cdot W_V^U(\eta),p,q}(\omega)$$

where the factor $c$ depends on $\eta$ but not on $\omega$. This identity implies the Markov property. □

A direct consequence of the Markov property is that the restriction of any FK measure $\Phi$ in $\mathcal{FK}(p,q,V)$ to $\Omega_U$ is contained in the convex hull $c\mathcal{FK}(p,q,U)$. Another consequence of the Markov property is that, if we close or we open all the edges touching the outer boundary of a box, we recover the FK measure in the box with free or wired boundary conditions. More precisely, let $\Lambda \subset \Lambda'$ be two boxes such that $\partial^{in} \Lambda' \cap \Lambda = \emptyset$. We define

$$\partial^{out} \Lambda = \big\{ x \in \mathbb{Z}^d \setminus \Lambda : \exists y \in \Lambda \quad \{x,y\} \in \mathbb{E}^d \big\}\,.$$

Let $E$ be the set of the edges having one endpoint in $\partial^{in} \Lambda$ and one endpoint in $\partial^{out} \Lambda$. Let $F$ be the set of the edges having both endpoints in $\partial^{out} \Lambda$. Then

$$\Phi_{\Lambda'}^{\pi,p,q}\Big(\cdot \,\Big|\, \text{all the edges of } E \text{ are closed}\Big) = \Phi_\Lambda^{f,p,q}(\,\cdot\,)\,,$$
$$\Phi_{\Lambda'}^{\pi,p,q}\Big(\cdot \,\Big|\, \text{all the edges of } E \cup F \text{ are open}\Big) = \Phi_\Lambda^{w,p,q}(\,\cdot\,)\,.$$

Finally, the FKG inequality and the Markov property together yield the following domination inequalities: if $V \subset W$ are finite subsets of $\mathbb{Z}^d$, then

$$\Phi_V^{f,p,q} \preceq \Phi_W^{f,p,q}\,, \qquad \Phi_W^{w,p,q} \preceq \Phi_V^{w,p,q}\,.$$

The second inequality implies that $\Phi_\Lambda^{w,p,q}(0 \longleftrightarrow \partial^{in} \Lambda)$ decreases as $\Lambda$ increases, so that the limit defining $\theta^w$ exists (see section 4.2).

# Part III

## Main results

# 5

# The Wulff crystal

We present here the core results of these notes. Under certain conditions, a large droplet of one phase floats in the other phase and the Wulff crystal emerges. Three variants are exposed in our three basic models: the Ising model, Bernoulli percolation, FK percolation. The history of the subject is summarized and the corresponding references are listed in section 5.5. The proofs we present in these notes are simplified compared to the original research papers. We provide full proofs in dimensions $d \geq 3$ and we indicate the modifications necessary to handle the two dimensional case, which is special.

## 5.1 Ising model

We shall mimic mathematically the initial experiment of section 1.1 with the help of the Ising model. Let us consider a box $\Lambda(n)$ of diameter $n$ full of pluses. We take $n$ very large, of the order of the Avogadro number $6.02 \times 10^{23}$. We start deleting pluses and replacing them by minuses, first a small quantity of minuses. It is possible to build a stochastic dynamics in the box which is conservative (i.e., the total numbers of minuses and pluses remain unchanged or equivalently the empirical magnetization $n^{-d} \sum_{x \in \Lambda(n)} \sigma(x)$ remains constant) and whose final equilibrium is the Gibbs measure $\mu^{+}_{\Lambda(n),T}$ conditioned to have the initial fixed magnetization. The simplest such dynamics is the so-called Kawasaki dynamics: at random exponential times, a pair of neighbouring particles might be exchanged according to a simple local probabilistic rule. As long as the empirical magnetization is larger than $m^*(T)$, the configuration in $\Lambda(n)$ at equilibrium is expected to be spatially homogeneous. If we keep pouring minuses into the box and removing pluses, we soon reach the value $m^*(T)$, and at this point we obtain the saturated pure phase $\mu^{+}_{T}$, i.e., the configuration in $\Lambda(n)$ looks like a finite sample of the infinite volume Gibbs measure $\mu^{+}_{T}$. We finally add some more minuses and we cross the threshold $m^*(T)$. We wish to understand the response of the system and the most likely configurations inside the box when there is an excess of minuses.

It turns out that this simple model indeed confirms the prediction of the phenomenological theory. At equilibrium, with probability tending to 1 as $n$ goes to $\infty$, one sees inside the box $\Lambda(n)$ a region where the configuration statistically looks like the minus phase $\mu_T^-$, surrounded by a region filled with the plus phase $\mu_T^+$. When rescaled by a factor $n$, the shape of this region converges as $n$ goes to $\infty$ towards a deterministic shape, called the Wulff crystal of the Ising model. This crystal is convex, it depends on the temperature and on the initial lattice $\mathbb{Z}^d$; it bears the name of Wulff, who studied it one century ago [131].

In order to detect conveniently the Wulff region, we rescale the box $\Lambda(n)$ by a factor $n$ and we send it onto the $d$–dimensional unit cube $[-1/2, 1/2]^d$. Let $\sigma$ be a spin configuration in $\Lambda(n)$. To $\sigma$ we associate a measure $\sigma_n$ on $[-1/2, 1/2]^d$ by setting

$$\sigma_n = \frac{1}{n^d} \sum_{x \in \Lambda(n)} \sigma(x)\, \delta_{x/n}$$

where $\delta_{x/n}$ is the Dirac mass at $x/n$. We call $\sigma_n$ the empirical magnetization. The expectation $b_n$ of $\sigma_n$ is

$$b_n = \frac{1}{n^{d+1}} \sum_{x \in \Lambda(n)} \sigma(x) x\,.$$

We denote by $\mathcal{L}^d$ or simply by $dx$ the $d$–dimensional Lebesgue measure.

**Theorem 5.1.** *Let $d \geq 2$ and let $T < T_c(d)$. There exists a bounded, closed, convex set $\mathcal{W}$ containing 0 in its interior, called the Wulff crystal of the Ising model such that the following holds.*
*Let $m < m^*$ be close enough to $m^*$ so that the rescaled Wulff crystal*

$$\mathcal{W}(m) = \left(\frac{m^* - m}{2m^*}\right)^{\frac{1}{d}} \frac{\mathcal{W}}{\mathcal{L}^d(\mathcal{W})^{\frac{1}{d}}}$$

*fits into the unit cube $[-1/2, 1/2]^d$. Let $w_n$ be the random measure defined by*

$$w_n(x)\, dx = \Big(1_{[-\frac{1}{2},\frac{1}{2}]^d}(x) - 2 \cdot 1_{\mathcal{W}(m)}\big(\frac{b_n}{m^* - m} + x\big)\Big) m^*\, dx\,.$$

*This is the measure having density $-m^*$ on $-b_n/(m^* - m) + \mathcal{W}(m)$ and $m^*$ on the complement. Under the conditional probability*

$$\mu_n(\cdot) = \mu^+_{\Lambda(n),T}\Big(\cdot \,\Big|\, \frac{1}{n^d} \sum_{x \in \Lambda(n)} \sigma(x) \leq m\Big)$$

*the difference between the random measures $\sigma_n$ and $w_n$ converges weakly in probability towards 0, i.e., for any continuous function $f : [-1/2, 1/2]^d \to \mathbb{R}$,*

$$\forall \varepsilon > 0 \qquad \lim_{n \to \infty} \mu_n\big(|\sigma_n(f) - w_n(f)| \geq \varepsilon\big) = 0\,.$$

*The probabilities of the deviations are of order* $\exp -cn^{d-1}$.

The last sentence of the theorem means the following. For any continuous function $f : [-1/2, 1/2]^d \to \mathbb{R}$, any $\varepsilon > 0$, there exist positive constants $b, c$ depending on $d, T, f, \varepsilon$ such that

$$\mu_n\left(\left|\frac{1}{n^d}\sum_{x\in\Lambda(n)}\sigma(x)\,f\Big(\frac{x}{n}\Big)+\int_{\mathcal{W}(m)} 2m^*\,f(-\frac{b_n}{m^*-m}+x)\,dx-\int_{[-\frac{1}{2},\frac{1}{2}]^d} m^* f(x)\,dx\right|>\varepsilon\right)$$
$$\leq b\exp(-cn^{d-1})\,.$$

The main assertion of the theorem is that the left–hand quantity goes to 0 for any continuous function $f$ and $\varepsilon > 0$. The objects appearing in the statement, namely the spontaneous magnetization $m^*$ and the Wulff crystal $\mathcal{W}$, are built as the thermodynamic limit of finite volume quantities (see section 2.2 for $m^*$ and section 11.2 for the surface tension). These objects can equivalently be defined with the help of the infinite volume Gibbs measure $\mu_T^+$. The way to prove theorem 5.1 is rather long, but it is likely that substantial simplifications will appear in the future. For instance, an intermediate step is to compute the asymptotics of the conditioning event.

**Theorem 5.2.** *Let $d \geq 2$ and let $T < T_c(d)$. For $m < m^*$ close enough to $m^*$, so that the rescaled Wulff crystal $\mathcal{W}(m)$ fits into the unit cube $[-1/2, 1/2]^d$, we have*

$$\lim_{n\to\infty}\frac{1}{n^{d-1}}\ln\mu^+_{\Lambda(n),T}\Big(\frac{1}{n^d}\sum_{x\in\Lambda(n)}\sigma(x)\leq m\Big) = -d\left(\frac{m^*-m}{2m^*}\right)^{\frac{d-1}{d}}\mathcal{L}^d(\mathcal{W})^{\frac{1}{d}}\,.$$

Theorem 5.2 and the volume large deviation principle of theorem 8.14 provide a complete picture of the large deviation behavior of the average magnetization. The rate function $I$ of the volume large deviation principle stated in theorem 8.14 vanishes on $]-m^*, m^*[$ because on this interval the large deviations are of surface order. As we are in the phase coexistence regime, a new large deviation principle on the surface scale emerges.

Theorems 5.1 and 5.2 in dimensions 2 are consequences of the much finer results of Dobrushin, Kotecký, Shlosman [63] and Pfister [109] for low temperatures and Ioffe and Schonmann [90, 91, 92] for all subcritical temperatures. In dimensions 3 and higher, they were proven by Bodineau [23] for low temperatures and by Cerf and Pisztora [43] until the slab percolation threshold for temperatures such that the associated infinite volume FK measure is unique. Recently, Bodineau proved that this slab percolation threshold coincides with the true critical point [26] and that for any subcritical temperature, the associated infinite volume FK measure is indeed unique [27]. See section 5.5 for a more detailed account of the history of these results.

In two dimensions the Wulff droplet can be identified with a random region surrounded by a minus spin cluster. Its external boundary is therefore a large contour separating plus and minus spins which follows closely the boundary of the Wulff crystal in the sense of the Hausdorff metric [63, 92]. In dimensions

$d \geq 3$, it is widely believed that for low temperatures, the Wulff droplet can still be defined by a microscopic contour. However for temperatures close to $T_c$, a fundamentally new situation is expected. The dominant minus spin cluster of the Wulff droplet should percolate all the way to the boundary of the box. More precisely, there should exist two big spin clusters, one of pluses and one of minuses, and they should both be omnipresent in the entire box; the densities of these clusters should undergo an abrupt change at the boundary of the Wulff droplet. In this case the phase boundaries cannot be described directly with contours.

*Proof of theorems 5.1 and 5.2.* Theorems 5.1 and 5.2 are consequences of the large deviation principle stated in theorem 7.1 and the Wulff theorem 15.1. Let $T, m$ be as in the statement of the theorem and let us define

$$\mathbb{M}(m) = \big\{ \nu \in \mathcal{M}(Q) : \nu(Q) \leq m \big\} .$$

For the notation, see section 7.2. We apply the large deviation principle stated in theorem 7.1:

$$-\inf \{ \mathcal{J}(\nu) : \nu \in \overset{\circ}{\mathbb{M}}(m) \} \leq \liminf_{n \to \infty} \frac{1}{n^{d-1}} \ln \mu^+_{\Lambda(n),T}\big(\sigma_n \in \mathbb{M}(m)\big)$$
$$\leq \limsup_{n \to \infty} \frac{1}{n^{d-1}} \ln \mu^+_{\Lambda(n),T}\big(\sigma_n \in \mathbb{M}(m)\big) \leq -\inf \{ \mathcal{J}(\nu) : \nu \in \overline{\mathbb{M}}(m) \} .$$

From the very definition of the rate function $\mathcal{J}$, we see that both infima appearing in the above inequalities are equal to

$$\inf \Big\{ \mathcal{I}(E) : E \in \mathcal{B}(Q),\, \mathcal{L}^d(E) \geq \frac{m^* - m}{2m^*} \Big\} .$$

By theorem 15.1, for $m$ close enough to $m^*$, the solution to this variational problem is given by the suitably rescaled Wulff crystal $\mathcal{W}(m)$ defined in theorem 5.1, so that

$$\lim_{n \to \infty} \frac{1}{n^{d-1}} \ln \mu^+_{\Lambda(n),T}\Big( \frac{1}{n^d} \sum_{x \in \Lambda(n)} \sigma(x) \leq m \Big) = -\mathcal{I}(\mathcal{W}(m))$$
$$= -\left( \frac{m^* - m}{2m^*} \right)^{\frac{d-1}{d}} \frac{\mathcal{I}(\mathcal{W})}{\mathcal{L}^d(\mathcal{W})^{\frac{d-1}{d}}} .$$

To get the formula of theorem 5.2, we use the Gauss–Green theorem (see sections 13.3, 13.3) and proposition 14.1 and we compute $\mathcal{I}(\mathcal{W})$ as follows:

$$d\mathcal{L}^d(\mathcal{W}) = \int_{\mathcal{W}} \operatorname{div} x \, d\mathcal{L}^d(x) = \int_{\partial^* \mathcal{W}} x \cdot \nu_{\mathcal{W}}(x) \, d\mathcal{H}^{d-1}(x)$$
$$= \int_{\partial^* \mathcal{W}} \tau(\nu_{\mathcal{W}}(x)) \, d\mathcal{H}^{d-1}(x) = \mathcal{I}(\mathcal{W}) .$$

We finally prove theorem 5.1. Let $f : [-1/2,1/2]^d \to \mathbb{R}$ be a continuous function, let $\varepsilon > 0$ and let

$$\mathbb{M}(m,f,\varepsilon) = \Big\{ \nu \in \mathcal{M}(Q) : \nu(Q) \leq m,$$
$$\Big| \nu(f) + \int_{\mathcal{W}(m)} 2m^* f\Big(-\frac{b(\nu)}{m^*-m} + x\Big)\, dx - \int_{[-\frac{1}{2},\frac{1}{2}]^d} m^* f(x)\, dx \Big| \geq \varepsilon \Big\}$$

where we set $b(\nu) = \int x\, d\nu(x)$ for $\nu \in \mathcal{M}(Q)$. By the large deviation principle of theorem 7.1,

$$\limsup_{n\to\infty} \frac{1}{n^{d-1}} \ln \mu^+_{\Lambda(n),T}\big(\sigma_n \in \mathbb{M}(m,f,\varepsilon) \,\big|\, \sigma_n \in \mathbb{M}(m)\big) \leq$$
$$-\inf\{\, \mathcal{J}(\nu) : \nu \in \overline{\mathbb{M}}(m,f,\varepsilon)\,\} + \inf\{\, \mathcal{J}(\nu) : \nu \in \mathbb{M}(m)\,\}\,.$$

The first infimum can be rewritten as a constrained isoperimetric problem, where the constraint prevents the translates of the Wulff crystal $\mathcal{W}(m)$ from being admissible candidates. The uniqueness part of the Wulff theorem 15.1 then implies that the first infimum is strictly larger than the second. □

We present next several pictures of the Wulff crystal of the two dimensional Ising model. The next page displays the exact deterministic shape of the Wulff crystal (see section 11.5 for the explicit formula). The following pages display pictures obtained by simulations at various temperatures. These simulations are performed with the program gising.

$T =$
0.5,
1.5,
2,
2.22

$T =$
2.22,
2.24,
2.26

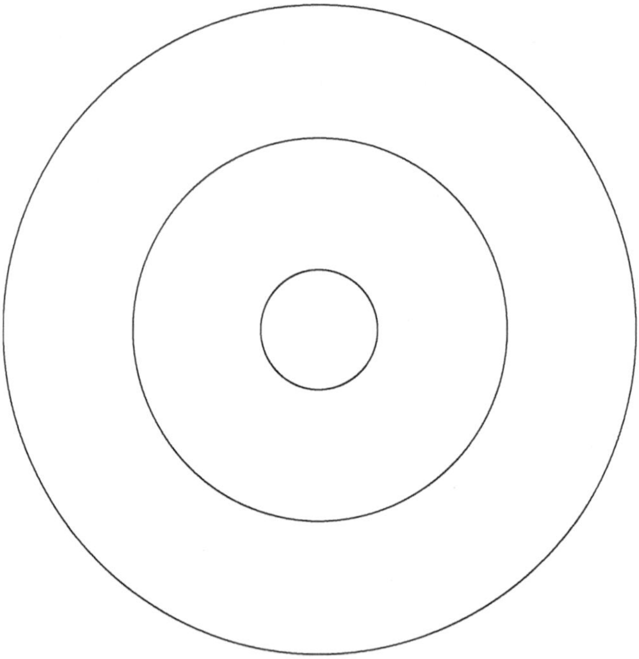

Theoretical shape of the Ising Wulff crystal
(computations done by Reda Messikh)

$T = 0.5$
12 days

$T = 1.5$
8 days

Simulation of the Ising Wulff crystal on a $1024 \times 1024$ grid with a 1 Ghz PC

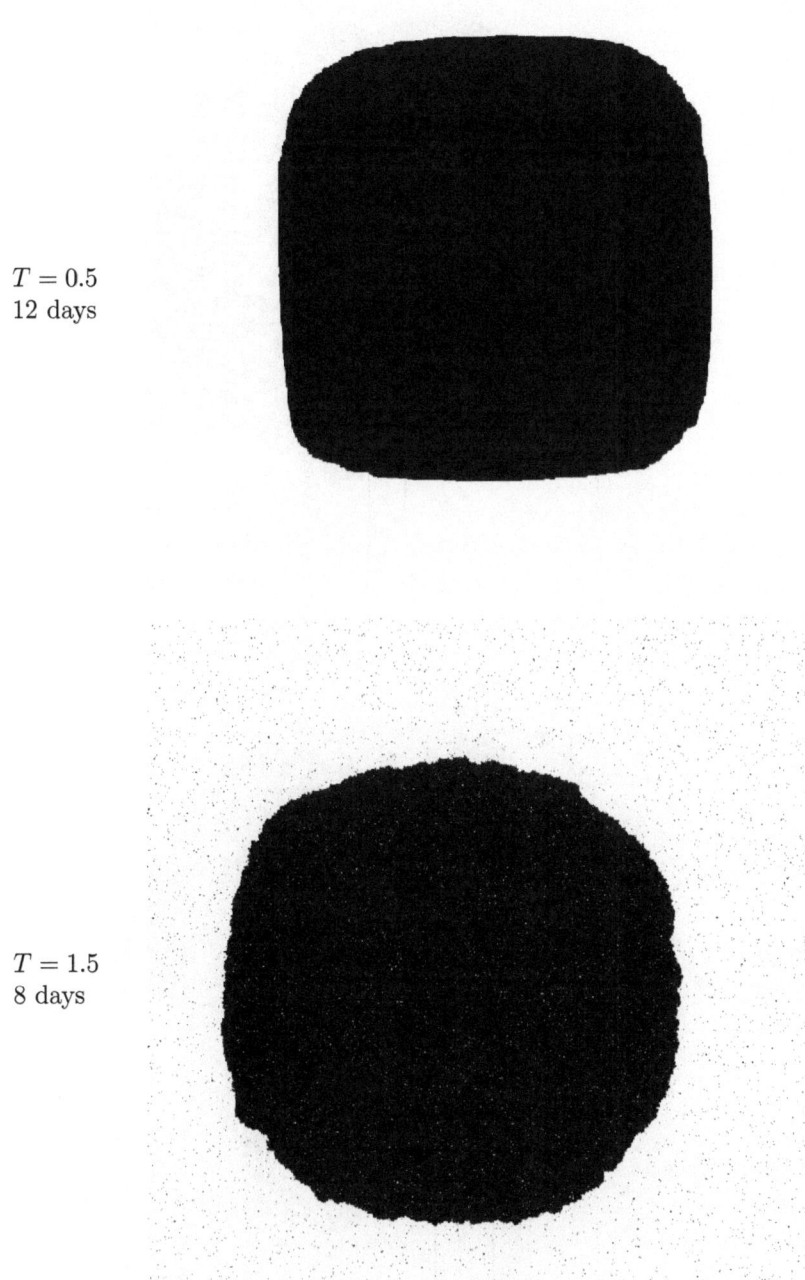

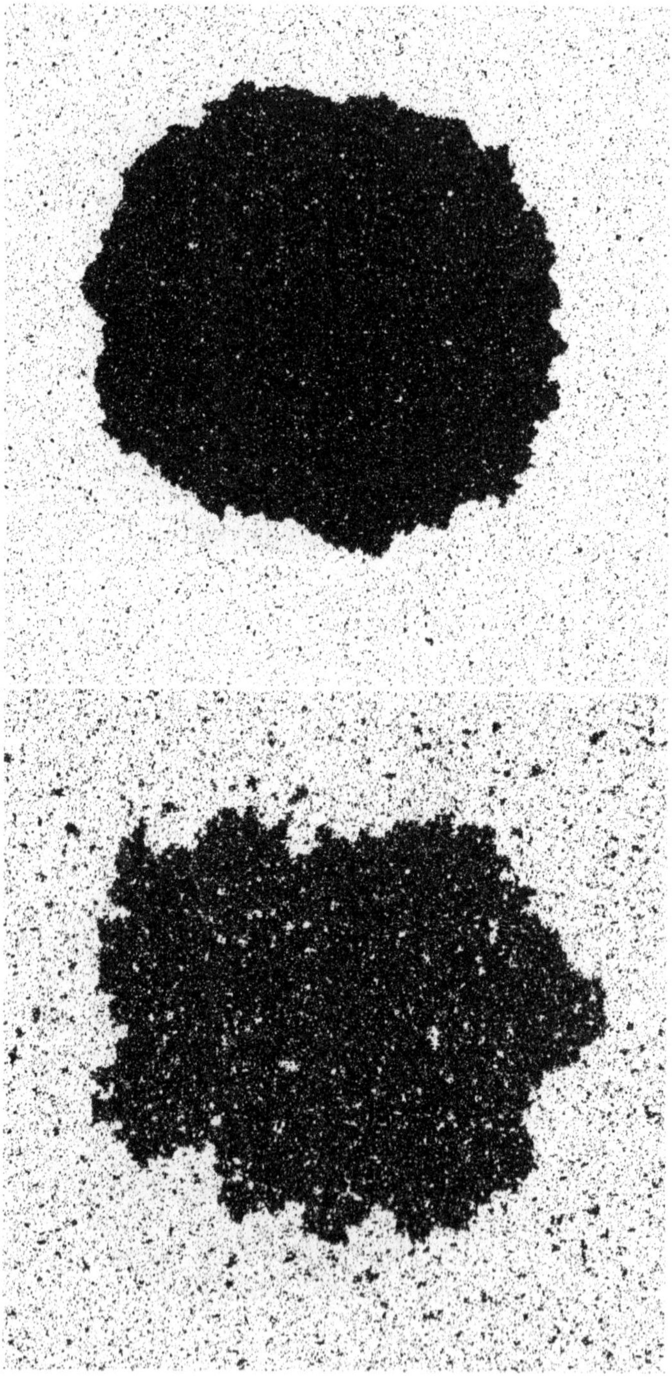

$T = 2$
8 days

$T = 2.22$
7 days

Simulation of the Ising Wulff crystal on a $1024 \times 1024$ grid with a 1 Ghz PC

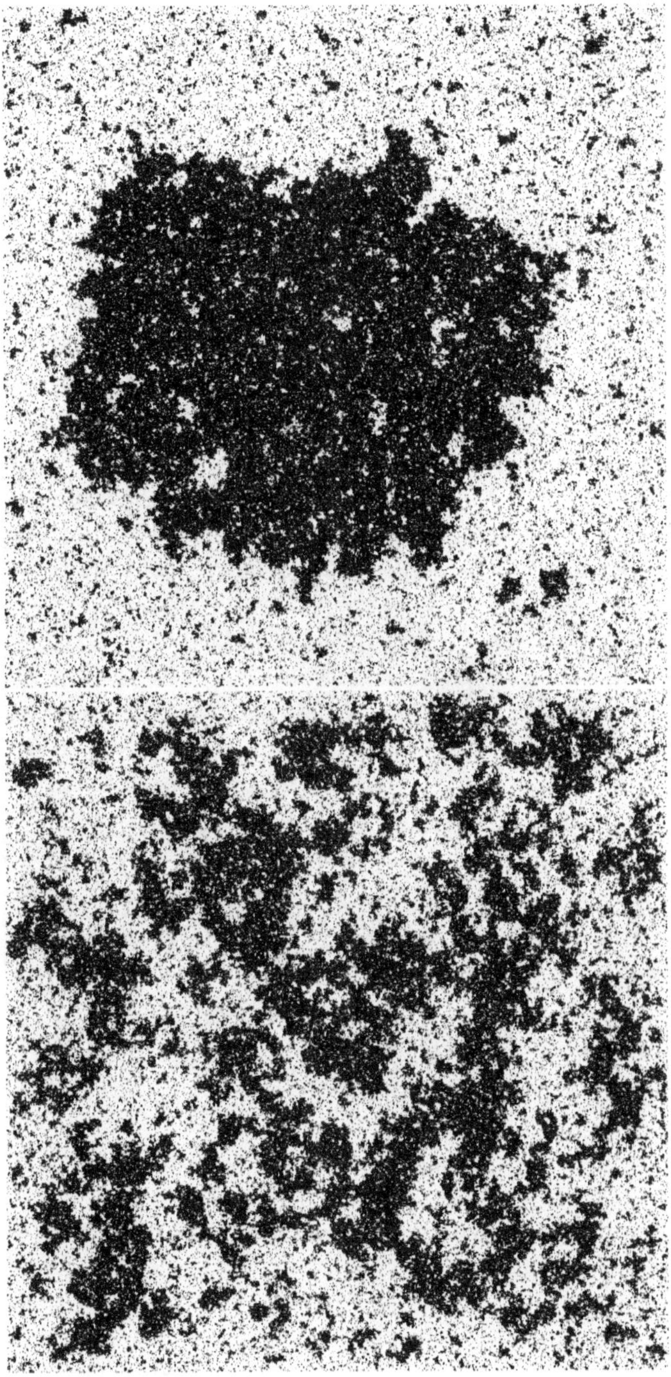

$T = 2.24$
69 days

$T = 2.3$
1 day

Simulation of the Ising Wulff crystal on a $1024 \times 1024$ grid with a 1 Ghz PC

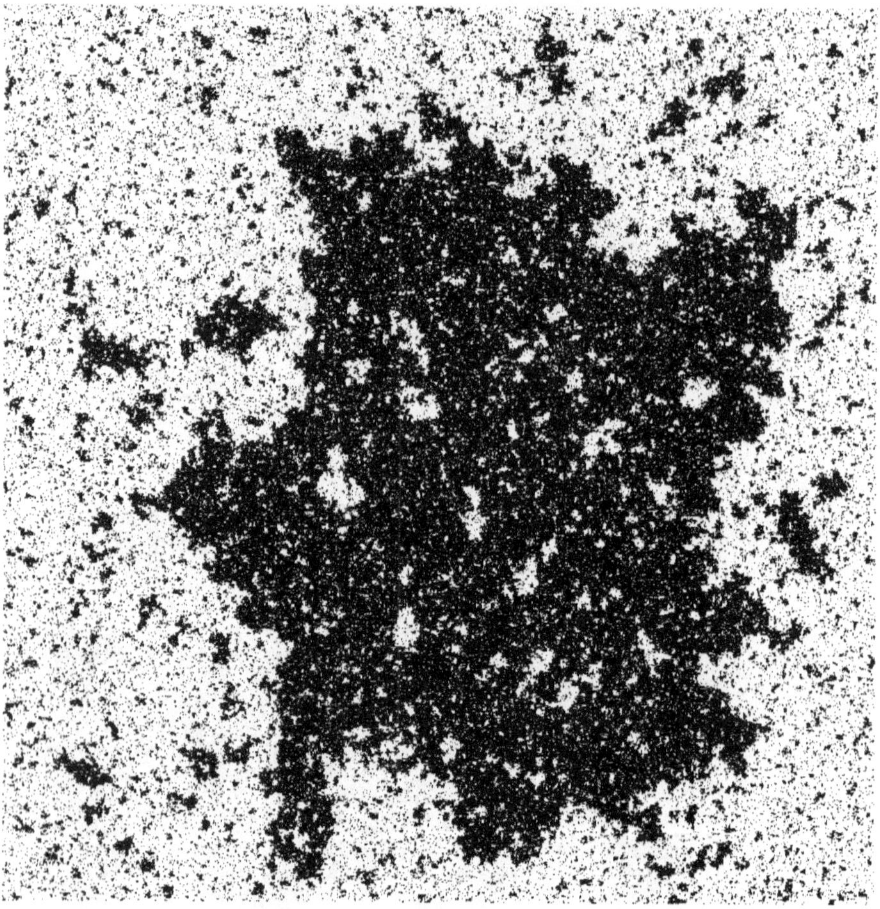

Simulation of the Ising Wulff crystal
on a $1024 \times 1024$ grid, $T = 2.26$
68 days of computation on a 1 Ghz PC

## 5.2 Bernoulli percolation

The counterpart of the phase coexistence regime in the percolation setting is the supercritical phase $p > p_c$. Indeed it is the regime where long range order occurs, due to the presence of an infinite open cluster. Aside from the infinite cluster, the configuration contains finite clusters of arbitrary large sizes. These large finite clusters can be thought of as droplets swimming in the infinite cluster. We wish to understand the typical structure of large finite clusters. The presence at a particular location of a large finite cluster is an event of low probability: for Bernoulli percolation in dimension $d$, for $p > p_c$, there exist two positive constants $c_1, c_2$ such that

$$\forall n \in \mathbb{N} \quad \exp(-c_1 n^{(d-1)/d}) \leq P(n \leq |C(0)| < \infty) \leq \exp(-c_2 n^{(d-1)/d}),$$

where $C(0)$ is the open cluster containing the origin. This is a result from Kesten and Zhang [93]. This estimate is based on the fact that the occurrence of a large finite cluster is due to a surface effect. Indeed at the frontier of a large open cluster, there is a set of closed edges, whose macroscopic components look like a large surface separating the sites of the cluster from the outside world. We first prove that one can put a sharp constant in the previous estimate, i.e., we compute the logarithmic asymptotics of the probability of a large finite cluster at the origin.

**Theorem 5.3.** *Let $d \geq 2$. For any $p > p_c$, the limit*

$$\lim_{n\to\infty} \frac{1}{n^{d-1}} \ln P\big(n^d \leq |C(0)| < \infty\big)$$

*exists and it is a finite strictly negative real number.*

The limit appearing in theorem 5.3 is a rather fascinating quantity. The proof of its existence requires an amazingly huge amount of work. Let us try to estimate the number of pages of research papers needed for a self–contained proof, which would require as a prerequisite for reading only a general knowledge of probability theory and functional analysis. In dimensions 2, it requires two third of [9] plus a geometric proof of the two dimensional Wulff theorem, as done in [53], that is around 50 pages. In dimensions 3, it requires pages 45–116 of [38], Pisztora's estimates [58, 112]; for the range of validity to be $p > p_c$, we must make appeal to [81]. Finally we need the Wulff geometric theorem [72]. In total, that is around 71+55+18+20=164 pages! Of course, if we extract only the minimal amount of information from these research papers needed for our purpose, the total would be much less, but still probably around 100 pages. Indeed, the proof presented here requires the paper [81] and the following sections and chapters of these notes: 8.1, 9, 11 (except 11.2), 12, 13, 14, 15.2, 16, 17. This state of the matter is not satisfactory at all. There must be simpler and more direct ways to obtain this result. Yet a better understanding of the percolation model seems necessary to improve the current outline of the proof.

In the next theorem, we describe the typical structure of the large finite clusters.

**Theorem 5.4.** *Let $d \geq 2$ and let $p > p_c$. There exists a bounded closed convex subset $W$ of $\mathbb{R}^d$ containing 0 in its interior, called the normalized Wulff crystal of the Bernoulli percolation model such that, under the conditional probability $P(\,\cdot\,|\,n^d \leq |C(0)| < \infty)$, as $n$ goes to $\infty$, the random measure*

$$\frac{1}{n^d} \sum_{x \in C(0)} \delta_{x/n}$$

*converges in probability with respect to the uniformly continuous bounded functions towards the set of measures*

$$\left\{ \theta \, 1_W(a+x)\, dx : a \in \mathbb{R}^d \right\}.$$

*The probabilities of the deviations are of order* $\exp -cn^{d-1}$.

Theorems 5.3 and 5.4 were proven in dimensions 2 by Alexander, Chayes and Chayes [9] and in dimensions 3 by Cerf [38]. We will provide here the proof in arbitrary dimensions $d$. In fact, the three dimensional proof goes through with very little changes. However, we present here a slightly different formulation of the results, and several key points have undergone substantial simplifications.

Roughly speaking, theorem 5.4 says that a large finite cluster of cardinality $n^d$ fills a whole region whose shape is close to $nW$; inside this region the density of the cluster is close to $\theta$, and outside it is vanishing. The precise meaning of the last sentence in the statement of theorem 5.4 is the following. Let $P_n$ be the conditional probability

$$P_n(\cdot) = P(\,\cdot\,|\,n^d \leq |C(0)| < \infty)\,.$$

For any $k \geq 1$, any uniformly continuous bounded function $f : \mathbb{R}^d \to \mathbb{R}^k$, any $\delta > 0$, there exist $b, c > 0$ depending on $d, p, f, \delta$ such that, for $n \geq 1$,

$$P_n\left( \exists a \in \mathbb{R}^d \quad \left| \frac{1}{n^d} \sum_{x \in C(0)} f\left(\frac{x}{n}\right) - \int_W f(a+x)\,\theta\, dx \right|_2 \leq \delta \right) \geq 1 - b\, e^{-cn^{d-1}}.$$

The first part of the theorem asserts that for any uniformly continuous bounded function $f$ and any $\delta > 0$, the left–hand quantity goes to 1 as $n$ goes to $\infty$.

*Proof of theorems 5.3 and 5.4.* We first deal with dimensions $d \geq 3$. Theorems 5.3 and 5.4 are consequences of the large deviation principle stated in theorem 7.3, the Wulff theorem 15.1 and the stability result corollary 15.4. The difference compared to theorems 5.1 and 5.2 is that we work in infinite volume and we use the enhanced upper bound of theorem 7.3. Let us define

$$\mathbb{M} = \left\{ \nu \in \mathcal{M}(\mathbb{R}^d) : 1 \leq \nu(\mathbb{R}^d) < +\infty \right\}$$

and let $\mathcal{E}_n$ be the event

$$\mathcal{E}_n = \Big\{ \frac{1}{n^d} \sum_{x\in C(0)} \delta_{x/n} \in \mathbb{M} \Big\}.$$

We apply the bounds given in theorem 7.3:

$$\begin{aligned} -\inf \{ \mathcal{J}(\nu) : \nu \in \overset{\circ}{\mathbb{M}} \} \leq \\ \liminf_{n\to\infty} \frac{1}{n^{d-1}} \ln P\big( \mathcal{E}_n \,\big|\, |C(0)| < \infty \big) \leq \limsup_{n\to\infty} \frac{1}{n^{d-1}} \ln P\big( \mathcal{E}_n \,\big|\, |C(0)| < \infty \big) \\ \leq -\sup_{f,\delta} \inf \big\{ \mathcal{J}(\varrho) : \varrho(\mathbb{R}^d) < +\infty,\, \exists \nu \in \mathbb{M}\ |\varrho(f) - \nu(f)| < \delta \big\} \end{aligned}$$

where the supremum is taken over $\delta > 0$ and the functions $f : \mathbb{R}^d \to \mathbb{R}$ which are bounded and uniformly continuous. Considering the constant function $f$ equal to 1 and using the very definition of the rate function $\mathcal{J}$, we see that the right–hand quantity is bounded above by

$$-\sup_{\delta>0} \inf \Big\{ \mathcal{I}(E) : E \in \mathcal{B}(\mathbb{R}^d),\ \frac{1-\delta}{\theta} \leq \mathcal{L}^d(E) < +\infty \Big\}.$$

For $\delta > 0$, we set $\lambda = ((1-\delta)/\theta)^{-1/d}$. For $E \in \mathcal{B}(\mathbb{R}^d)$ such that $\mathcal{L}^d(E) \geq (1-\delta)/\theta$, we have $\mathcal{L}^d(\lambda E) \geq 1$ and $\mathcal{I}(\lambda E) = \lambda^{d-1}\mathcal{I}(E)$. As $\delta$ goes to 0, the factor $\lambda$ goes to $1/\theta$, thus the above value is in fact equal to

$$-\inf \Big\{ \mathcal{I}(E) : E \in \mathcal{B}(\mathbb{R}^d),\ \frac{1}{\theta} \leq \mathcal{L}^d(E) < +\infty \Big\}.$$

Therefore the large deviation upper and lower bounds are both equal to this value so that the limit appearing in theorem 5.3 exists and, by theorem 15.1, it is equal to the energy of a suitably rescaled Wulff crystal $-\mathcal{I}(W)$.

We finally prove theorem 5.4. Let $f : \mathbb{R}^d \to \mathbb{R}^k$ be a uniformly continuous bounded function, let $\varepsilon > 0$ and let

$$\mathbb{M}(f,\varepsilon) = \Big\{ \nu \in \mathcal{M}(\mathbb{R}^d) : 1 \leq \nu(\mathbb{R}^d) < +\infty, \\ \forall a \in \mathbb{R}^d \quad \Big| \nu(f) - \int_W f(a+x)\,\theta\,dx \Big|_2 \geq \varepsilon \Big\}.$$

Let $\mathcal{F}_n$ be the event

$$\mathcal{F}_n = \Big\{ \frac{1}{n^d} \sum_{x\in C(0)} \delta_{x/n} \in \mathbb{M}(f,\varepsilon) \Big\}.$$

We apply the enhanced upper bound given in theorem 7.3:

$$\begin{aligned} \limsup_{n\to\infty} \frac{1}{n^{d-1}} \ln P\big( \mathcal{F}_n \,\big|\, |C(0)| < \infty \big) \\ \leq -\sup_{g,\delta} \inf \big\{ \mathcal{J}(\varrho) : \varrho(\mathbb{R}^d) < +\infty,\, \exists \nu \in \mathbb{M}(f,\varepsilon) \quad |\varrho(g) - \nu(g)|_2 < \delta \big\} \end{aligned}$$

where the supremum is taken over $\delta > 0$ and the functions $g : \mathbb{R}^d \to \mathbb{R}^{k+1}$ which are bounded and uniformly continuous. Considering the particular function $g = (f, 1)$ in this supremum, we see that the right–hand quantity is bounded above by

$$-\sup_{\delta>0} \inf \Big\{ \mathcal{J}(\varrho) : 1 - \delta \leq \varrho(\mathbb{R}^d) < +\infty, \qquad \forall a \in \mathbb{R}^d \quad \Big| \varrho(f) - \int_W f(a+x)\,\theta\,dx \Big|_2 \geq \varepsilon - \delta \Big\} .$$

If $\varrho$ is such that $\mathcal{J}(\varrho) < \infty$, then there exists a Borel subset $E$ of $\mathbb{R}^d$ such that $\varrho$ is the measure with density $\theta 1_E$; in this case we have

$$\forall a \in \mathbb{R}^d \qquad \Big| \varrho(f) - \int_W f(a+x)\,\theta\,dx \Big|_2 \leq \| |f|_2 \|_\infty \theta \mathcal{L}^d(E \Delta (a+W)) .$$

Therefore the previous quantity is bounded above by

$$-\sup_{\delta>0} \inf \Big\{ \mathcal{I}(E) : E \in \mathcal{B}(\mathbb{R}^d),\ \frac{1-\delta}{\theta} \leq \mathcal{L}^d(E) < +\infty \qquad \forall a \in \mathbb{R}^d \quad \mathcal{L}^d(E \Delta (a+W)) \geq \frac{\varepsilon - \delta}{\| |f|_2 \|_\infty \theta} \Big\} .$$

The infimum is a constrained isoperimetric problem, where the constraint prevents the translates of the Wulff crystal $W$ from being admissible candidates. Corollary 15.4 implies that this quantity is strictly less than $-\mathcal{I}(W)$. We conclude that

$$\limsup_{n\to\infty} \frac{1}{n^{d-1}} \ln P\big(\mathcal{F}_n \,\big|\, \mathcal{E}_n\big) < 0 .$$

To handle the two dimensional case with exactly the same strategy, one should rely on a different large deviation principle, as explained after theorem 7.3. However the bounds provided in theorem 7.4 are enough to prove theorem 5.3: the large deviation upper bound holds as before, and we simply apply the large deviation lower bound with the measure $\nu = \theta 1_{\overset{\circ}{W}}$. Theorem 5.4 is simpler in two dimensions, because the distance and the perimeter obey the same scaling: in particular, the probability to have a finite open cluster of diameter $n$ is of order $\exp -cn$, with $c > 0$, and it is unlikely that the Wulff crystal created by $C(0)$ is far from 0. □

## 5.3 FK percolation

The results for the FK model are valid when $p$ is larger than a threshold which is conjectured to coincide with the critical point. This threshold is defined differently in dimensions $d = 2$ and in dimensions $d \geq 3$.

In dimensions $d = 2$, the results hold as long as there is exponential decay for the dual connections. Let $q \geq 1$. As in [80], we define

$$p_g = \sup \left\{ p : \exists c > 0 \quad \forall x, y \in \mathbb{Z}^2 \quad \Phi^{p,q}_{\infty}(x \leftrightarrow y) \leq \exp(-c|x-y|) \right\}.$$

By the results of [82], it is known that exponential decay occurs as soon as the connectivities decay at a sufficient polynomial rate. Following [51], we introduce the point $\widehat{p}_g$ dual to $p_g$:

$$\widehat{p}_g = \frac{q(1-p_g)}{p_g + q(1-p_g)} \geq p_c(q).$$

The point $\widehat{p}_g$ is conjectured to agree with the critical point $p_c(q)$.

In dimensions $d \geq 3$, the results hold as long as there is percolation in sufficiently thick slabs. We introduce next the corresponding critical point. Let $q \geq 1$ and let $d \geq 3$. Let $S(n,L) = [-n,n]^{d-1} \times [1,L]$. Let $\Phi^{p,q}_{n,L}$ be the FK measure on $S(n,L)$ with parameters $p,q$ and free boundary conditions. Let $\Pi(p,L)$ be the property:

$$\exists \alpha > 0 \qquad \forall n \geq 1 \quad \forall x, y \in S(n,L) \quad \Phi^{p,q}_{n,L}(x \longleftrightarrow y) > \alpha.$$

This property is monotonic: if $p \leq p'$ and $L \leq L'$ then $\Pi(p,L) \Rightarrow \Pi(p',L')$. We define the slab percolation threshold

$$\widehat{p}_c(q) = \lim_{L \to \infty} \inf \left\{ p : \Pi(p,L) \text{ occurs} \right\}.$$

Certainly $p_c(q) \leq \widehat{p}_c(q) < 1$. In order to unify the notation, we set $\widehat{p}_c(q) = \widehat{p}_g$ in dimensions $d = 2$.

Next we state the FK percolation counterpart of theorem 5.2. Recall that for $A \subset \mathbb{Z}^d$, we denote by $\partial^{in} A$ the sites of $A$ which have a nearest neighbour in $\mathbb{Z}^d \setminus A$ and we define

$$\varrho(A) = \left| \{ x \in A : x \longleftrightarrow \partial^{in} A \} \right|.$$

For $q \geq 1$ and $p \in [0,1] \setminus \mathcal{U}(q)$, we denote by $\theta$ the percolation probability corresponding to the parameters $p,q$ (see theorem 4.2 and thereafter for the definitions of $\mathcal{U}(q)$ and $\theta$).

**Theorem 5.5.** *Let $d \geq 2$, let $q \geq 1$ and let $p \in ]\widehat{p}_c, 1[$ be such that $p \notin \mathcal{U}(q)$. There exists a bounded, closed, convex set $\mathcal{W}$ containing 0 in its interior, called the Wulff crystal of the FK model, such that the following holds. Let $\alpha < \theta$ be close enough to $\theta$ so that the rescaled Wulff crystal*

$$\mathcal{W}(\alpha) = \left( \frac{\theta - \alpha}{\theta} \right)^{\frac{1}{d}} \frac{\mathcal{W}}{\mathcal{L}^d(\mathcal{W})^{\frac{1}{d}}}$$

*fits into the unit cube $[-1/2, 1/2]^d$. Then*

$$\lim_{n \to \infty} \frac{1}{n^{d-1}} \ln \Phi^{w}_{\Lambda(n)} \Big( \frac{\varrho(\Lambda(n))}{|\Lambda(n)|} \leq \alpha \Big) = -d \left( \frac{\theta - \alpha}{\theta} \right)^{\frac{d-1}{d}} \mathcal{L}^d(\mathcal{W})^{\frac{1}{d}}.$$

The proof of this theorem rests on the large deviation principle of theorem 7.5 in dimensions $d \geq 3$ and the Wulff isoperimetric theorem 15.1; it is analogous to the proof of theorem 5.2. In two dimensions, the large deviation bounds provided in theorem 7.6 are enough to complete the proof. Theorem 5.5 and the volume large deviation principle of theorem 8.3 provide a complete picture of the large deviation behavior of $\varrho(\Lambda(n))/|\Lambda(n)|$. Theorems 5.1, 5.3, 5.4 have also their counterparts in the FK setting. In fact, theorems 5.3 and 5.4 hold for $p$ in the uniqueness region $]\widehat{p}_c, 1] \setminus \mathcal{U}(q)$, with the infinite volume FK measure $\Phi^{p,q}_\infty$ instead of $P$. The points of the proofs which require some adaptation are discussed in section 17.4. For instance, we have

$$\lim_{n\to\infty} \frac{1}{n^{d-1}} \ln \Phi^{p,q}_\infty\big(n^d \leq |C(0)| < \infty\big) = -d\,\mathcal{L}^d(\mathcal{W}) = -d\,\mathcal{L}^d(\mathcal{W})\,\theta^{-1+1/d}.$$

For theorem 5.1, the event under consideration should be interpreted in terms of percolation clusters in order to get a FK version.

## 5.4 What do we know about the Wulff crystal?

Very little is known about the Wulff crystal itself. For $q \geq 1$ and $p$ close to 1, the diameter of the Wulff crystal $\mathcal{W}$ is of order $-\ln(1-p)$, which tends to infinity as $p$ goes to 1. In dimensions 3, it is possible to obtain a first order expansion of the shape of the Wulff crystal of the FK model with the help of combinatorial computations involving domino tiling [41].

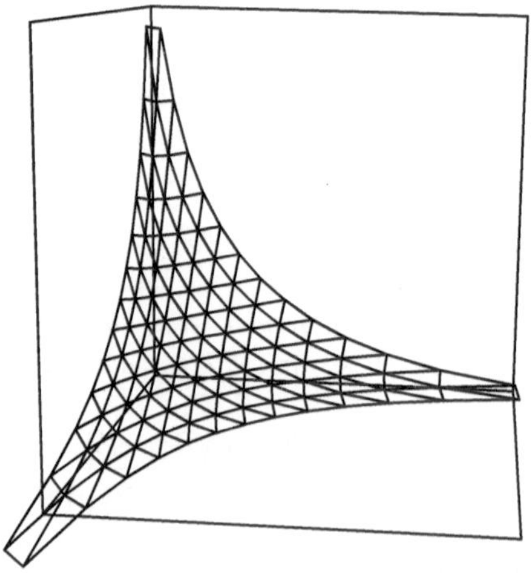

The limiting shape of the corner of the Wulff crystal as $p \to 1$

**Theorem 5.6.** *Let $\varepsilon = -1/\ln(1-p)$ and let $T_\varepsilon$ be the map*

$$T_\varepsilon(x) = \frac{1}{\varepsilon}(1,1,1) - x\,.$$

*As $p$ goes to 1, the boundary of the corner of $T_\varepsilon(\mathcal{W})$ converges to the surface*

$$\Big\{ \big(f(A,B,C) - \ln A, f(A,B,C) - \ln B, f(A,B,C) - \ln C\big) : A,B,C > 0 \Big\}\,,$$

*where for $A,B,C$ positive,*

$$f(A,B,C) = \frac{1}{4\pi^2}\int_{[0,2\pi]}\int_{[0,2\pi]} \ln|A + Be^{iu} + Ce^{iv}|\,du\,dv\,.$$

In dimensions $d = 2$, it is possible to obtain an explicit formula for the shape of the Wulff crystal (see section 11.5). As $T$ goes to $T_c$ from below, the Wulff crystal becomes a circle. In the joint limit $(n,T) \to (\infty, T_c)$, in an adequate regime, the Ising model creates then a circular droplet [42].

There are a lot of challenging open questions related to the surface tension and the Wulff crystal. A major problem is to understand the fluctuations of the interfaces. Dobrushin [61] proved that at low temperatures, interfaces perpendicular to the axis directions are localized (see also [83] for the extension to FK percolation). It is conjectured that in dimensions $d = 3$ there exists a temperature $T_R < T_c$ where a so–called "roughening" transition occurs: the nature of the fluctuations undergoes a fundamental change for $T > T_R$. As far as we know, there is still no proof or disproof for the existence of this transition in dimensions $d = 3$. It is also expected that the facets of the Wulff crystal should disappear at $T_R$ (see [106]). Bodineau, Schonmann and Shlosman investigate the question of the flatness of the Wulff crystal [31]. Note that theorem 5.6 does not give any control on the size of the facets of the true crystal. This circle of questions is also linked to the delicate problem of the sharp large deviations: can we obtain a more precise expansion of the probability of a deviation for the empirical magnetization? Refined estimates are available in dimensions two for the Ising model [62, 63, 89] and for Bernoulli percolation [5]. Recently, the correct exponent describing the fluctuations of the random curve around the Wulff crystal in the FK model has been obtained rigorously [6, 10].

Another very interesting topic is of course the dynamics. One should try to understand the dynamical mechanism leading to the creation of a Wulff crystal. It is naturally expected that several reasonable choices of microscopic dynamics induce the macroscopic dynamics associated to the motion by mean curvature. For instance it seems to be the case for the non–conservative Glauber dynamics, which is relevant for the beautiful theory of metastability (see [117, 120]). In his last years, De Giorgi became interested in the problem of the geometric evolution of Caccioppoli sets and he introduced the concept of minimal barriers. Another available tool is the method of the auxiliary function and the use of viscosity solutions (see [52]). Besides the Wulff crystal, there are numerous issues of the phenomenological theory of phase coexistence that one should try to analyze from a microscopic point of view (see for instance the book of Visintin [128]).

## 5.5 Bibliographical comments

The first mathematical proof of phase separation was achieved by Dobrushin, Kotecký and Shlosman [63] in the context of the two dimensional Ising model fifteen years ago. The main tool of their work is the cluster expansion, which, on the one hand allowed the derivation of results much finer than necessary to verify the Wulff construction, but on the other hand restricted the validity of the results to two dimensions and to low temperatures. Pfister [109] simplified the proof with the help of duality arguments. Alexander, Chayes and Chayes [9] have proved the Wulff construction in the entire supercritical phase of two dimensional Bernoulli percolation. Alexander [5] subsequently refined the probabilistic estimates. By using Pfister's approach and certain coarse-graining techniques from [112], Ioffe [90, 91] extended the basic large deviation principle for the magnetization up to the critical temperature. Finally, Ioffe and Schonmann [92] extended the results of [63] up to $T_c$. Several other works investigated the phase separation phenomenon in dimension 2: [45, 62, 89, 111, 118, 119, 120]. There was a bundle of important results for various models in dimensions higher than 3 which indicated the occurrence of a phase separation phenomenon [4, 16, 17, 18, 58, 93, 112].

As far as we understand, the 2D results on the Ising model relied at some point on the strict convexity of the Wulff shape, which at that time could be proved only with the help of Onsager's famous exact computation. No such ingredient was used in the percolation setting [9]. This fact, plus the results of Kesten–Zhang [93] and Pisztora [112] were conclusive clues indicating that the problem was to be solved first in the percolation setting, rather than directly in the Ising model. The work [37] was a preliminary try before attacking the problem in dimension three: the results for 2D percolation were reformulated in a weak form with the help of a large deviation principle. This approach prompted further 2D geometric investigations [39].

The first precise study of phase separation for a truly microscopic model in dimension 3 was achieved for the Bernoulli percolation model [38]. It was then quickly extended to the Ising model in two parallel partially overlapping works: on the one hand in [23] which relies on [17, 38, 112] and is limited to sufficiently low temperatures, and on the other hand in [43] which relies on [38, 112] and is valid until the slab percolation threshold $\widehat{p}_c(2)$ for temperatures such that the associated infinite volume FK measure is unique. Whenever $p > \widehat{p}_c(q)$ in the FK model with parameter $q$, slab arguments can be successfully put into action. A lot of results describing the supercritical phase of Bernoulli percolation were first derived under this assumption. Grimmett and Marstrand [81] proved that $\widehat{p}_c(1) = p_c(1)$. Recently, Bodineau [26, 27] has proved that $\widehat{p}_c(2) = p_c(2)$ and that for $p > p_c(2)$, there exists a unique infinite volume FK measure corresponding to the parameters $p, q = 2$. Therefore the results of [43] hold for any subcritical temperature. A challenging conjecture is to prove the equality $\widehat{p}_c(q) = p_c(q)$ for any $q \geq 1$, as well as the uniqueness of the infinite volume FK measure for $p \neq p_c(q)$. A fundamental contribution of Pisztora's work [112] is to transfer the slab technology from Bernoulli percolation to the FK model and hence to the Ising and Potts models via the coupling described in section 4.3. All the results relying on Pisztora's coarse graining estimates will be valid in the region $q \geq 1$, $p > \widehat{p}_c(q)$, $p \notin \mathcal{U}(q)$. The slab technology is useful only in dimensions larger than or equal to 3. Specific arguments can be developed to get similar results in two dimensions. See [51] for the 2D analogue of Pisztora's coarse graining results.

In a subsequent work [44], the phenomenon of phase separation was handled in a more general framework, namely, general large deviation principles and weak laws of large numbers for the Ising, Potts and FK percolation models were proven in a rather general domain with general boundary conditions. The Kac Ising model was further investigated in [24, 25]. Further developments appeared in [29, 30].

The analysis of the phase coexistence phenomenon in dimensions higher than 3 requires a completely new strategy compared to the two dimensional proofs. First, a change of topology in the space of the shapes is needed in order to formulate adequately the result itself. Second, the two dimensional proofs [9, 63, 90, 91, 92, 109] were all relying on a polygonalization procedure (called the skeleton) to approximate the shapes of the interfaces, together with a combinatorial bound on the number of admissible skeletons. Nobody has succeeded yet in designing an analogous procedure in dimension 3. The leading idea of the work [38] was to embed from the very beginning the analysis in a continuous world in order to replace the combinatorial argument, whose purpose is to control the explosion of the possible shapes for the droplet, by a compactness argument. In other terms, one should restate the problem in the correct variational framework.

Although the solution of the classical isoperimetric problem has been known since ancient times, it was formulated in a very satisfactory setting only fifty years ago: here, by satisfactory, we mean a setting where the space of the candidate solutions is endowed with a convenient topology having the required closure and semi–continuity properties so that the direct methods of the calculus of variations can be applied [76]. To that aim, it is necessary to extend the notion of perimeter or surface area to non regular sets. This was achieved by Caccioppoli [35, 36] and developed afterwards by De Giorgi [54, 55, 56]. Hence, at the geometric level, we use the nowadays standard tools of geometric measure theory in codimension 1 [67, 72, 73, 77, 100, 124].

Let us point out that the idea to use these tools is also present in a series of works in the context of the Ising model with Kac potentials (a model which is intermediate between the nearest–neighbour Ising model and a mean–field model). Alberti, Bellettini, Cassandro, Presutti [4, 16] developed the general philosophy of embedding the problem in a continuous setting in order to use the BV framework (the equivalent functional formulation of the sets of finite perimeter). The idea to look at the locally averaged magnetization was originally introduced in [4]. Enhanced results were obtained in [17, 18], but still in some limit where the range of interactions goes to infinity.

The proofs we present in these notes are simplified compared to the research papers [23, 28, 38, 43, 44]. Indeed, this sequence of works relies directly on the coarse–graining estimates of Pisztora, which describe accurately the typical configurations on a scale intermediate between the microscopic and the macroscopic scale. These coarse–graining estimates are done with the help of the FK representation of the Ising model and a general renormalization procedure [112]. This renormalization scheme uses a powerful stochastic domination inequality which later has been generalized [98], and the final estimates of Pisztora are obtained thanks to a subtle discrete isoperimetric inequality [58]. Yet the above quoted research papers [23, 28, 38, 43, 44] include themselves a second renormalization step. On the whole, these proofs perform two renormalization steps, which is a bit odd. The argument we present here requires only a single renormalization step and it does not make appeal to the stochastic domination inequality used by Pisztora, instead simple Peierls counting arguments enable us to do the job. Moreover the initial coarse–graining estimates are simpler and weaker than those of Pisztora.

Let us finish by mentioning the most recent works around the Wulff crystal. The low temperature expansion of the 3D Wulff crystal is computed in [41]. Alexander succeeded recently in deriving cube root fluctuations of the random curve around the Wulff crystal in the FK model in two dimensions [6, 10]. Couronné and Messikh provide a two dimensional version of Pisztora's coarse graining estimates [51]. Messikh analyzes the phase coexistence phenomenon in the 2D Ising model close to criticality: the Wulff crystal becomes then a circle [104, 42] (see also [105] for an application to image segmentation). Couronné has shown that the statistical repartition of the large finite clusters in the FK percolation model can be approximated by a Poisson process [50]. He has also studied the Wulff crystal for oriented percolation in dimensions $d \geq 3$ [48]. In this model, the Wulff crystal has a singular point. Bodineau, Schonmann and Shlosman investigate the question of the flatness of the Wulff crystal [31]. Biskup, Chayes and Kotecky study the formation/dissolution of equilibrium droplets in the context of the 2D Ising model [20, 21] and they derive the Gibbs–Thomson formula in the droplet formation regime [22]. Alexander, Biskup and Chayes devise an Ising–based model of a solvent–solute system [7, 8] and they study the associated phase separation phenomenon.

# Part IV

# Large deviation principles

# 6

# Large deviation theory

The connection between the probabilistic world and the realm of the calculus of variations is done through the language of large deviations. In this chapter, we give the basic definitions and some general results from the theory of large deviations.

## 6.1 Main definitions

Let $(X, \mathcal{O})$ be a regular topological space; regular means that for any $x \in X$ and any closed subset $F$ of $X$ such that $x \notin F$, there exist two disjoint open sets $U, V$ such that $x \in U$ and $F \subset V$. We suppose that $X$ is also endowed with a $\sigma$–field $\mathcal{B}$ satisfying the following hypothesis.

**Hypothesis.** Each element $x \in X$ admits a basis of neighbourhoods consisting of open measurable sets:

$$\forall x \in X \quad \forall O \in \mathcal{O} \quad x \in O \quad \Longrightarrow \quad \exists V \in \mathcal{B} \cap \mathcal{O} \quad x \in V \subset O\,.$$

**Definition 6.1.** *A rate function on $(X, \mathcal{O})$ is a lower semicontinuous map $I : X \to \mathbb{R}^+ \cup \{ +\infty \}$. A rate function $I$ is said to be good or coercive if its level sets $\{ x \in X : I(x) \leq \lambda \}$, $\lambda \in \mathbb{R}^+$, are compact.*

Let $(\mu_n)_{n\geq 1}$ be a sequence of probability measures defined on the $\sigma$–field $\mathcal{B}$. The sequence $(\mu_n)_{n\geq 1}$ satisfies a large deviation principle with speed $n$ governed by the rate function $I$ and with respect to the topology $\mathcal{O}$ if for any $A \in \mathcal{B}$

$$-\inf \{ I(x) : x \in \mathring{A} \} \leq \liminf_{n\to\infty} \frac{1}{n} \ln \mu_n(A) \leq \limsup_{n\to\infty} \frac{1}{n} \ln \mu_n(A) \leq -\inf \{ I(x) : x \in \overline{A} \}\,.$$

A large deviation principle provides asymptotic estimates for the probability of certain events. Let $(X_n)_{n\geq 1}$ be a sequence of random variables taking their values in $(X, \mathcal{B})$. We say that the sequence $(X_n)_{n\geq 1}$ satisfies a large deviation principle if the sequence of their laws does so. Suppose that $(X_n)_{n\geq 1}$ satisfies a large deviation principle governed by a rate function $I$ admitting a unique global minimum at $x^*$. Necessarily, $I(x^*) = 0$. For any measurable neighbourhood $U$ of $x^*$, we have $\inf_{X\setminus U} I > 0$ and the large deviation principle implies then

$$\lim_{n\to\infty} P(X_n \in U) = 1\,,$$

which is a genuine law of large numbers. Let us point out that a large deviation principle depends heavily upon the topology: the same sequence of probability measures might satisfy several large deviation principles associated to different topologies. The finer the topology is, the more informative is the large deviation principle.

A set $A \in \mathcal{B}$ is said to be a continuity set for $I$ if

$$\inf\{\, I(x) : x \in \overset{\circ}{A}\,\} = \inf\{\, I(x) : x \in \overline{A}\,\}\,.$$

For such a set, we have a true limit:

$$\lim_{n\to\infty} \frac{1}{n} \ln \mu_n(A) = -\inf\{\, I(x) : x \in A\,\}\,.$$

The main interest of this abstract formulation of large deviation principles lies in its generality and robustness. Indeed, in a lot of very diverse problems, asymptotic estimates can be expressed conveniently with the help of a large deviation principle. Often the proof of a large deviation principle is done in two distinct steps. One proves the lower bound on the one hand and the upper bound on the other hand, usually with very different arguments. In addition it is advisable to prove first the upper bound for compact sets.

**Definition 6.2.** *A sequence of probability measures $(\mu_n)_{n\geq 1}$ defined on the $\sigma$–field $\mathcal{B}$ satisfies a weak large deviation principle with speed $n$ governed by the rate function $I$ and with respect to the topology $\mathcal{O}$ if*

- $\forall A \in \mathcal{B} \qquad -\inf\{\, I(x) : x \in \overset{o}{A}\,\} \leq \liminf_{n\to\infty} \dfrac{1}{n} \ln \mu_n(A)$
- $\forall A \in \mathcal{B}$, $\overline{A}$ *compact,* $\qquad \limsup_{n\to\infty} \dfrac{1}{n} \ln \mu_n(A) \leq -\inf\{\, I(x) : x \in \overline{A}\,\}\,.$

## 6.2 $I$–tightness

Once a weak large deviation principle is proved, the main problem to get the full large deviation principle is to prove that, on the correct exponential scale, all the probability mass concentrates on the neighbourhood of compact sets.

We suppose throughout the remainder of this section that the space $X$ is a topological vector space. This means that $X$ is a vector space endowed with a topology such that every point of $X$ is a closed set and the vector space operations are continuous [115]. The topology is then invariant by translation and it is conveniently described by a basis of neighbourhoods of the origin. We denote such a basis by $\mathcal{U}$. If $x \in X$, the collection $x+U$, $U \in \mathcal{U}$, is a basis of neighbourhoods for $x$. The large deviation lower bound is equivalent to the following statement:

$$\forall x \in X \quad \forall U \in \mathcal{U} \qquad \liminf_{n\to\infty} \frac{1}{n} \ln P\big(X_n \in x+U\big) \geq -I(x).$$

We always use the above statement to prove the large deviation lower bound. If $A, U$ are two subsets of $X$, their sum $A+U$ is the set $\{\, a+u : a \in A,\ u \in U\,\}$.

**Definition 6.3.** *We say that the sequence of random variables $(X_n)_{n\in\mathbb{N}}$ is $I$–tight if there exist positive constants $c, \lambda_0$ such that*

$$\forall \lambda \geq \lambda_0 \quad \forall U \in \mathcal{U} \qquad \limsup_{n\to\infty} \frac{1}{n} \ln P\big(X_n \notin I^{-1}([0,\lambda]) + U\big) \leq -c\lambda.$$

Our proofs of the large deviation upper bound rely on the following formulation.

**Lemma 6.4.** *Suppose that the rate function $I$ is good. The sequence of random variables $(X_n)_{n\in\mathbb{N}}$ satisfies the large deviation upper bound if and only if it is $I$–tight and it satisfies the local estimate*

$$\forall x \in X, \quad I(x) < \infty, \quad \forall \varepsilon > 0, \quad \exists U = U(x,\varepsilon) \in \mathcal{U}$$
$$\limsup_{n\to\infty} \frac{1}{n} \ln P\big(X_n \in x+U\big) \leq -I(x)(1-\varepsilon).$$

*Remark 6.5.* In the situation where the rate function is not good (that is, it is only lower semicontinuous), we have the following result: If the sequence of random variables $(X_n)_{n\in\mathbb{N}}$ is $I$–tight and satisfies the local estimate, then the weak large deviation upper bound holds.

*Proof.* The conditions stated in the lemma are direct consequences of the large deviation upper bound, should it hold. We prove next that these conditions imply the large deviation upper bound. Let $A$ be a closed subset of $X$. Let $\varepsilon > 0$ and let $V$ be a symmetric neighbourhood of the origin (that is, $V = -V$). Let $\lambda \geq \lambda_0$. Since $I$ is a good rate function, then the set $I^{-1}([0,\lambda])$ is compact. To each $x$ in $I^{-1}([0,\lambda])$ we associate a neighbourhood $U = U(x,\varepsilon)$ of the origin as given by the local estimate. We can also impose that $U(x,\varepsilon) \subset V$. The collection of sets

$$x + U(x,\varepsilon), \quad x \in I^{-1}([0,\lambda])$$

is an open covering of $I^{-1}([0,\lambda])$, from which we can extract a finite subcover associated to a finite number of points $x_i$, $1 \leq i \leq r$. The set

$$\bigcup_{1\le i\le r} x_i + U(x_i,\varepsilon)$$

is an open set containing the compact set $I^{-1}([0,\lambda])$. We will use next the following classical result of the theory of topological vector spaces (see for instance [115], Theorem 1.10).

**Lemma 6.6.** *Let $K$ and $C$ be two subsets of $X$, such that $K$ is compact and $C$ is closed. If $K\cap C=\emptyset$, then there exists a neighbourhood $U$ of $0$ such that $(K+U)\cap(C+U)=\emptyset$.*

*Proof.* For completeness, we include the proof. Let $V\in\mathcal{U}$. From the continuity of the vector addition, there exist $U_1,U_2\in\mathcal{U}$ such that $U_1+U_2\subset V$. Setting $U=U_1\cap U_2\cap(-U_1)\cap(-U_2)$ we get a symmetric neighbourhood of the origin (i.e., $U=-U$) such that $U+U\subset V$. Applying this result again, we get $U\in\mathcal{U}$ symmetric such that $U+U+U+U\subset V$. Thus, for any $x\in K$, there exists $U_x\in\mathcal{U}$ symmetric such that $x+U_x+U_x+U_x\subset X\setminus C$, whence $(x+U_x+U_x)\cap(C+U_x)=\emptyset$. The collection of sets $x+U_x$, $x\in K$, is an open covering of $K$, from which we can extract a finite subcover associated to a finite number of points $x_i$, $1\le i\le r$. Let $U=\bigcap_{1\le i\le r}U_{x_i}$. Then

$$K+U\subset\bigcup_{1\le i\le r}(x_i+U_{x_i}+U)\subset\bigcup_{1\le i\le r}(x_i+U_{x_i}+U_{x_i})$$

and the last union does not intersect $C+U$. □

From lemma 6.6, we know that there exists $U\in\mathcal{U}$ such that

$$I^{-1}([0,\lambda])+U\subset\bigcup_{1\le i\le r}x_i+U(x_i,\varepsilon)\,.$$

We write

$$\begin{aligned}P(X_n\in A)&\le P\big(X_n\in A\cap(I^{-1}([0,\lambda])+U)\big)+P\big(X_n\notin I^{-1}([0,\lambda])+U\big)\\ &\le\sum_{1\le i\le r}P\big(X_n\in A\cap(x_i+U(x_i,\varepsilon))\big)+P\big(X_n\notin I^{-1}([0,\lambda])+U\big)\\ &\le\sum_{\substack{1\le i\le r\\ x_i\in A-V}}P\big(X_n\in x_i+U(x_i,\varepsilon)\big)+P\big(X_n\notin I^{-1}([0,\lambda])+U\big)\,.\end{aligned}$$

Since $V$ is symmetric, then $A-V=A+V$. By the definition of the sets $U(x_i,\varepsilon)$, $1\le i\le r$, using the local estimate, the $I$–tightness of the sequence and the previous inequalities, we get

$$\begin{aligned}\limsup_{n\to\infty}\frac{1}{n}\ln P\big(X_n\in A\big)&\le-\min\Big(\min_{\substack{1\le i\le r\\ x_i\in A+V}}I(x_i)(1-\varepsilon)\,,\;c\lambda\Big)\\ &\le-\min\Big(\inf_{x\in A+V}I(x)(1-\varepsilon)\,,\;c\lambda\Big)\,.\end{aligned}$$

We send successively $\lambda$ to $\infty$, $\varepsilon$ to 0 and we optimize over $V$; we get

$$\limsup_{n\to\infty} \frac{1}{n} \ln P\big(X_n \in A\big) \leq -\sup_{V\in\mathcal{V}} \inf\{\,I(x) : x \in A+V\,\},$$

where $\mathcal{V}$ is the collection of the symmetric neighborhoods of 0. Let us set

$$\alpha = \inf\{\,I(x) : x \in \overline{A}\,\}$$

and let $\beta < \alpha$. Then $\{\,x \in X : I(x) \leq \beta\,\} \cap \overline{A} = \emptyset$. Yet $\overline{A} = \bigcap_{V\in\mathcal{V}} \overline{A+V}$. Since $I$ is a good rate function, there exists $V \in \mathcal{V}$ such that

$$\{\,x \in X : I(x) \leq \beta\,\} \cap \overline{A+V} = \emptyset,$$

whence $\inf\{\,I(x) : x \in A+V\,\} \geq \beta$. This being true for any $\beta < \alpha$, we conclude that

$$\sup_{V\in\mathcal{V}} \inf\{\,I(x) : x \in A+V\,\} \geq \alpha.$$

The large deviation upper bound follows from the previous inequalities. □

In the last steps of the previous proof, we have used the following little lemma, which appears repeatedly in large deviation arguments.

**Lemma 6.7.** *Let $f_1(\varepsilon), \dots, f_r(\varepsilon)$ be $r$ non–negative functions defined on $]0,1[$. Then*

$$\limsup_{\varepsilon\to 0} \varepsilon \ln\Big(\sum_{1\leq i\leq r} f_i(\varepsilon)\Big) = \max_{1\leq i\leq r} \limsup_{\varepsilon\to 0} \varepsilon \ln f_i(\varepsilon).$$

*Proof.* Obviously,

$$\max_{1\leq i\leq r} f_i(\varepsilon) \leq \sum_{1\leq i\leq r} f_i(\varepsilon) \leq r \max_{1\leq i\leq r} f_i(\varepsilon).$$

Taking ln, multiplying by $\varepsilon$ and sending $\varepsilon$ to 0, we see that

$$\begin{aligned}
\limsup_{\varepsilon\to 0} \varepsilon \ln\Big(\sum_{1\leq i\leq r} f_i(\varepsilon)\Big) &= \limsup_{\varepsilon\to 0} \varepsilon \ln \max_{1\leq i\leq r} f_i(\varepsilon) \\
&= \inf_{\varepsilon>0} \sup_{\delta\in]0,\varepsilon[} \delta \ln \max_{1\leq i\leq r} f_i(\delta) = \inf_{\varepsilon>0} \max_{1\leq i\leq r} \sup_{\delta\in]0,\varepsilon[} \delta \ln f_i(\delta) \\
&= \max_{1\leq i\leq r} \inf_{\varepsilon>0} \sup_{\delta\in]0,\varepsilon[} \delta \ln f_i(\delta) = \max_{1\leq i\leq r} \limsup_{\varepsilon\to 0} \varepsilon \ln f_i(\varepsilon).
\end{aligned}$$

We can exchange inf and max because we optimize over a finite set. □

## 6.3 Contraction principle

Large deviation principles possess a fundamental stability property: they are preserved by continuous functions. Let $(X, \mathcal{O}_X)$ and $(Y, \mathcal{O}_Y)$ be two regular topological spaces. We suppose also that $X$ and $Y$ are both endowed with two $\sigma$–fields $\mathcal{B}_X$ and $\mathcal{B}_Y$ satisfying our usual assumption (see beginning of section 6.1). Let $f : X \to Y$ be a continuous map, measurable with respect to the $\sigma$–fields $\mathcal{B}_X$ and $\mathcal{B}_Y$. Let $(\mu_n)_{n\geq 1}$ be a sequence of probability measures on $\mathcal{B}_X$ and for $n \geq 1$, let us denote by $\nu_n$ the image measure of $\mu_n$ through $f$, that is the measure on $\mathcal{B}_Y$ defined by

$$\forall B \in \mathcal{B}_Y \quad \nu_n(B) = \mu_n\big(f^{-1}(B)\big)\,.$$

**Proposition 6.8.** *If the sequence $(\mu_n)_{n\geq 1}$ satisfies a large deviation principle governed by the good rate function $I$ and with respect to the topology $\mathcal{O}_X$, then the sequence $(\nu_n)_{n\geq 1}$ satisfies a large deviation principle with respect to the topology $\mathcal{O}_\mathcal{Y}$, with the same speed, governed by the good rate function $J$ given by*

$$\forall y \in Y \qquad J(y) \,=\, \inf\big\{\, I(x) : x \in X,\, y = f(x)\,\big\}\,.$$

*Proof.* The map $J$ takes its values in $[0, +\infty]$. Let $t \geq 0$. Let $y \in Y$ be such that $J(y) \leq t$. Since $I$ has compact level sets, it attains its minimum over the closed set $\{\, x \in X : f(x) = y\,\}$ so that there exists $x \in X$ such that $y = f(x)$ and $J(y) = I(x)$. Therefore

$$\{\, y \in Y : J(y) \leq t\,\} \,=\, f\big(\{\, x \in X : I(x) \leq t\,\}\big)\,,$$

and $J$ has also compact level sets. Next we prove the large deviation lower bound. Let $B \in \mathcal{B}_Y$. Then

$$\liminf_{n\to\infty} \frac{1}{n} \ln \nu_n(B) = \liminf_{n\to\infty} \frac{1}{n} \ln \mu_n\big(f^{-1}(B)\big) \geq -\inf\big\{\, I(x) : x \in \overset{\circ}{\overbrace{f^{-1}(B)}}\,\big\}.$$

Yet $f^{-1}(\overset{\circ}{B}) \subset \overset{\circ}{\overbrace{f^{-1}(B)}}$, whence

$$\begin{aligned}
-\inf\big\{\, I(x) : x \in \overset{\circ}{\overbrace{f^{-1}(B)}}\,\big\} &\geq -\inf\big\{\, I(x) : x \in f^{-1}(\overset{\circ}{B})\,\big\}\\
= -\inf\Big\{\, \inf\big\{\, I(x) : x \in X,\, f(x) = y\,\big\} : y \in \overset{\circ}{B}\Big\} &= -\inf\big\{\, J(y) : y \in \overset{\circ}{B}\,\big\}\,.
\end{aligned}$$

We go now for the large deviation upper bound:

$$\limsup_{n\to\infty} \frac{1}{n} \ln \nu_n(B) = \limsup_{n\to\infty} \frac{1}{n} \ln \mu_n\big(f^{-1}(B)\big) \leq -\inf\big\{\, I(x) : x \in \overline{f^{-1}(B)}\,\big\}.$$

Yet $\overline{f^{-1}(B)} \subset f^{-1}(\overline{B})$, whence

$$\begin{aligned}
-\inf\big\{\, I(x) : x \in \overline{f^{-1}(B)}\,\big\} &\leq -\inf\big\{\, I(x) : x \in f^{-1}(\overline{B})\,\big\}\\
= -\inf\Big\{\, \inf\big\{\, I(x) : x \in X,\, f(x) = y\,\big\} : y \in \overline{B}\Big\} &= -\inf\big\{\, J(y) : y \in \overline{B}\,\big\}\,.
\end{aligned}$$

In the last step we simply use the definition of $J$ and we interchange the order of the infima. □

## 6.4 Varadhan's lemma

We end this chapter with the functional formulation of a large deviation principle, classically known as Varadhan's lemma.

**Theorem 6.9.** *Let $(\mu_n)_{n\geq 1}$ be a sequence of probability measures on $(X,\mathcal{B})$ which satisfies a large deviation principle with speed $n$, governed by the good rate function $I$. For any bounded continuous function $f : X \to \mathbb{R}$, we have*

$$\lim_{n\to\infty} \frac{1}{n} \ln \int_X \exp\big(nf(x)\big)\, d\mu_n(x) \;=\; \sup_{x\in X} \big(f(x) - I(x)\big)\,.$$

*Proof.* Let $x \in X$ and let $V$ be a neighbourhood of $x$. We have

$$\frac{1}{n} \ln \int_X \exp\big(nf(x)\big)\, d\mu_n(x) \,\geq\, \frac{1}{n} \ln \Big( \exp\big(n \inf_V f\big) \mu_n(V)\Big)\,,$$

whence, using the large deviation lower bound,

$$\liminf_{n\to\infty} \frac{1}{n} \ln \int_X \exp\big(nf(x)\big)\, d\mu_n(x) \,\geq\, \inf_V f - I(x)\,.$$

Since $f$ is continuous at $x$, then $\inf_V f$ converges towards $f(x)$ when $V$ shrinks to $x$. Taking the supremum over $x \in X$, we get

$$\liminf_{n\to\infty} \frac{1}{n} \ln \int_X \exp\big(nf(x)\big)\, d\mu_n(x) \,\geq\, \sup_{x\in X} \big(f(x) - I(x)\big)\,.$$

Let us prove the converse inequality. Let $\lambda, \delta > 0$. By hypothesis, the level set $I^{-1}([0,\lambda]) = \{\, x \in X : I(x) \leq \lambda \,\}$ is compact. For any $x \in I^{-1}([0,\lambda])$, there exists an open measurable neighbourhood $V_x$ of $x$ such that

$$\sup\{\, f(x) : x \in V_x \,\} \,\leq\, f(x) + \delta\,, \qquad \inf\{\, I(x) : x \in \overline{V}_x \,\} \,\geq\, I(x) - \delta\,.$$

Notice that we use here the fact that the topological space $(X, \tau)$ is regular, so that for any open set $V$ containing $x$, there exists an open set $U$ containing $x$ such that $\overline{U} \subset V$. The collection $V_x$, $x \in I^{-1}([0,\lambda])$, is an open cover of the compact set $I^{-1}([0,\lambda])$. Let $(x_j, V_{x_j})$, $j \in J$, be a finite subcover extracted from this covering. Setting $A = \bigcup_{j\in J} V_{x_j}$, we have

$$\begin{aligned}
&\int_X \exp\big(nf(x)\big)\, d\mu_n(x) \,\leq \\
&\qquad \sum_{j\in J} \int_{V_{x_j}} \exp\big(nf(x)\big)\, d\mu_n(x) + \int_{X\setminus A} \exp\big(nf(x)\big)\, d\mu_n(x) \\
&\qquad\qquad \leq \sum_{j\in J} \exp(nf(x_j) + n\delta)\, \mu_n(V_{x_j}) + \exp(n \sup_X |f|)\, \mu_n(X \setminus A)\,.
\end{aligned}$$

Therefore, using the large deviation upper bound and lemma 6.7,

$$\limsup_{n\to\infty} \frac{1}{n} \ln \int_X \exp\big(nf(x)\big)\, d\mu_n(x)$$
$$\leq \max\Big( \max_{j\in J} \big(f(x_j) - I(x_j) + 2\delta\big), \sup_X |f| - \lambda \Big)$$
$$\leq \max\Big( \sup_{x\in X} \big(f(x) - I(x) + 2\delta\big), \sup_X |f| - \lambda \Big).$$

We conclude by letting successively $\delta$ go to 0 and then $\lambda$ go to $\infty$. □

# 7

# Surface large deviation principles

The key to prove the laws of large numbers stated in theorems 5.1, 5.2, 5.3, 5.4, 5.5 is to prove first the adequate large deviation principles. In this chapter, we state these large deviation principles. Their proofs use a lot of common tools, namely, the coarse–graining results of chapter 9, the geometric results on covering and approximation of chapter 14, the interface lemma 12.1 and also the decoupling lemma 10.10. The proofs of theorems 7.1, 7.3, 7.5 are completed in chapters 16, 18, 19 respectively. These final parts of the proofs are globally similar. Instead of including the three proofs in a general scheme, we choose to present them separately. Indeed, the case of Bernoulli percolation is easier to understand, and the complexity of the renormalization arguments increases slightly in the FK setting and notably in the Ising case. So we would advise the courageous reader to study first the Bernoulli percolation proof, then the FK case and finally the Ising case.

## 7.1 Surface energy

For each model, the basic quantity to build the large deviation principle is the surface tension. The surface tension $\tau$ is a map from the $(d-1)$–dimensional unit sphere $S^{d-1}$ of $\mathbb{R}^d$ to $\mathbb{R}^+$ which depends on the microscopic definition of the model. Each model has its own adequate definition of surface tension. For a unit vector $\nu$, let $A$ be a unit hypersquare orthogonal to $\nu$. We sum up next the various definitions of $\tau(\nu)$.

*Bernoulli percolation.* Let $\operatorname{cyl} A$ be the cylinder $A+\mathbb{R}\nu$. The surface tension $\tau(\nu)$ is equal to the limit

$$\lim_{n\to\infty} -\frac{1}{n^{d-1}} \ln P\left(\begin{array}{l}\text{inside } n\operatorname{cyl} A \text{ there exists a finite set of closed edges } E\\ \text{cutting } n\operatorname{cyl} A \text{ in at least two unbounded components,}\\ \text{the edges of } E \text{ at distance less than } 2d \text{ from the boun–}\\ \text{–dary of } n\operatorname{cyl} A \text{ are at a distance less than } 2d \text{ from } nA\end{array}\right).$$

A subadditive argument yields the existence of this limit: given a large square, we tile it into smaller squares and, supposing that the corresponding event for each small square occurs, we glue together the associated sets of closed edges in order to obtain a set of edges realizing the above event for the large square. We apply then the FKG inequality to get an almost subadditive inequality. The condition on the localization of the set of closed edges $E$ near the boundary of the cylinder $n \operatorname{cyl} A$ is crucial to perform the glueing without deteriorating the probabilistic estimates. This subadditive argument goes through for any probability measure on edge configurations which is translation invariant, ergodic and satisfies the FKG property. In particular, it applies to the infinite volume FK measure $\Phi_\infty$. If we take $q = 2$, we get the surface tension of the Ising model at temperature $T = -2/\ln(1-p)$.

*FK percolation.* Let $D_n$ be the cylinder $D_n = \{\, na + t\nu : a \in A,\ |t| \leq n \,\}$. The set $D_n \setminus nA$ has two connected components, which we denote by $C_1^n$ and $C_2^n$. For $i = 1, 2$ and $n \in \mathbb{N}$, let $A_i^n$ be the set of the points of $C_i^n \cap \mathbb{Z}^d$ which have a nearest neighbour in $\mathbb{Z}^d \setminus D_n$:

$$A_i^n = \{\, x \in C_i^n \cap \mathbb{Z}^d : \exists y \in \mathbb{Z}^d \setminus D_n \quad |x - y| = 1 \,\}.$$

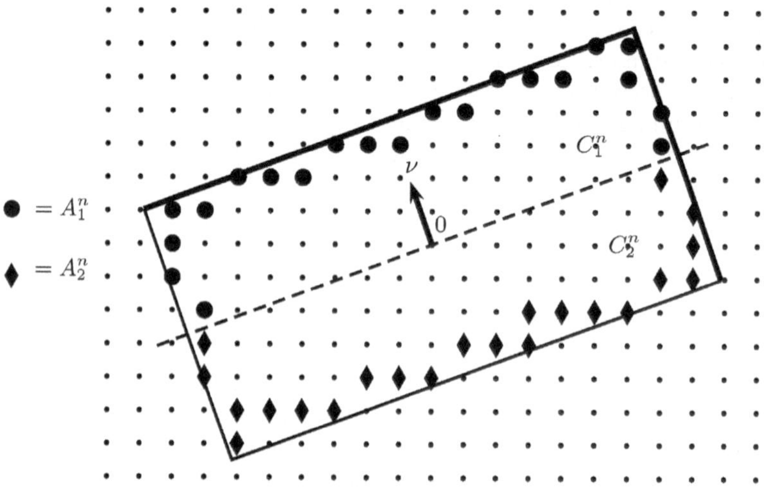


the cylinder $D_n$ and the sets $A_1^n$, $A_2^n$

Let $q \geq 1$ and $p \in ]0,1[$. The limit

$$\lim_{n \to \infty} -\frac{1}{n^{d-1}} \ln \Phi_{D_n}^{w,p,q} \begin{pmatrix} \text{there is no open path} \\ \text{in } D_n \text{ from } A_1^n \text{ to } A_2^n \end{pmatrix}$$

exists in $[0, \infty]$. In the uniqueness region $p \in ]\widehat{p}_c, 1] \setminus \mathcal{U}(q)$, this limit is equal to the surface tension defined by the formula used in Bernoulli percolation (with $\Phi_\infty^{p,q}$ in place of $P$).

*Ising model.* We finally reformulate the definition of the surface tension in the spin setting. We define the sets $D_n$, $A_1^n$, $A_2^n$ as above. The Ising Hamiltonian in $D_n$ is

$$\forall \sigma \in \{-,+\}^{D_n} \qquad H_n(\sigma) \,=\, -\frac{1}{2} \sum_{\substack{x,y \in D_n \\ |x-y|=1}} \sigma(x)\sigma(y)\,.$$

Let $\mathcal{E}_n$ be the set of the spin configurations $\sigma$ inside $D_n$ such that $\sigma(x) = +$ for $x \in A_1^n \cup A_2^n$. Let $\mathcal{F}_n$ be the set of the spin configurations $\sigma$ inside $D_n$ such that $\sigma(x) = -$ for $x \in A_1^n$ and $\sigma(x) = +$ for $x \in A_2^n$. The partition functions $Z_n^+$, $Z_n^{-,+}$ corresponding to pure $+$ and mixed $-,+$ boundary conditions at temperature $T$ are

$$Z_n^+ \,=\, \sum_{\sigma \in \mathcal{E}_n} \exp -\frac{H_n(\sigma)}{T}\,, \qquad Z_n^{-,+} \,=\, \sum_{\sigma \in \mathcal{F}_n} \exp -\frac{H_n(\sigma)}{T}\,.$$

Let $T > 0$. The limit

$$\lim_{n\to\infty} \; -\frac{1}{n^{d-1}} \ln \frac{Z_n^{-,+}}{Z_n^+}$$

exists in $[0,\infty]$. It coincides with the limit defining the surface tension in the FK model with $q = 2$ and $p = 1 - \exp(-2/T)$.

The function $\tau$ satisfies the weak simplex inequality, it is continuous and invariant under the isometries which leave $\mathbb{Z}^d$ invariant (see chapter 11 for details). Moreover $\tau$ is positive in the regime $p > \widehat{p}_c$. For $q = 1, 2$, it is known that $\widehat{p}_c = p_c$ [81, 26]. This equality is conjectured to hold for any $q \geq 1$. We denote by $\mathcal{W}_\tau$ the Wulff shape associated to the surface tension $\tau$, called also the crystal of $\tau$,

$$\mathcal{W}_\tau \,=\, \big\{\, x \in \mathbb{R}^d : x \cdot w \leq \tau(w) \text{ for all } w \text{ in } S^{d-1} \,\big\}\,.$$

Since $\tau$ is continuous and bounded away from 0, its crystal $\mathcal{W}_\tau$ is convex, closed, bounded and contains the origin 0 in its interior (see proposition 14.1). With the help of the Wulff crystal, we define the surface energy $\mathcal{I}(A)$ of a Borel set $A$ as

$$\mathcal{I}(A) \,=\, \sup \,\Big\{ \int_A \operatorname{div} f(x)\, dx : f \in C_c^1(\mathbb{R}^d, \mathcal{W}_\tau) \,\Big\}$$

where $C_c^1(\mathbb{R}^d, \mathcal{W}_\tau)$ is the set of the $C^1$ vector functions defined on $\mathbb{R}^d$ with values in $\mathcal{W}_\tau$ having compact support and div is the usual divergence operator. As a consequence of the classical Gauss–Green theorem, for a set $A$ having a smooth boundary $\partial A$, the surface energy $\mathcal{I}(A)$ is the surface integral of $\tau$ on the boundary of $A$, that is

$$\mathcal{I}(A) \,=\, \int_{\partial A} \tau(\nu_A(x))\, d\mathcal{H}^{d-1}(x)\,,$$

where $\nu_A(x)$ is the exterior normal vector to $\partial A$ at $x$ and $\mathcal{H}^{d-1}$ is the $d-1$ dimensional Hausdorff measure in $\mathbb{R}^d$ (see section 13.1 for the definition).

## 7.2 The empirical magnetization

Let $d \geq 2$ and let $T < T_c(d)$. Let $\tau$ be the surface tension of the Ising model at temperature $T$. Let $\mathcal{W}_\tau$ be the associated Wulff crystal and let $\mathcal{I}$ be the corresponding surface energy, as defined in section 7.1. Let

$$Q = [-1/2, 1/2]^d$$

be the $d$–dimensional unit cube. We denote by $\mathcal{M}(Q)$ the vector space of the finite signed Borel measures on $Q$. We equip $\mathcal{M}(Q)$ with the weak topology, that is the coarsest topology for which the linear functionals

$$\nu \in \mathcal{M}(Q) \mapsto \int f \, d\nu \,, \qquad f \in C_c(\mathbb{R}^d, \mathbb{R})$$

are continuous, where $C_c(\mathbb{R}^d, \mathbb{R})$ is the set of the continuous functions from $\mathbb{R}^d$ to $\mathbb{R}$ having compact support. We define a map $\mathcal{J} : \mathcal{M}(Q) \to [0, +\infty]$ by setting $\mathcal{J}(\nu) = \mathcal{I}(A)$ if $\nu$ is the measure having density $-m^* 1_A + m^* 1_{Q \setminus A}$ with respect to the Lebesgue measure, where $A$ is a Borel subset of $Q$, and $\mathcal{J}(\nu) = +\infty$ otherwise.

Let $\sigma$ be a spin configuration in $\Lambda(n)$. To $\sigma$ we associate a measure $\sigma_n$ on $Q$ by setting

$$\sigma_n = \frac{1}{n^d} \sum_{x \in \Lambda(n)} \sigma(x) \, \delta_{x/n} \,.$$

**Theorem 7.1.** *Let $d \geq 2$ and let $T < T_c(d)$. Under $\mu^+_{\Lambda(n),T}$, the random measure $\sigma_n$ satisfies a large deviation principle in $\mathcal{M}(Q)$, with respect to the weak topology, with speed $n^{d-1}$ and governed by the rate function $\mathcal{J}$, i.e., for any Borel subset $\mathbb{M}$ of $\mathcal{M}(Q)$,*

$$\begin{aligned} -\inf \{ \mathcal{J}(\nu) : \nu \in \overset{o}{\mathbb{M}} \} &\leq \liminf_{n \to \infty} \frac{1}{n^{d-1}} \ln \mu^+_{\Lambda(n),T}\big(\sigma_n \in \mathbb{M}\big) \\ &\leq \limsup_{n \to \infty} \frac{1}{n^{d-1}} \ln \mu^+_{\Lambda(n),T}\big(\sigma_n \in \mathbb{M}\big) \leq -\inf \{ \mathcal{J}(\nu) : \nu \in \overline{\mathbb{M}} \} \,. \end{aligned}$$

This large deviation principle ensures a weak law of large numbers. If we impose specific boundary conditions or volume constraints to the system, then, with probability going to 1 as $n$ goes to $\infty$, the random measure $\sigma_n$ will be close to an admissible measure whose surface energy is minimal. A general compactness argument implies the existence of at least one such minimizer. However, in most examples one cannot say much about the minimizers themselves. One notable exception is the Wulff problem. The difficulty stems from the fact that the surface tension $\tau$ is anisotropic and almost no quantitative information about its magnitude is available. Results on the regularity of $\tau$ would yield results on the local regularity of minimizing configurations [12].

## 7.3 Minimal surfaces

Once the link between the models of statistical mechanics and the geometric measure theory has been established, it can be extended to various situations. Several complicated geometric functionals emerge naturally from relatively simple stochastic models, which provide potentially useful simulation tools. We present here two examples: the minimal surfaces and the Mumford–Shah functional.

Consider a bounded open region $\Omega$ in $\mathbb{R}^d$ whose boundary $\Gamma$ has a Lipschitz parametrization. Note that this hypothesis is satisfied when $\Omega$ is a bounded open set with a $C^1$ boundary, when $\Omega$ is a polyhedral domain, or when $\Omega$ is a convex set (a convex function restricted to a compact subset of its domain is Lipschitz, see [113], theorem 10.4). Let $\Gamma^-$ and $\Gamma^+$ be two disjoint and relatively open subsets of $\Gamma$ such that the relative boundary of $\Gamma \setminus \Gamma^- \setminus \Gamma^+$ in $\Gamma$ has zero $\mathcal{H}^{d-1}$ measure (where $\mathcal{H}^{d-1}$ is the $d-1$ dimensional Hausdorff measure). We will study the Ising model in the region $\Omega$ with boundary conditions equal to $-$ on $\Gamma^-$ and to $+$ on $\Gamma^+$. Let $|\cdot|_\infty$ be the usual supremum norm. To obtain a discrete version of the region $\Omega$, we define for $n \in \mathbb{N}$,

- the rescaled lattice:

$$\mathbb{Z}^d_n = \mathbb{Z}^d/n$$

- the discrete counterpart of $\Omega$:

$$\Omega_n = \big\{\, x \in \mathbb{Z}^d_n : \exists y \in \Omega \quad |x-y|_\infty < 1/n \,\big\}$$

- the inner vertex boundary of $\Omega_n$:

$$\Gamma_n = \big\{\, x \in \Omega_n : \exists y \in \mathbb{Z}^d_n \setminus \Omega_n \quad |x-y| = 1/n \,\big\}$$

- the minus boundary conditions on $\Omega_n$:

$$\Gamma^-_n = \big\{\, x \in \Gamma_n : \exists y \in \Gamma^- \quad |x-y|_\infty < 1/n \,\big\}$$

- the plus boundary conditions on $\Omega_n$:

$$\Gamma^+_n = \big\{\, x \in \Gamma_n : \exists y \in \Gamma^+ \quad |x-y|_\infty < 1/n,\ \forall y \in \Gamma^- \quad |x-y|_\infty \geq 1/n \,\big\}.$$

We use the sets $\Gamma^-_n, \Gamma^+_n$ to specify the boundary conditions; namely we put pluses on $\Gamma^+_n$ and minuses on $\Gamma^-_n$. More precisely, we define the Hamiltonian or energy $H_n(\sigma)$ of a spin configuration $\sigma$ in $\Omega_n$ by

$$H_n(\sigma) = -\frac{1}{2}\sum_{\substack{x,y\in\Omega_n\\|x-y|=1}} \sigma(x)\sigma(y) - \sum_{\substack{x\in\Gamma^+_n,\, y\in\Omega_n\\|x-y|=1}} \sigma(y) - \sum_{\substack{x\in\Gamma^-_n,\, y\in\Omega_n\\|x-y|=1}} -\sigma(y)\,.$$

The Ising Gibbs measure $\mu_n$ in $\Omega_n$ with plus boundary conditions on $\Gamma^+_n$ and minus boundary conditions on $\Gamma^-_n$ and at temperature $T > 0$ is the probability measure on $\{-,+\}^{\Omega_n}$ defined by:

$$\forall \sigma \in \Omega_n \qquad \mu_n(\sigma) = \frac{1}{Z_n} \exp -\frac{H_n(\sigma)}{T}$$

where $Z_n$ is a normalizing factor called the partition function.

We recall that the perimeter $\mathcal{P}(E)$ of a Borel subset $E$ of $\mathbb{R}^d$ is

$$\mathcal{P}(E) = \sup \Big\{ \int_E \operatorname{div} f(x)\, dx : f \in C_c^\infty(\mathbb{R}^d, B(1)) \Big\}$$

where $C_c^\infty(\mathbb{R}^d, B(1))$ is the set of the compactly supported $C^\infty$ vector functions from $\mathbb{R}^d$ to the unit ball $B(1)$ and div is the usual divergence operator. If $E$ is a set of finite perimeter (i.e., $\mathcal{P}(E) < \infty$), one can define its reduced boundary $\partial^* E$; it is a subset of the topological boundary which can be used to extend several formulas of the classical differential calculus on surfaces (it is exactly the topological boundary for a $C^1$ set). At each point $x$ in the reduced boundary $\partial^* E$ of a set $E$ having finite perimeter, it is possible to define a measure theoretic exterior normal vector $\nu_E(x)$ to $E$ at $x$ which plays the role of the normal vector for a smooth surface (see sections 13.3, 13.3).

We define a surface energy $\mathcal{I}_\Gamma$ on the collection $\mathcal{B}(\Omega)$ of the Borel subsets of $\Omega$ as follows. For any $A \in \mathcal{B}(\Omega)$ having finite perimeter we set

$$\begin{aligned}\mathcal{I}_\Gamma(A) = &\int_{\partial^* A \cap \Omega} \tau(\nu_A(x))\, d\mathcal{H}^{d-1}(x) \\ &+ \int_{\partial^* A \cap \Gamma^+} \tau(\nu_A(x))\, d\mathcal{H}^{d-1}(x) + \int_{\partial^* (\Omega \setminus A) \cap \Gamma^-} \tau(\nu_{\Omega \setminus A}(x))\, d\mathcal{H}^{d-1}(x) .\end{aligned}$$

We set $I_\Gamma(A) = +\infty$ if $A$ does not have finite perimeter. The set $A$ represents the minus phase and its complement $\Omega \setminus A$ the plus phase. The first integral is the surface energy of the interfaces inside $\Omega$, the second integral corresponds to the contacts between $A$ and $\Gamma^+$, the third to the contacts between $\Omega \setminus A$ and $\Gamma^-$. We denote as before by $\mathcal{M}(\Omega)$ the vector space of the finite signed Borel measures on $\Omega$. We equip $\mathcal{M}(\Omega)$ with the weak topology, that is the coarsest topology for which the linear functionals

$$\nu \in \mathcal{M}(\Omega) \mapsto \int f\, d\nu , \qquad f \in C_c(\mathbb{R}^d, \mathbb{R})$$

are continuous, where $C_c(\mathbb{R}^d, \mathbb{R})$ is the set of the continuous maps from $\mathbb{R}^d$ to $\mathbb{R}$ having compact support. We define a map $\mathcal{J} : \mathcal{M}(\Omega) \to [0, +\infty]$ by setting $\mathcal{J}_\Gamma(\nu) = \mathcal{I}_\Gamma(A)$ if $\nu$ is the measure with density $m^* 1_{\Omega \setminus A} - m^* 1_A$ with respect to the Lebesgue measure, where $A$ is a Borel subset of $\Omega$, and $\mathcal{J}_\Gamma(\nu) = +\infty$ otherwise.

Let $\sigma$ be a spin configuration in $\Omega_n$. To $\sigma$ we associate a measure $\sigma_n$ on $\Omega$ by setting

$$\sigma_n = \frac{1}{n^d} \sum_{x \in \Omega_n} \sigma(x)\, \delta_x .$$

The results of [44] imply that, for $d \geq 3$ and $T < T_c$, under $\mu_n$, the random measure $\sigma_n$ satisfies a large deviation principle in $\mathcal{M}(\Omega)$, with respect to the weak topology, with speed $n^{d-1}$ and governed by the rate function $\mathcal{J}_\Gamma$. The treatment of the boundary conditions is technically delicate but it leads to interesting variational questions. The Winterbottom problem [130] was handled in the framework of the Ising model in [29]. For several applications, it is simpler to consider the Ising model in finite sets which approximate the Wulff crystal: the advantage is that the complicated constraint on the value of $m$ in theorems 5.1 and 5.2 disappears. The following result is a consequence of [44] and the Wulff theorem 15.1. We denote by $\mu^+_{n\mathcal{W},T}$ the Ising Gibbs measure in $nW$ at temperature $T$ with plus boundary conditions (with the preceding notation, we take $\Omega = \overset{\circ}{\mathcal{W}}$, $\Gamma^+ = \partial\mathcal{W}$).

**Corollary 7.2.** *Let $d \geq 2$ and let $T < T_c(d)$. For $m \in ]-m^*, m^*[$, we have*

$$\lim_{n\to\infty} \frac{1}{n^{d-1}} \ln \mu^+_{n\mathcal{W},T}\Big( \sum_{x\in n\mathcal{W}} \sigma(x) \leq m\mathcal{L}^d(n\mathcal{W})\Big) = -d\left(\frac{m^*-m}{2m^*}\right)^{\frac{d-1}{d}} \mathcal{L}^d(\mathcal{W})^{\frac{1}{d}} .$$

The study of the phase boundaries leads naturally to the theory of minimal surfaces, and for instance to the Plateau problem corresponding to anisotropic surface measures. Let $\Omega$ be a bounded open set in $\mathbb{R}^3$ with smooth boundary and let $\gamma$ be a Jordan curve drawn on $\partial\Omega$ which separates $\partial\Omega$ into two disjoint relatively open sets $\Gamma^+$ and $\Gamma^-$. Typical configurations in the Ising model on a fine grid in $\Omega$ with boundary conditions plus on $\Gamma^+$ and minus on $\Gamma^-$ will exhibit two phases separated by an interface close to a minimal surface which is a global solution to the following Plateau type problem:

$$\text{minimize} \int_S \tau(\nu_S(x))\, d\mathcal{H}^2(x) \ :\ S \text{ is a surface in } \Omega \text{ spanned by } \gamma$$

where $\nu_S(x)$ is the normal vector to $S$ at $x$. It is conjectured that, as the temperature approaches $T_c$ from below, the surface tension $\tau$ becomes more and more isotropic, thus the solution of the above minimization problem should converge to the solution of the classical Plateau problem.

Models with more than two coexisting phases like the Potts model lead to very hard variational problems [44]. In the Potts model with parameter $q = 3$, one can easily put constraints on the system so that a double bubble should appear. However, the corresponding variational problems are extremely hard even in the isotropic case. The famous conjecture related to the symmetric double bubble in the three dimensional case with isotropic surface energy has only been resolved recently [85, 87, 88] and the asymmetric case (with two bubbles having different volumes) remains unresolved even in the isotropic case.

## 7.4 The cluster shapes

We consider the Bernoulli bond percolation model on $\mathbb{Z}^d$ with parameter $p$, where $p$ is a fixed value in the supercritical regime $p_c < p < 1$. Let $\tau$ be the surface tension of the percolation model corresponding to the parameter $p$, let $\mathcal{W}_\tau$ be the associated Wulff crystal and let $\mathcal{I}$ be the corresponding surface energy, as defined in section 7.1.

We denote by $\mathcal{M}(\mathbb{R}^d)$ the vector space of the finite signed Borel measures on $\mathbb{R}^d$. We equip $\mathcal{M}(\mathbb{R}^d)$ with the weak topology, that is the coarsest topology for which the linear functionals

$$\nu \in \mathcal{M}(\mathbb{R}^d) \mapsto \int f\, d\nu\,, \qquad f \in C_c(\mathbb{R}^d, \mathbb{R})$$

are continuous, where $C_c(\mathbb{R}^d, \mathbb{R})$ is the set of the continuous maps from $\mathbb{R}^d$ to $\mathbb{R}$ having compact support.

We define a map $\mathcal{J} : \mathcal{M}(\mathbb{R}^d) \to [0, +\infty]$ by setting $\mathcal{J}(\nu) = \mathcal{I}(A)$ if $\nu$ is the measure with density $\theta\, 1_A$ with respect to the Lebesgue measure, where $A$ is a Borel subset of $\mathbb{R}^d$, and $\mathcal{J}(\nu) = +\infty$ otherwise. For the definition of $\mathcal{I}$, see section 7.1.

**Theorem 7.3.** *Let $d \geq 3$ and let $p > p_c$. The sequence of random measures*

$$\mathcal{C}_n = \frac{1}{n^d} \sum_{x \in C(0)} \delta_{x/n}$$

*satisfies a large deviation principle in $\mathcal{M}(\mathbb{R}^d)$ with speed $n^{d-1}$ and rate function $\mathcal{J}$, i.e., for any Borel subset $\mathbb{M}$ of $\mathcal{M}(\mathbb{R}^d)$,*

$$-\inf\{\, \mathcal{J}(\nu) : \nu \in \overset{o}{\mathbb{M}} \} \leq \liminf_{n\to\infty} \frac{1}{n^{d-1}} \ln P\big(\mathcal{C}_n \in \mathbb{M}\big)$$
$$\leq \limsup_{n\to\infty} \frac{1}{n^{d-1}} \ln P\big(\mathcal{C}_n \in \mathbb{M}\big) \leq -\inf\{\, \mathcal{J}(\nu) : \nu \in \overline{\mathbb{M}}\,\}\,.$$

*Under the conditional probability $\widehat{P}(\cdot) = P\big(\cdot \,\big|\, |C(0)| < \infty\big)$ we have the enhanced large deviation upper bound: for any Borel subset $\mathbb{M}$ of $\mathcal{M}(\mathbb{R}^d)$,*

$$\limsup_{n\to\infty} \frac{1}{n^{d-1}} \ln \widehat{P}\big(\mathcal{C}_n \in \mathbb{M}\big) \leq$$
$$- \sup_{f,\delta} \inf \big\{\, \mathcal{J}(\varrho) : \varrho(\mathbb{R}^d) < +\infty,\ \exists \nu \in \mathbb{M} \quad |\varrho(f) - \nu(f)|_2 < \delta \,\big\}$$

*where the supremum is taken over $\delta > 0$ and the functions $f : \mathbb{R}^d \to \mathbb{R}^k$ which are bounded and uniformly continuous (the dimension $k$ is let free as well; $\varrho(f) - \nu(f)$ is a vector in $\mathbb{R}^k$ and $|\cdot|_2$ is the Euclidean norm).*

In two dimensions, one should proceed differently to define $\mathcal{J}$. The main difference with dimensions $d \geq 3$ is that a single connection between two distant sites creates some surface energy. For instance, if $A$ is the union of two balls, then $\mathcal{I}(A)$ should take into account the surface tension corresponding to a minimal path connecting the two balls. Thus the rate function should be increased to incorporate the cost of such connections. To build correctly the rate function, a possibility would be to use the Hausdorff lower semicontinuous envelope of the length in the plane, as studied in [39]. The surface energy $\mathcal{I}$ should then be the anisotropic version of this quantity. With an adequate definition of the surface energy, a similar large deviation principle holds in dimensions two as well [37]. Nevertheless, the strategy used in higher dimension yields without any further adaptation the following large deviation bounds.

**Theorem 7.4.** *Let $d = 2$ and let $p > p_c$. The sequence of random measures*

$$\mathcal{C}_n = \frac{1}{n^2} \sum_{x \in C(0)} \delta_{x/n}$$

*satisfies the following large deviation bounds:*
*For any Borel subset $\mathbb{M}$ of $\mathcal{M}(\mathbb{R}^2)$,*

$$\limsup_{n \to \infty} \frac{1}{n} \ln P\big(\mathcal{C}_n \in \mathbb{M}\big) \leq -\inf\{\,\mathcal{J}(\nu) : \nu \in \overline{\mathbb{M}}\,\}\,.$$

*Let $\nu \in \mathcal{M}(\mathbb{R}^2)$ be of the form $\nu = \theta 1_A$, where $A$ is an open connected set containing $0$. For any weak neighbourhood $\mathcal{U}$ of $\nu$ in $\mathcal{M}(\mathbb{R}^2)$, we have*

$$\liminf_{n \to \infty} \frac{1}{n} \ln P\big(\mathcal{C}_n \in \mathcal{U}\big) \geq -\mathcal{J}(\nu)\,.$$

This is not a complete large deviation principle because we do not have the lower bound for any $\nu \in \mathcal{M}(\mathbb{R}^2)$, and, as explained above, we should also modify the rate function in order to match the lower bound and the upper bound. However the estimates of theorem 7.4 are enough to prove theorem 5.3.

## 7.5 FK percolation

A possible road to prove the large deviation principle of theorem 7.1 consists in first proving a large deviation principle for the clusters of the FK model and then transferring it to the Ising model via the random colouring. In this section, we state a variant of this large deviation principle which is more natural in the FK setting. Let $d \geq 2$, $q \geq 1$ and let $p > \widehat{p}_c(d)$, $p \notin \mathcal{U}(q)$. Let $\theta = \theta(p,q)$ be the percolation probability (see section 4.2). Let $\tau$ be the surface tension of the FK model corresponding to the parameters $p, q$. Let $\mathcal{W}_\tau$ be the associated Wulff crystal and let $\mathcal{I}$ be the corresponding surface energy, as defined in section 7.1. Let

$$Q = [-1/2, 1/2]^d$$

be the $d$–dimensional unit cube. We denote by $\mathcal{M}(Q)$ the vector space of the finite signed Borel measures on $Q$. We equip $\mathcal{M}(Q)$ with the weak topology, that is the coarsest topology for which the linear functionals

$$\nu \in \mathcal{M}(Q) \mapsto \int f\, d\nu\,, \qquad f \in C_c(\mathbb{R}^d, \mathbb{R})$$

are continuous, where $C_c(\mathbb{R}^d, \mathbb{R})$ is the set of the continuous maps from $\mathbb{R}^d$ to $\mathbb{R}$ having compact support. We define a map $\mathcal{J} : \mathcal{M}(Q) \to [0, +\infty]$ by setting $\mathcal{J}(\nu) = \mathcal{I}(A)$ if $\nu$ is the measure with density $\theta\, 1_{Q\setminus A}$ with respect to the Lebesgue measure, where $A$ is a Borel subset of $Q$, and $\mathcal{J}(\nu) = +\infty$ otherwise. For the definition of $\mathcal{I}$, see section 7.1.

**Theorem 7.5.** *Let $d \geq 3$, $q \geq 1$ and $p > \widehat{p}_c(d)$, $p \notin \mathcal{U}(q)$. Under $\Phi^{w,p,q}_{\Lambda(n)}$, the random measure*

$$\mathcal{C}_n = \frac{1}{n^d} \sum_{\substack{x \in \Lambda(n) \\ x \longleftrightarrow \partial^{in} \Lambda(n)}} \delta_{x/n}$$

*satisfies a large deviation principle in $\mathcal{M}(Q)$, with respect to the weak topology, with speed $n^{d-1}$ and governed by the rate function $\mathcal{J}$, i.e., for any Borel subset $\mathbb{M}$ of $\mathcal{M}(Q)$,*

$$-\inf\{\, \mathcal{J}(\nu) : \nu \in \overset{o}{\mathbb{M}} \} \leq \liminf_{n\to\infty} \frac{1}{n^{d-1}} \ln \Phi^{w,p,q}_{\Lambda(n)}(\mathcal{C}_n \in \mathbb{M})$$
$$\leq \limsup_{n\to\infty} \frac{1}{n^{d-1}} \ln \Phi^{w,p,q}_{\Lambda(n)}(\mathcal{C}_n \in \mathbb{M}) \leq -\inf\{\, \mathcal{J}(\nu) : \nu \in \overline{\mathbb{M}} \}\,.$$

As in the case of Bernoulli percolation, a similar large deviation principle should hold in two dimensions, but with a slightly more complicated rate function (see the comment after theorem 7.3). However, we obtain without additional effort the following large deviation bounds.

**Theorem 7.6.** *Let $d = 2$, $Q = [-1/2, 1/2]^2$ and let $q \geq 1$, $p > \widehat{p}_c(2) = \widehat{p}_g$. Under $\Phi^{w,p,q}_{\Lambda(n)}$, the random measure*

$$\mathcal{C}_n = \frac{1}{n^2} \sum_{\substack{x \in \Lambda(n) \\ x \longleftrightarrow \partial^{in} \Lambda(n)}} \delta_{x/n}$$

*satisfies the following large deviation bounds:*
*For any Borel subset $\mathbb{M}$ of $\mathcal{M}(Q)$,*

$$\limsup_{n\to\infty} \frac{1}{n} \ln \Phi^{w,p,q}_{\Lambda(n)}(\mathcal{C}_n \in \mathbb{M}) \leq -\inf\{\, \mathcal{J}(\nu) : \nu \in \overline{\mathbb{M}} \}\,.$$

*Let $\nu \in \mathcal{M}(Q)$ be of the form $\nu = \theta 1_{Q\setminus A}$, where $A$ is an open connected subset of $Q$. For any weak neighbourhood $\mathcal{U}$ of $\nu$ in $\mathcal{M}(Q)$, we have*

$$\liminf_{n\to\infty} \frac{1}{n} \ln \Phi^{w,p,q}_{\Lambda(n)}(\mathcal{C}_n \in \mathcal{U}) \geq -\mathcal{J}(\nu)\,.$$

8

# Volume large deviations

This chapter is devoted to volume large deviation principles, as opposed to surface large deviation principles. While the former are reminiscent of the case of i.i.d. random variables and Cramér's theorem, the latter principles are caused by the phase coexistence phenomenon. The results of this chapter will be used later on in the proofs of the coarse graining estimates of section 9.

## 8.1 Bernoulli percolation

For $A \subset \mathbb{Z}^d$, we denote by $\partial^{in} A$ the sites of $A$ which have a nearest neighbour in $\mathbb{Z}^d \setminus A$ and we define

$$\varrho(A) = \big|\{ x \in A : x \longleftrightarrow \partial^{in} A \}\big| .$$

The map $\varrho$ is subadditive, i.e.,

$$\forall A, B \subset \mathbb{Z}^d, \quad A \cap B = \emptyset \quad \Rightarrow \quad \varrho(A \cup B) \leq \varrho(A) + \varrho(B) .$$

Indeed, if $x \in A$ is connected by an open path to $\partial^{in}(A \cup B)$, then this path must cross $\partial^{in} A$, so that $\{ x \longleftrightarrow \partial^{in} A \}$ occurs.

**Proposition 8.1.** *For any $p \in [0,1]$, we have with probability one*

$$\lim_{n \to \infty} \frac{\varrho(\Lambda(n))}{|\Lambda(n)|} = \theta(p) .$$

*Remark 8.2.* The multiparameter subadditive ergodic theorem (see [1], theorem 2.4 or [122], theorem 1.1) implies that the above limit exists almost surely. The proof below relies only on the ergodic theorem.

*Proof.* Let $m \geq 2n$ be two integers. For $x \in \Lambda(n)$, we have $\Lambda(n) \subset x + \Lambda(m)$, whence

$$\varrho(\Lambda(n)) \geq \sum_{x\in\Lambda(n)} 1_{\{x \longleftrightarrow x+\partial^{in}\Lambda(m)\}},$$

and, sending $m$ to $\infty$,

$$\frac{\varrho(\Lambda(n))}{|\Lambda(n)|} \geq \frac{1}{|\Lambda(n)|}\sum_{x\in\Lambda(n)} 1_{\{x \longleftrightarrow \infty\}} = \frac{|C_\infty \cap \Lambda(n)|}{|\Lambda(n)|}.$$

Sending now $n$ to $\infty$ and applying the ergodic theorem, we obtain

$$\liminf_{n\to\infty} \frac{\varrho(\Lambda(n))}{|\Lambda(n)|} \geq P(0 \longleftrightarrow \infty).$$

Conversely, for $m \leq n$,

$$\begin{aligned}\varrho(\Lambda(n)) &= \sum_{x\in\Lambda(n)} 1_{\{x \longleftrightarrow \partial^{in}\Lambda(n)\}}\\ &\leq |\Lambda(n)\setminus\Lambda(n-m)| + \sum_{x\in\Lambda(n-m)} 1_{\{x \longleftrightarrow x+\partial^{in}\Lambda(m)\}}\end{aligned}$$

whence

$$\frac{\varrho(\Lambda(n))}{|\Lambda(n)|} \leq \frac{|\Lambda(n)\setminus\Lambda(n-m)|}{|\Lambda(n)|} + \frac{1}{|\Lambda(n-m)|}\sum_{x\in\Lambda(n-m)} 1_{\{x \longleftrightarrow x+\partial^{in}\Lambda(m)\}}.$$

Sending $n$ to $\infty$, the ergodic theorem yields

$$\lim_{n\to\infty} \frac{1}{|\Lambda(n-m)|}\sum_{x\in\Lambda(n-m)} 1_{\{x \longleftrightarrow x+\partial^{in}\Lambda(m)\}} = P\big(0 \longleftrightarrow \partial^{in}\Lambda(m)\big)$$

so that

$$\limsup_{n\to\infty} \frac{\varrho(\Lambda(n))}{|\Lambda(n)|} \leq P\big(0 \longleftrightarrow \partial^{in}\Lambda(m)\big).$$

We conclude by sending $m$ to $\infty$. □

Our goal is to analyze the large deviations from the law of large numbers we have just proved. The large deviations from below are handled in theorem 5.5. We state next the result concerning the large deviations of $\varrho(\Lambda(n))$ from above.

**Theorem 8.3.** *Let $p \in ]0,1[$. For any $\alpha \in [0,1]$, the limit*

$$J(\alpha) = \lim_{n\to\infty} -\frac{1}{n^d}\ln P\Big(\frac{\varrho(\Lambda(n))}{|\Lambda(n)|} \geq \alpha\Big)$$

*exists and is finite. The map $\alpha \in [0,1] \mapsto J(\alpha) \in \mathbb{R}^+$ is convex continuous. It vanishes on $[0,\theta(p)]$ and it is strictly positive on $]\theta(p),1]$.*

*Proof.* Let $m \leq n$ be two integers. Let $n = mq+r$ be the Euclidean division of $n$ by $m$. We partition $\Lambda(n)$ into $q^d$ disjoint translates of $\Lambda(m)$, denoted by $\Lambda_i$, $1 \leq i \leq q^d$, which cover $\Lambda(qm)$, and a remaining region $\Lambda_{q^d+1} = \Lambda(n) \setminus \Lambda(qm)$. Let $E$ be the event

$$E = \Big\{ \text{all the edges having an endpoint in } \Lambda_{q^d+1} \cup \bigcup_{1 \leq i \leq q^d} \partial^{in} \Lambda_i \text{ are open} \Big\}.$$

On the event $E$, all the points belonging to the boundaries $\partial^{in} \Lambda_i$, $1 \leq i \leq q^d$, are connected together and they are also connected to $\partial^{in} \Lambda(n)$. Thus, for $\alpha \in [0,1]$,

$$E \cap \bigcap_{1 \leq i \leq q^d} \{ \varrho(\Lambda_i) \geq \alpha |\Lambda_i| \} \subset \{ \varrho(\Lambda(n)) \geq \alpha |\Lambda(n)| \},$$

whence by the FKG inequality (or simply by independence) and the translation invariance of $P$,

$$P\big(\varrho(\Lambda(n)) \geq \alpha |\Lambda(n)|\big) \geq P(E) P\big(\varrho(\Lambda(m)) \geq \alpha |\Lambda(m)|\big)^{q^d}.$$

The number of edges involved in the definition of the event $E$ is at most $2dq^d |\partial^{in} \Lambda(m)| + 2d |\partial^{in} \Lambda(n)| m$. Taking logarithms and sending $n$ to $\infty$, we get

$$\liminf_{n \to \infty} \frac{1}{|\Lambda(n)|} \ln P\Big( \frac{\varrho(\Lambda(n))}{|\Lambda(n)|} \geq \alpha \Big) \geq \\ \frac{1}{|\Lambda(m)|} \ln P\Big( \frac{\varrho(\Lambda(m))}{|\Lambda(m)|} \geq \alpha \Big) + 2d \frac{|\partial^{in} \Lambda(m)|}{|\Lambda(m)|} \ln p.$$

Sending now $m$ to $\infty$, we see that the limit defining $J(\alpha)$ in the statement of the theorem exists. Moreover

$$P\Big( \frac{\varrho(\Lambda(n))}{|\Lambda(n)|} \geq \alpha \Big) \geq P\big(\text{all the edges of } \Lambda(n) \text{ are open}\big) \geq p^{d|\Lambda(n)|}$$

so that $J(\alpha) \leq -d \ln p$ is finite. Next we show that the map $J$ is convex. Let $\lambda, \alpha_1, \alpha_2 \in [0,1]$ with $\alpha_1 < \alpha_2$. Let also $m \leq n$ be two integers. We use the same partition of $\Lambda(n)$ as before, and we set $r = \lfloor \lambda q^d \rfloor$, the largest integer less than or equal to $\lambda q^d$. On the event

$$E \cap \bigcap_{1 \leq i \leq r} \{ \varrho(\Lambda_i) \geq \alpha_1 |\Lambda_i| \} \cap \bigcap_{r < i \leq q^d} \{ \varrho(\Lambda_i) \geq \alpha_2 |\Lambda_i| \}$$

the following inequality holds (recall that $\alpha_1 < \alpha_2$):

$$\varrho(\Lambda(n)) \geq \big(r\alpha_1 + (q^d - r)\alpha_2\big) |\Lambda(m)| + |\Lambda_{q^d+1}| \\ \geq \big(\lambda\alpha_1 + (1-\lambda)\alpha_2\big) |\Lambda(n)|.$$

Therefore

$$P\Big(\varrho(\Lambda(n)) \geq (\lambda\alpha_1 + (1-\lambda)\alpha_2)|\Lambda(n)|\Big) \geq$$
$$P(E)\, P\Big(\varrho(\Lambda(m)) \geq \alpha_1|\Lambda(m)|\Big)^r\, P\Big(\varrho(\Lambda(m)) \geq \alpha_2|\Lambda(m)|\Big)^{q^d - r}$$

and, taking limits in the usual way,

$$-J\big(\lambda\alpha_1 + (1-\lambda)\alpha_2\big) \geq -\lambda J(\alpha_1) - (1-\lambda)J(\alpha_2)\,.$$

Since $n^{-d}\varrho(\Lambda(n))$ converges towards $\theta$ with probability 1 by proposition 8.1, then $J$ is equal to 0 on $[0,\theta[$. The map $J$ is convex and finite on $[0,1]$, hence it is continuous on $]0,1[$ and thus $J(\theta)=0$. We prove now that $J$ is continuous at 1. Let $m \leq n$ be two integers. We use the same partition $(\Lambda_i,\, 1 \leq i \leq q^d+1)$ of $\Lambda(n)$ as before. Let $\varepsilon, \delta$ be such that $0 < \varepsilon < \delta < 1$ and let us define

$$I(\delta) = \big\{\, i : 1 \leq i \leq q^d,\ \varrho(\Lambda_i) \geq (1-\delta)|\Lambda_i|\,\big\}\,.$$

We have then

$$\begin{aligned}\varrho(\Lambda(n)) &\leq \sum_{i \notin I(\delta)} \varrho(\Lambda_i) + \sum_{i \in I(\delta)} \varrho(\Lambda_i) + \varrho(\Lambda_{q^d+1})\\ &\leq \big(q^d - |I(\delta)|\big)(1-\delta)|\Lambda(m)| + |I(\delta)||\Lambda(m)| + |\Lambda(n)| - q^d|\Lambda(m)|\\ &= \delta|I(\delta)||\Lambda(m)| + |\Lambda(n)| - \delta q^d|\Lambda(m)|\,.\end{aligned}$$

The event $\varrho(\Lambda(n)) \geq (1-\varepsilon)|\Lambda(n)|$ implies that

$$q^d - |I(\delta)| \leq \left\lceil \frac{\varepsilon n^d}{\delta m^d} \right\rceil = N\,.$$

Decomposing this event over the possible choices of the set $I(\delta)$, we get for $\varepsilon$ small and $n$ large

$$\begin{aligned}P\big(\varrho(\Lambda(n)) \geq (1-\varepsilon)|\Lambda(n)|\big) &\leq \sum_{k=0}^{N} \binom{q^d}{k} P\big(\varrho(\Lambda(m)) \geq (1-\delta)|\Lambda(m)|\big)^{q^d-k}\\ &\leq (N+1)\binom{q^d}{N} P\big(\varrho(\Lambda(m)) \geq (1-\delta)|\Lambda(m)|\big)^{q^d-N}.\end{aligned}$$

Taking logarithms, sending $n$ to $+\infty$, we get with the help of a standard expansion of the binomial coefficient (see lemma 1.4)

$$\begin{aligned}J(1-\varepsilon) \geq\ & \frac{\varepsilon}{\delta}\ln\frac{\varepsilon}{\delta} + \Big(1-\frac{\varepsilon}{\delta}\Big)\ln\Big(1-\frac{\varepsilon}{\delta}\Big)\\ & -\Big(1-\frac{\varepsilon}{\delta}\Big)\frac{1}{m^d}\ln P\big(\varrho(\Lambda(m)) \geq (1-\delta)|\Lambda(m)|\big)\,.\end{aligned}$$

Sending first $\varepsilon$ to 0, we have

$$\lim_{\substack{\varepsilon \to 0 \\ \varepsilon > 0}} J(1-\varepsilon) \geq -\frac{1}{m^d} \ln P\big(\varrho(\Lambda(m)) \geq (1-\delta)|\Lambda(m)|\big) .$$

Sending now $\delta$ to 0 and then $m$ to $\infty$, we obtain

$$\lim_{\substack{\varepsilon \to 0 \\ \varepsilon > 0}} J(1-\varepsilon) \geq J(1)$$

hence $J$, being non–decreasing, is continuous at 1. We finally prove that $J$ is positive on $]\theta, 1]$. Let $\alpha \in ]\theta, 1[$. There exists an integer $m \geq 1$ such that

$$E\Big(\frac{\varrho(\Lambda(m))}{|\Lambda(m)|}\Big) < \frac{\alpha+\theta}{2} .$$

Let $n$ be an integer larger than $m$. Using again the same partition of $\Lambda(n)$ as before, by the subadditivity of $\varrho$, we have

$$\varrho(\Lambda(n)) \leq \sum_{1 \leq i \leq q^d+1} \varrho(\Lambda_i) \leq \sum_{1 \leq i \leq q^d} \varrho(\Lambda_i) + |\Lambda(n) \setminus \Lambda(qm)| .$$

Let $n$ be large enough so that

$$\alpha - \frac{|\Lambda(n) \setminus \Lambda(qm)|}{|\Lambda(n)|} \geq \frac{\alpha+\theta}{2} .$$

Combining the previous inequalities, we see that

$$P\Big(\frac{\varrho(\Lambda(n))}{|\Lambda(n)|} \geq \alpha\Big) \leq P\bigg(\frac{1}{q^d} \sum_{1 \leq i \leq q^d} \frac{\varrho(\Lambda_i)}{|\Lambda_i|} \geq \frac{\alpha+\theta}{2}\bigg) . \qquad (\circ)$$

Yet the random variables $\varrho(\Lambda_i)/|\Lambda_i|$, $1 \leq i \leq q^d$, are i.i.d., bounded by 1, and their common expectation is strictly less than $(\alpha+\theta)/2$. By the classical Cramér theorem in $\mathbb{R}$ (see theorems 1.3 and 1.5), there exists $c > 0$ such that the right–hand quantity of inequality $(\circ)$ is bounded by $\exp(-cq^d)$, whence

$$\limsup_{n \to \infty} \frac{1}{|\Lambda(n)|} \ln P\Big(\frac{\varrho(\Lambda(n))}{|\Lambda(n)|} \geq \alpha\Big) \leq -\frac{c}{|\Lambda(m)|} < 0$$

and we conclude that $J(\alpha) > 0$. □

**Proposition 8.4.** *Under the probability $P$, the sequence $n^{-d}\varrho(\Lambda(n))$, $n \geq 1$, satisfies a large deviation principle with speed $n^d$ and governed by the good rate function $I = J$.*

The proof of the large deviation lower bound in the interval $[0, \theta]$ is very instructive. It consists in creating an event whereby a special box is disconnected from $\partial^{in}\Lambda(n)$, the volume of which is the adequate proportion of the volume of $\Lambda(n)$. Outside of this box, the situation is typical and an approximate density $\theta$ of vertices is connected to $\partial^{in}\Lambda(n)$. The probability of such an event is larger than $\exp(-c|\partial^{in}\Lambda(n)|)$ for some $c > 0$, because we only need to impose a constraint on the edges touching the boundary of the special box. Thus we get even a lower bound of surface order, which is the correct order.

*Proof.* We set $I = J$. That $I$ is a good rate function is a consequence of the convexity of $J$ on $[0,1]$ and the continuity of $J$ proved in theorem 8.3. The large deviation upper bound follows from the definition of $J$, because $J$ is non–decreasing on $[0,1]$. Indeed, let $F$ be a closed subset of $[0,1]$ and let $a = \inf F$. Then

$$P\Big(\frac{\varrho(\Lambda(n))}{|\Lambda(n)|} \in F\Big) \leq P\Big(\frac{\varrho(\Lambda(n))}{|\Lambda(n)|} \geq a\Big)$$

whence

$$\limsup_{n\to\infty} \frac{1}{n^d} \ln P\Big(\frac{\varrho(\Lambda(n))}{|\Lambda(n)|} \in F\Big) \leq -I(a) = \inf_F I\,.$$

To obtain the large deviation lower bound, we will prove the following local estimate:

$$\forall \alpha \in [0,1] \quad \forall \varepsilon > 0 \quad \liminf_{n\to\infty} \frac{1}{n^d} \ln P\Big(\frac{\varrho(\Lambda(n))}{|\Lambda(n)|} \in ]\alpha-\varepsilon, \alpha+\varepsilon[\Big) \geq -I(\alpha)\,.$$

We consider separately the cases where $\alpha \geq \theta$ and $\alpha < \theta$. Let first $\alpha \in [\theta, 1]$ and let $\varepsilon > 0$. We write

$$P\Big(\frac{\varrho(\Lambda(n))}{|\Lambda(n)|} \in ]\alpha-\varepsilon, \alpha+\varepsilon[\Big) \geq P\Big(\frac{\varrho(\Lambda(n))}{|\Lambda(n)|} \geq \alpha\Big) - P\Big(\frac{\varrho(\Lambda(n))}{|\Lambda(n)|} \geq \alpha+\varepsilon\Big)\,.$$

The map $J$ is convex and vanishes only on $[0,\theta]$, thus it is increasing on $[\theta,1]$ and $J(\alpha+\varepsilon) > J(\alpha)$. By the identity defining $J$ in theorem 8.3, the second term on the right–hand side is negligible compared to the first, whence

$$\liminf_{n\to\infty} \frac{1}{n^d} \ln P\Big(\frac{\varrho(\Lambda(n))}{|\Lambda(n)|} \in ]\alpha-\varepsilon, \alpha+\varepsilon[\Big) \geq -J(\alpha) = -I(\alpha)\,.$$

Now suppose that $\theta > 0$. Let $\alpha \in [0,\theta[$ and let $\varepsilon, \eta > 0$. There exists an integer $m \geq 1$ such that

$$P\Big(\frac{\varrho(\Lambda(m))}{|\Lambda(m)|} \in ]\theta(1-\varepsilon), \theta(1+\varepsilon)[\Big) \geq 1-\eta\,.$$

Let $n$ be an integer larger than $m$. Let $s = \lfloor (1-\alpha/\theta)^{1/d} n \rfloor$ and let

$$q_1 = \Big\lfloor \frac{n}{2m} \Big\rfloor, \quad q_2 = \Big\lfloor \frac{s}{2m} \Big\rfloor + 1\,.$$

We partition $\Lambda(2mq_1) \setminus \Lambda(2mq_2)$ into $N = 2^d q_1^d - 2^d q_2^d$ disjoint translates of $\Lambda(m)$, denoted by $\Lambda_i$, $1 \leq i \leq N$. Let $E, F$ be the events

$$E = \Big\{ \text{all the edges having at least one endpoint in } (\Lambda(n) \setminus \Lambda(2mq_1)) \cup \bigcup_{1\leq i\leq N} \partial^{in} \Lambda_i \text{ are open} \Big\}\,,$$

$$F = \Big\{ \text{all the edges having one endpoint in } \partial^{in} \Lambda(s-2) \text{ are closed} \Big\}\,.$$

On the event

$$E \cap F \cap \bigcap_{1 \le i \le N} \left\{ \frac{\varrho(\Lambda_i)}{|\Lambda_i|} \in ]\theta(1-\varepsilon), \theta(1+\varepsilon)[ \right\}$$

we have on the one hand

$$\varrho(\Lambda(n)) \ge \varrho\big(\Lambda(n) \setminus \Lambda(2mq_1)\big) + \sum_{1 \le i \le N} \varrho(\Lambda_i) \ge \theta(1-\varepsilon)\big|\Lambda(n) \setminus \Lambda(2mq_2)\big|$$

and on the other hand

$$\varrho(\Lambda(n)) \le \sum_{1 \le i \le N} \varrho(\Lambda_i) + \varrho\big(\big(\Lambda(n) \setminus \Lambda(2mq_1)\big) \cup \big(\Lambda(2mq_2) \setminus \Lambda(s-2)\big)\big)$$

$$\le \theta(1+\varepsilon)\big|\Lambda(2mq_1) \setminus \Lambda(2mq_2)\big| + \big|\big(\Lambda(n) \setminus \Lambda(2mq_1)\big) \cup \big(\Lambda(2mq_2) \setminus \Lambda(s-2)\big)\big| .$$

For $n$ large enough, these inequalities imply that

$$(\alpha - 2\varepsilon)|\Lambda(n)| \le \varrho(\Lambda(n)) \le (\alpha + 2\varepsilon)|\Lambda(n)|$$

whence, using the independence of $E$, $F$, and the variables $\varrho(\Lambda_i)$, $1 \le i \le N$,

$$P\Big(\frac{\varrho(\Lambda(n))}{|\Lambda(n)|} \in ]\alpha-2\varepsilon, \alpha+2\varepsilon[\Big) \ge P(E)\, P(F)\, P\Big(\frac{\varrho(\Lambda(m))}{|\Lambda(m)|} \in ]\theta(1-\varepsilon), \theta(1+\varepsilon)[\Big)^N .$$

The number of edges involved in the definition of the events $E, F$ is negligible compared to $|\Lambda(n)|$. Taking logarithms and sending $n$ to infinity, we get

$$\liminf_{n \to \infty} \frac{1}{|\Lambda(n)|} \ln P\Big(\frac{\varrho(\Lambda(n))}{|\Lambda(n)|} \in ]\alpha - 2\varepsilon, \alpha + 2\varepsilon[\Big) \ge \frac{\alpha}{\theta m^d} \ln(1-\eta) .$$

Sending $\eta$ to 0 yields the local large deviation estimate we need. □

**Corollary 8.5.** *For any $t \in \mathbb{R}$, the limit*

$$p(t) = \lim_{n \to \infty} \frac{1}{n^d} \ln E\Big( \exp\big(t\varrho(\Lambda(n))\big)\Big)$$

*exists. The function $p(t)$ is called the pressure. It is convex and lower semicontinuous. The rate function $I$ is the Fenchel–Legendre transform of $p(t)$:*

$$\forall \alpha \in [0,1] \qquad I(\alpha) = \sup_{t \in \mathbb{R}} \big(t\alpha - p(t)\big) .$$

*Proof.* A direct application of Varadhan's lemma (see theorem 6.9) yields the existence of the limit stated in the corollary and also that

$$\forall t \in \mathbb{R} \qquad p(t) = \sup_{\alpha \in \mathbb{R}} \big(t\alpha - I(\alpha)\big) .$$

Since $I$ is convex lower semicontinuous, the Fenchel–Legendre transform of its Fenchel–Legendre transform $p(t)$ is $I$ itself. □

## 8.2 FK percolation

We will now extend the volume large deviation result of section 8.1 to FK percolation. To this end, we follow the proof for the independent case and we try to bypass the steps relying on independence, typically by using correlation inequalities. We define $\varrho$ exactly as in the case of Bernoulli percolation and we wish to study the asymptotic behavior of $n^{-d}\varrho(\Lambda(n))$. Yet the situation is more involved in the FK setting, because we have to consider finite volume FK measures as well. Only the case of wired boundary conditions can be handled directly in a satisfactory way. The next lemma is the key to extend the volume large deviation results from Bernoulli percolation to FK percolation.

**Lemma 8.6.** *Let $p \in [0,1]$ and $q \in [1,+\infty[$. Then*

$$\lim_{n\to\infty} \Phi^w_{\Lambda(n)}\Big(\frac{\varrho(\Lambda(n))}{|\Lambda(n)|}\Big) = \theta^w(p,q)\,.$$

*Remark 8.7.* Since the measure $\Phi^w_\infty$ is ergodic and translation invariant [79], the multiparameter subadditive ergodic theorem ([1], theorem 2.4 or [122], theorem 1.1) yields

$$\theta^w(p,q) = \lim_{n\to\infty} \frac{\varrho(\Lambda(n))}{|\Lambda(n)|} \qquad \Phi^w_\infty \text{ almost surely}\,.$$

However we choose to prove a simpler result, which is enough for our purpose.

*Proof.* For $n \geq 1$, we have

$$\Phi^w_{\Lambda(n)}\big(\varrho(\Lambda(n))\big) \geq \Phi^w_{\Lambda(3n)}\big(\varrho(\Lambda(n)\big) \geq \sum_{x\in\Lambda(n)} \Phi^w_{\Lambda(3n)}\big(x \longleftrightarrow x+\partial^{in}\Lambda(2n)\big)$$

$$\geq \sum_{x\in\Lambda(n)} \Phi^w_{x+\Lambda(4n)}\big(x \longleftrightarrow x+\partial^{in}\Lambda(2n)\big) \geq |\Lambda(n)|\,\Phi^w_{\Lambda(4n)}\big(0 \longleftrightarrow \partial^{in}\Lambda(4n)\big)\,.$$

The first inequality holds because $\Phi^w_{\Lambda(n)}$ dominates $\Phi^w_{\Lambda(3n)}$ (see section 4.4). The third inequality holds because $\Lambda(3n) \subset x + \Lambda(4n)$ for any $x \in \Lambda(n)$, thus $\Phi^w_{x+\Lambda(4n)}$ dominates $\Phi^w_{\Lambda(3n)}$. Sending $n$ to $\infty$, we get

$$\liminf_{n\to\infty} \Phi^w_{\Lambda(n)}\Big(\frac{\varrho(\Lambda(n))}{|\Lambda(n)|}\Big) \geq \theta^w(p,q)\,.$$

Conversely, let $m \leq n$. Writing

$$\varrho(\Lambda(n)) \leq \sum_{x\in\Lambda(n-m)} 1_{\{x \longleftrightarrow \partial^{in}\Lambda(n)\}} + |\Lambda(n)\setminus\Lambda(n-m)|$$

and because $\Phi^w_{x+\Lambda(m)}$ dominates $\Phi^w_{\Lambda(n)}$ for $x \in \Lambda(n-m)$ (see section 4.4), we get

$$\Phi^w_{\Lambda(n)}\big(\varrho(\Lambda(n))\big) \leq \sum_{x\in\Lambda(n-m)} \Phi^w_{x+\Lambda(m)}\big(x \longleftrightarrow x+\partial^{in}\Lambda(m)\big) + |\Lambda(n)\setminus\Lambda(n-m)|\,.$$

By translation invariance, sending $n$ to $\infty$, we obtain

$$\limsup_{n\to\infty}\ \Phi^w_{\Lambda(n)}\Big(\frac{\varrho(\Lambda(n))}{|\Lambda(n)|}\Big) \leq \Phi^w_{\Lambda(m)}\big(0 \longleftrightarrow \partial^{in}\Lambda(m)\big)\,.$$

Sending $m$ to $\infty$ gives the desired inequality. □

Next we extend our volume large deviation estimate to FK percolation.

**Theorem 8.8.** *Let $p \in ]0,1[$ and $q \in [1,+\infty[$. The sequence of the laws of $n^{-d}\varrho(\Lambda(n))$ under the probability $\Phi^w_{\Lambda(n)}$, $n \geq 1$, satisfies a large deviation principle with speed $n^d$ and governed by the good rate function $I$ given by*

$$\forall \alpha\in[0,1] \qquad I(\alpha) = \sup_{t\in\mathbb{R}}\Big(t\alpha - \lim_{n\to\infty}\frac{1}{n^d}\ln\Phi^w_{\Lambda(n)}\Big(\exp\big(t\varrho(\Lambda(n))\big)\Big)\Big)\,.$$

*This rate function vanishes on $[0,\theta^w(p,q)]$ and is positive on $]\theta^w(p,q),1]$.*

*Proof.* All the arguments in the proof of theorem 8.3 and proposition 8.4 are valid for FK percolation (only replace $P$ by $\Phi^w_{\Lambda(n)}$ and $\theta$ by $\theta^w$ in the proof), except for proving that $I$ vanishes on $[0,\theta^w]$, is positive on $]\theta^w,1]$ and continuous at 1. Let us prove that $I$ vanishes on $[0,\theta^w]$. For $\alpha\geq 0$, we have

$$\Phi^w_{\Lambda(n)}\Big(\frac{\varrho(\Lambda(n))}{|\Lambda(n)|}\Big) \leq \alpha + \Phi^w_{\Lambda(n)}\Big(\frac{\varrho(\Lambda(n))}{|\Lambda(n)|}\geq\alpha\Big)\,,$$

whence, using lemma 8.6,

$$\liminf_{n\to\infty}\ \Phi^w_{\Lambda(n)}\Big(\frac{\varrho(\Lambda(n))}{|\Lambda(n)|}\geq\alpha\Big) \geq \theta^w-\alpha\,.$$

Therefore $I$ vanishes on $[0,\theta^w[$, and by continuity, it vanished also at $\theta^w$.

A slight modification has to be done to control the right–hand side of inequality (∘) (before proposition 8.4). We will rely on a conditioning. Let $\alpha>\theta^w$. For $m\geq 1$, we define $\Lambda'(m)=\Lambda(m)\setminus\partial^{in}\Lambda(m)$. By lemma 8.6, there exists an integer $m\geq 1$ such that

$$\Phi^w_{\Lambda'(m)}\Big(\frac{\varrho(\Lambda'(m))}{|\Lambda'(m)|}\Big) < \frac{\alpha+\theta^w}{2}\,.$$

Let $n$ be an integer larger than $m$. Let $n=mq+r$ be the Euclidean division of $n$ by $m$. We partition $\Lambda(n)$ into $q^d$ disjoint translates of $\Lambda(m)$, denoted by $\Lambda_i$, $1\leq i\leq q^d$, which cover $\Lambda(qm)$, and a remaining region $\Lambda_{q^d+1}=\Lambda(n)\setminus\Lambda(qm)$. For $i\in\{1\cdots q^d\}$, we set $\Lambda'_i=\Lambda_i\setminus\partial^{in}\Lambda_i$. By the subadditivity of $\varrho$, we have

$$\varrho(\Lambda(n)) \leq \sum_{1\leq i\leq q^d} \left(\varrho(\Lambda_i') + \varrho(\partial^{in}\Lambda_i)\right) + |\Lambda(n)\setminus\Lambda(qm)|\,.$$

We suppose that $m, n$ are large enough so that

$$\alpha - \frac{|\Lambda(n)\setminus\Lambda(qm)|}{|\Lambda(n)|} - q^d\frac{|\partial^{in}\Lambda(m)|}{|\Lambda(n)|} \geq \frac{\alpha+\theta^w}{2}\,.$$

Combining the previous inequalities, we see that

$$\Phi^w_{\Lambda(n)}\left(\frac{\varrho(\Lambda(n))}{|\Lambda(n)|} \geq \alpha\right) \leq \Phi^w_{\Lambda(n)}\left(\frac{1}{q^d}\sum_{1\leq i\leq q^d}\frac{\varrho(\Lambda_i')}{|\Lambda_i'|} \geq \frac{\alpha+\theta^w}{2}\right).$$

Let $E$ be the event

$$E = \Big\{\text{all the edges having one endpoint in } \bigcup_{1\leq i\leq q^d}\partial^{in}\Lambda_i \text{ are open}\Big\}.$$

By the FKG property (see section 4.4),

$$\Phi^w_{\Lambda(n)}\left(\frac{1}{q^d}\sum_{1\leq i\leq q^d}\frac{\varrho(\Lambda_i')}{|\Lambda_i'|} \geq \frac{\alpha+\theta^w}{2}\right) \leq \Phi^w_{\Lambda(n)}\left(\frac{1}{q^d}\sum_{1\leq i\leq q^d}\frac{\varrho(\Lambda_i')}{|\Lambda_i'|} \geq \frac{\alpha+\theta^w}{2}\,\Big|\, E\right)$$

Conditionally on $E$, all the vertices of $\bigcup_{1\leq i\leq q^d}\partial^{in}\Lambda_i$ are wired together, so that by the Markov property (see section 4.4), events occurring in different boxes $\Lambda_i'$ are independent. In particular, conditionally on $E$, the random variables $\varrho(\Lambda_i')/|\Lambda_i'|$, $1\leq i\leq q^d$, are i.i.d., their common law being the law of $\varrho(\Lambda'(m))/|\Lambda'(m)|$ under $\Phi^w_{\Lambda'(m)}$. Hence their common expectation is strictly less than $(\alpha+\theta^w)/2$ and we conclude with the help of the classical Cramér theorem in $\mathbb{R}$ (see theorems 1.3 and 1.5) as in the proof of theorem 8.3. Let us do the beginning of the proof of the continuity of $I$ at 1. Let $m\leq n$ be two integers. We use the same partition $(\Lambda_i,\ 1\leq i\leq q^d+1)$ of $\Lambda(n)$ as before. For $i\in\{\,1\cdots q^d\,\}$, we set $\Lambda_i'=\Lambda_i\setminus\partial^{in}\Lambda_i$. Let $\varepsilon,\delta$ be such that $0<\varepsilon<\delta<1$ and let us define

$$I(\delta) = \left\{\, i : 1\leq i\leq q^d,\ \varrho(\Lambda_i') \geq (1-\delta)|\Lambda_i'|\,\right\}.$$

We have then

$$\begin{aligned}\varrho(\Lambda(n)) &\leq \left(q^d - |I(\delta)|\right)(1-\delta)|\Lambda'(m)| + |I(\delta)||\Lambda'(m)| + |\Lambda(n)| - q^d|\Lambda'(m)|\\ &= \delta|I(\delta)||\Lambda'(m)| + |\Lambda(n)| - \delta q^d|\Lambda'(m)|\,.\end{aligned}$$

The event $\varrho(\Lambda(n)) \geq (1-\varepsilon)|\Lambda(n)|$ implies that

$$q^d - |I(\delta)| \leq \left\lceil\frac{\varepsilon|\Lambda(n)|}{\delta|\Lambda'(m)|}\right\rceil = N\,.$$

Decomposing this event over the possible choices of the set $I(\delta)$, we get with the help of the FKG inequality and the Markov property (see section 4.4),

$$\Phi^w_{\Lambda(n)}\big(\varrho(\Lambda(n)) \geq (1-\varepsilon)|\Lambda(n)|\big) \leq \sum_{\substack{q^d-N\leq k\leq q^d\\ 1\leq i_1<\cdots<i_k\leq q^d}} \Phi^w_{\Lambda(n)}\big(I(\delta)=\{i_1,\dots,i_k\}\big) \leq$$

$$\sum_{\substack{q^d-N\leq k\leq q^d\\ 1\leq i_1<\cdots<i_k\leq q^d}} \Phi^w_{\Lambda(n)}\left(\begin{matrix}\forall j\in\{1\cdots k\}\\ \varrho(\Lambda'_{i_j})\geq(1-\delta)|\Lambda'_{i_j}|\end{matrix}\;\middle|\;\begin{matrix}\text{the edges having at least one vertex}\\ \text{in } \partial^{in}\Lambda_{i_1}\cup\cdots\cup\partial^{in}\Lambda_{i_k} \text{ are open}\end{matrix}\right)$$

$$\leq \sum_{q^d-N\leq k\leq q^d} \binom{q^d}{k} \Phi^w_{\Lambda'(m)}\big(\varrho(\Lambda'(m))\geq(1-\delta)|\Lambda'(m)|\big)^k .$$

The end of the proof is then the same as in theorem 8.3. The expression of $I$ given in the statement of the theorem is a consequence of corollary 8.5. To prove the large deviation lower bound, we also need the following law of large numbers:

$$\forall\varepsilon>0\qquad \lim_{n\to\infty}\Phi^w_{\Lambda(n)}\Big(\Big|\frac{\varrho(\Lambda(n))}{|\Lambda(n)|}-\theta^w\Big|>\varepsilon\Big) = 0.$$

This result is an immediate consequence of lemma 8.6 and the fact that $J$ is positive on $]\theta^w,1]$. The argument is detailed in the case of the Ising model (see corollary 8.13). □

The volume large deviation estimates are very robust with respect to boundary conditions. Indeed, the FK measures possess the following decoupling property.

**Supersurface decoupling.** Let $\Lambda$ be a box. For any boundary conditions $\pi_1$, $\pi_2$ on $\partial^{in}\Lambda$ (see section 4.4 or [112] for the definition), any event $A$ depending on the edges in $\Lambda$,

$$\Phi^{\pi_1}_\Lambda(A) \leq q^{2|\partial^{in}\Lambda|}\Phi^{\pi_2}_\Lambda(A).$$

The proof is straightforward: changing the boundary conditions can only affect the counting factor $cl^\pi$ by $|\partial^{in}\Lambda|$. More precisely, for any configuration $\omega$ in $\Lambda$, we have $|cl^{\pi_1}(\omega)-cl^{\pi_2}(\omega)|\leq|\partial^{in}\Lambda|$, hence

$$\Phi^{\pi_1}_\Lambda(A) = \frac{\sum_{\omega\in A}\big(\prod_{e\in\mathbb{E}^d(\Lambda)}p^{\omega(e)}(1-p)^{1-\omega(e)}\big)\,q^{cl^{\pi_1}(\omega)}}{\sum_{\omega\in\Omega_\Lambda}\big(\prod_{e\in\mathbb{E}^d(\Lambda)}p^{\omega(e)}(1-p)^{1-\omega(e)}\big)\,q^{cl^{\pi_1}(\omega)}}$$

$$\leq q^{2|\partial^{in}\Lambda|}\frac{\sum_{\omega\in A}\big(\prod_{e\in\mathbb{E}^d(\Lambda)}p^{\omega(e)}(1-p)^{1-\omega(e)}\big)\,q^{cl^{\pi_2}(\omega)}}{\sum_{\omega\in\Omega_\Lambda}\big(\prod_{e\in\mathbb{E}^d(\Lambda)}p^{\omega(e)}(1-p)^{1-\omega(e)}\big)\,q^{cl^{\pi_2}(\omega)}} = q^{2|\partial^{in}\Lambda|}\Phi^{\pi_2}_\Lambda(A).$$

The supersurface decoupling readily implies that all the finite volume FK measures satisfy the same volume large deviation principle, and so does $\Phi_\infty$.

**Corollary 8.9.** *Let $p \in ]0,1[$ and $q \in [1,+\infty[$. Let $\pi(n)$, $n \geq 1$, be a sequence of boundary conditions on $\partial^{in}\Lambda(n)$. The sequence of the laws of $n^{-d}\varrho(\Lambda(n))$ under $\Phi^{\pi(n)}_{\Lambda(n)}$, $n \geq 1$, satisfies the same large deviation principle as in theorem 8.8.*

The supersurface decoupling implies that spatially disjoint events are decoupled up to supersurface order. More precisely, let $D_i$, $i \in I$, be a finite collection of disjoint compact subsets of $[-1/2,1/2]^d$. For $i \in I$, let $(S^i_n)_{n \geq 1}$ be a sequence of events such that $S^i_n$ depends only on the edges in $nD_i \cap \Lambda(n)$. Then

$$\forall \alpha > d-1 \qquad \limsup_{n\to\infty} \frac{1}{n^\alpha} \ln \Phi^w_{\Lambda(n)}\Big(\bigcap_{i\in I} S^i_n\Big) = \sum_{i\in I} \limsup_{n\to\infty} \frac{1}{n^\alpha} \ln \Phi^w_{\Lambda(n)}(S^i_n)\,.$$

The same equality is valid when lim sup is replaced by lim inf. To prove the surface large deviation principle, we will need decoupling until surface order (see section 10).

## 8.3 Ising model

The proof of the volume large deviation principle for the Ising model is very similar in spirit to the FK percolation case. However it does not seem obvious to deduce directly this large deviation principle from the corresponding result in FK percolation. A potential problem is that the probabilistic estimates can be transferred up to (and not including) volume large deviations. Indeed, the colouring operation induces also random fluctuations, with large deviations occurring on the volume level. Starting from a FK configuration, the Ising configuration associated in the coupling might present an excess of pluses either because there are too many points connected to $\partial^{in}\Lambda(n)$ or because too many small clusters are colored +. Therefore the link between the rate functions of theorems 8.8 and 8.12 is not even elucidated and one has to do a full independent proof of theorem 8.14. However, we will omit some details which are slight variants of the percolation case.

For $A$ a finite subset of $\mathbb{Z}^d$ and $\sigma$ a spin configuration in $A$, we define

$$M(A) = \sum_{x\in A} \sigma(x)\,,$$

i.e., $M(A)$ is simply the sum of the spins in $A$. The map $M$ is additive:

$$\forall A, B \subset \mathbb{Z}^d, \quad A \cap B = \emptyset \quad \Rightarrow \quad M(A \cup B) = M(A) + M(B)\,.$$

**Lemma 8.10.** *For any $T > 0$, we have*

$$\lim_{n\to\infty} \mu^+_{\Lambda(n),T}\Big(\frac{M(\Lambda(n))}{|\Lambda(n)|}\Big) = m^*(T)\,.$$

*Remark 8.11.* Since the infinite volume measure $\mu_T^+$ is ergodic and translation invariant [66], the multiparameter subadditive ergodic theorem ([1], theorem 2.4 or [122], theorem 1.1) yields

$$\lim_{n\to\infty} \frac{M(\Lambda(n))}{|\Lambda(n)|} = m^*(T) = \mu_T^+(\sigma(0)) \qquad \mu_T^+ \text{ almost surely}.$$

However we prefer to avoid the use of the delicate subadditive ergodic theorem, and we prefer also to work with finite volume Gibbs measures.

*Proof.* We could write a proof similar to the proof of lemma 8.6. However we prefer to deduce this result from lemma 8.6 with the help of the coupling between the FK and Ising models. Indeed, for any box $\Lambda$ and for $T > 0$, we have

$$\forall x \in \Lambda \qquad \mu_{\Lambda,T}^+(\sigma(x)) = \Phi_\Lambda^{w,p,2}(x \longleftrightarrow \partial^{in}\Lambda),$$

where $p = 1 - \exp(-2/T)$ (see section 4.3). Summing over $x \in \Lambda$, we get

$$\mu_{\Lambda,T}^+(M(\Lambda)) = \Phi_\Lambda^{w,p,2}(\varrho(\Lambda)).$$

Applying this identity to $\Lambda(n)$ for $n \geq 1$, dividing by $|\Lambda(n)|$ and letting $n$ go to $\infty$, we obtain with the help of lemma 8.6 that

$$\lim_{n\to\infty} \mu_{\Lambda(n),T}^+\Big(\frac{M(\Lambda(n))}{|\Lambda(n)|}\Big) = \lim_{n\to\infty} \Phi_{\Lambda(n)}^{w,p,2}\Big(\frac{\varrho(\Lambda(n))}{|\Lambda(n)|}\Big) = \theta^w(p,2).$$

Since $\theta^w(p,2) = m^*(T)$, the desired conclusion follows. □

**Theorem 8.12.** *Let $T > 0$. For any $\alpha \in [-1,1]$, the limit*

$$J(\alpha) = \lim_{n\to\infty} -\frac{1}{n^d} \ln \mu_{\Lambda(n),T}^+\Big(\frac{M(\Lambda(n))}{|\Lambda(n)|} \geq \alpha\Big)$$

*exists and is finite. The map $\alpha \in [-1,1] \mapsto J(\alpha) \in \mathbb{R}^+$ is convex continuous. It vanishes on $[-1, m^*(T)]$ and it is strictly positive on $]m^*(T), 1]$.*

*Proof.* Since the temperature $T$ is fixed throughout the proof, we remove it from the notation when possible, writing for instance $\mu_\Lambda^\eta$ instead of $\mu_{\Lambda,T}^\eta$. Let $m \leq n$ be two integers. Let $n = mq+r$ be the Euclidean division of $n$ by $m$. We partition $\Lambda(n)$ into $q^d$ disjoint translates of $\Lambda(m)$, denoted by $\Lambda_i$, $1 \leq i \leq q^d$, which cover $\Lambda(qm)$, and a remaining region $\Lambda_{q^d+1} = \Lambda(n) \setminus \Lambda(qm)$. Let $E$ be the event

$$E = \Big\{ \text{all the sites in } \Lambda_{q^d+1} \cup \bigcup_{1\leq i\leq q^d} \partial^{in}\Lambda_i \text{ are set to } + \Big\}.$$

Using the additivity of $M$, we have

$$E \cap \bigcap_{1\leq i\leq q^d} \{ M(\Lambda_i) \geq \alpha|\Lambda_i| \} \subset \{ M(\Lambda(n)) \geq \alpha|\Lambda(n)| \},$$

whence by the FKG inequality and the Markov property (see section 2.4),

$$\mu^+_{\Lambda(n)}\Big(\frac{M(\Lambda(n))}{|\Lambda(n)|}\geq\alpha\Big)\geq\mu^+_{\Lambda(n)}(E)\,\mu^+_{\Lambda(m)}\Big(\frac{M(\Lambda(m))}{|\Lambda(m)|}\geq\alpha\Big)^{q^d}.$$

The number of sites involved in the definition of the event $E$ is at most $q^d|\partial^{in}\Lambda(m)|+|\partial^{in}\Lambda(n)|m$. Taking logarithms and sending $n$ to $\infty$, we get

$$\liminf_{n\to\infty}\frac{1}{|\Lambda(n)|}\ln\mu^+_{\Lambda(n)}\Big(\frac{M(\Lambda(n))}{|\Lambda(n)|}\geq\alpha\Big)\geq$$
$$\frac{1}{|\Lambda(m)|}\ln\mu^+_{\Lambda(m)}\Big(\frac{M(\Lambda(m))}{|\Lambda(m)|}\geq\alpha\Big)-4d\frac{|\partial^{in}\Lambda(m)|}{T|\Lambda(m)|}.$$

Sending now $m$ to $\infty$, we see that the limit defining $J(\alpha)$ in the statement of the theorem exists. Moreover

$$\mu^+_{\Lambda(n)}\Big(\frac{M(\Lambda(n))}{|\Lambda(n)|}\geq\alpha\Big)\geq\mu^+_{\Lambda(n)}\begin{pmatrix}\text{all the sites of }\Lambda(n)\\ \text{are set to plus}\end{pmatrix}\geq\exp-\frac{4d|\Lambda(n)|}{T},$$

so that $J(\alpha)\leq 4d/T$ is finite. To prove that $J$ is convex and continuous, we proceed exactly as in theorem 8.3, using the event $E$ defined at the beginning of this proof. To prove the continuity at 1, some little extra care is needed; the argument is detailed in the case of the FK percolation model (see theorem 8.8). We finally prove that $J$ is positive on $]m^*,1]$. Let $\alpha\in]m^*,1]$. There exists an integer $m\geq 1$ such that

$$\mu^+_{\Lambda(m)}\Big(\frac{M(\Lambda(m))}{|\Lambda(m)|}\Big)<\frac{\alpha+m^*}{2}.$$

Let $n$ be an integer larger than $m$. Using again the same partition of $\Lambda(n)$ as before, by the subadditivity of $M$, we have

$$M(\Lambda(n))\leq\sum_{1\leq i\leq q^d+1}M(\Lambda_i)\leq\sum_{1\leq i\leq q^d}M(\Lambda_i)+|\Lambda(n)\setminus\Lambda(qm)|.$$

Let $n$ be large enough so that

$$\alpha-\frac{|\Lambda(n)\setminus\Lambda(qm)|}{|\Lambda(n)|}\geq\frac{\alpha+m^*}{2}.$$

Let $E$ be the event defined at the beginning of this proof. Applying the FKG inequality and combining the previous inequalities, we see that

$$\mu^+_{\Lambda(n)}\Big(\frac{M(\Lambda(n))}{|\Lambda(n)|}\geq\alpha\Big)\leq\mu^+_{\Lambda(n)}\Big(\frac{M(\Lambda(n))}{|\Lambda(n)|}\geq\alpha\,\Big|\,E\Big)$$
$$\leq\mu^+_{\Lambda(n)}\Big(\frac{1}{q^d}\sum_{1\leq i\leq q^d}\frac{M(\Lambda_i)}{|\Lambda_i|}\geq\frac{\alpha+m^*}{2}\,\Big|\,E\Big).$$

By the Markov property (see section 2.4), conditionally on $E$, the random variables $M(\Lambda_i)/|\Lambda_i|$, $1 \leq i \leq q^d$, are i.i.d., their common law being the law of $M(\Lambda(m))/|\Lambda(m)|$ under $\mu^+_{\Lambda(m)}$. Hence their common expectation is strictly less than $(\alpha+m^*)/2$. By the classical Cramér theorem in $\mathbb{R}$ (see theorems 1.3 and 1.5), there exists $c > 0$ such that the right–hand quantity of the above inequality is bounded by $\exp(-cq^d)$, whence

$$\limsup_{n\to\infty} \frac{1}{|\Lambda(n)|} \ln \mu^+_{\Lambda(n)}\Big(\frac{M(\Lambda(n))}{|\Lambda(n)|} \geq \alpha\Big) \leq -\frac{c}{|\Lambda(m)|} < 0$$

and we conclude that $J(\alpha) > 0$. □

Lemma 8.10 and theorem 8.12 imply the following weak law of large numbers.

**Corollary 8.13.** *For any $T > 0$, we have*

$$\forall \varepsilon > 0 \qquad \lim_{n\to\infty} \mu^+_{\Lambda(n),T}\Big(\Big|\frac{M(\Lambda(n))}{|\Lambda(n)|} - m^*\Big| > \varepsilon\Big) = 0\,.$$

*Proof.* Let $\varepsilon > 0$. Theorem 8.12 readily implies that

$$\lim_{n\to\infty} \mu^+_{\Lambda(n)}\Big(\frac{M(\Lambda(n))}{|\Lambda(n)|} > m^* + \varepsilon\Big) = 0\,.$$

Let $\delta > 0$. By decomposing the expectation over the intervals $[-1, m^* - \delta[$, $[m^* - \delta, m^* + \varepsilon]$, $]m^* + \varepsilon, 1]$, we have

$$\begin{aligned}
\mu^+_{\Lambda(n)}\Big(\frac{M(\Lambda(n))}{|\Lambda(n)|}\Big) &\leq (m^* - \delta)\mu^+_{\Lambda(n)}\Big(\frac{M(\Lambda(n))}{|\Lambda(n)|} < m^* - \delta\Big) \\
&\quad +(m^* + \varepsilon)\mu^+_{\Lambda(n)}\Big(\frac{M(\Lambda(n))}{|\Lambda(n)|} \geq m^* - \delta\Big) + \mu^+_{\Lambda(n)}\Big(\frac{M(\Lambda(n))}{|\Lambda(n)|} > m^* + \varepsilon\Big) \\
&= \mu^+_{\Lambda(n)}\Big(\frac{M(\Lambda(n))}{|\Lambda(n)|} < m^* - \delta\Big)(-\delta - \varepsilon) + m^* + \varepsilon + \mu^+_{\Lambda(n)}\Big(\frac{M(\Lambda(n))}{|\Lambda(n)|} > m^* + \varepsilon\Big).
\end{aligned}$$

Sending $n$ to $\infty$ and using lemma 8.10, we get

$$0 \leq -(\delta + \varepsilon) \limsup_{n\to\infty} \mu^+_{\Lambda(n)}\Big(\frac{M(\Lambda(n))}{|\Lambda(n)|} < m^* - \delta\Big) + \varepsilon\,.$$

Sending now $\varepsilon$ to 0, we conclude that

$$\lim_{n\to\infty} \mu^+_{\Lambda(n)}\Big(\frac{M(\Lambda(n))}{|\Lambda(n)|} < m^* - \delta\Big) = 0$$

as required. □

**Theorem 8.14.** *Let $T > 0$. Let $\eta \in \{-,+\}^{\mathbb{Z}^d}$. The sequence of the laws of $n^{-d}\sum_{x\in\Lambda(n)} \sigma(x)$ under the finite volume Gibbs measures $\mu^\eta_{\Lambda(n),T}$, $n \geq 1$, satisfies a large deviation principle with speed $n^d$ and governed by the good rate function $I$ given by $I(x) = J(|x|)$ for $x \in [-1,1]$, where $J$ is the function defined in theorem 8.12. This rate function does not depend on $\eta$. It vanishes on $[-m^*, m^*]$ and it is positive on $[-1, -m^*[\cup]m^*, 1]$.*

*Remark 8.15.* Let $\mu_T$ be any infinite volume Gibbs measure corresponding to the temperature $T$. Under $\mu_T$, the sequence $n^{-d}\sum_{x\in\Lambda(n)}\sigma(x)$, $n\geq 1$, satisfies the same large deviation principle.

*Remark 8.16.* As in corollary 8.5, for any $t\in\mathbb{R}$, the limit

$$p(t) = \lim_{n\to\infty}\frac{1}{n^d}\ln\mu^+_{\Lambda(n),T}\Big(\exp\big(tM(\Lambda(n))\big)\Big)$$

exists. The function $p(t)$ is called the pressure. The rate function $I$ of theorem 8.14 is the Fenchel–Legendre transform of $p(t)$.

*Proof.* Since the temperature $T$ is fixed throughout the proof, we remove it from the notation when possible, writing for instance $\mu^\eta_\Lambda$ instead of $\mu^\eta_{\Lambda,T}$. To deal with the boundary conditions, we rely on the following decoupling inequality, whose proof is similar to the proof of the supersurface decoupling inequality for FK percolation.

**Supersurface decoupling.** For any box $\Lambda$, any boundary conditions $\eta$, $\xi$ on $\partial^{in}\Lambda$, any event $A$ depending on the sites in $\Lambda\setminus\partial^{in}\Lambda$,

$$\mu^\eta_\Lambda(A)\leq\exp(4d|\partial^{in}\Lambda|/T)\,\mu^\xi_\Lambda(A)\,.$$

That $I$ is a good rate function is a consequence of the convexity of $J$ on $[0,1]$ and the continuity of $J$ proved in theorem 8.12. The large deviation upper bound follows from the definition of $J$, because $J$ is non–decreasing on $[0,1]$. Indeed, let $F$ be a closed subset of $[-1,1]$ and let

$$a = \sup F\cap[-1,0]\,,\qquad b = \inf F\cap[0,1]\,.$$

Then

$$\mu^\eta_{\Lambda(n)}\Big(\frac{M(\Lambda(n))}{|\Lambda(n)|}\in F\Big)\leq\mu^\eta_{\Lambda(n)}\Big(\frac{M(\Lambda(n))}{|\Lambda(n)|}\leq a\Big)+\mu^\eta_{\Lambda(n)}\Big(\frac{M(\Lambda(n))}{|\Lambda(n)|}\geq b\Big)\,.$$

Using supersurface decoupling and a change of sign, we have

$$\limsup_{n\to\infty}\frac{1}{n^d}\ln\mu^\eta_{\Lambda(n)}\Big(\frac{M(\Lambda(n))}{|\Lambda(n)|}\leq a\Big) = \limsup_{n\to\infty}\frac{1}{n^d}\ln\mu^-_{\Lambda(n)}\Big(-\frac{M(\Lambda(n))}{|\Lambda(n)|}\geq -a\Big) = J(-a)\,.$$

Indeed, the law of $-M(\Lambda(n))$ under $\mu^-_{\Lambda(n)}$ is equal to the law of $M(\Lambda(n))$ under $\mu^+_{\Lambda(n)}$. Using lemma 6.7, we conclude that

$$\limsup_{n\to\infty}\frac{1}{n^d}\ln\mu^\eta_{\Lambda(n)}\Big(\frac{M(\Lambda(n))}{|\Lambda(n)|}\in F\Big)\leq-\min(J(-a),J(b)) = -\inf_F I\,.$$

To obtain the large deviation lower bound, we will prove the following local estimate: for any $\alpha\in[-1,1]$,

$$\forall \varepsilon > 0 \qquad \liminf_{n\to\infty} \frac{1}{n^d} \ln \mu^+_{\Lambda(n)}\Big(\Big|\frac{M(\Lambda(n))}{|\Lambda(n)|} - \alpha\Big| < \varepsilon\Big) \geq -I(\alpha)\,.$$

We consider separately the cases where $|\alpha| \geq m^*$ and $|\alpha| < m^*$. By symmetry, we need only to examine the situation where $\alpha \geq 0$. Let first $\alpha \in [m^*, 1]$ and let $\varepsilon > 0$. We write

$$\mu^+_{\Lambda(n)}\Big(\Big|\frac{M(\Lambda(n))}{|\Lambda(n)|} - \alpha\Big| < \varepsilon\Big) \geq \mu^+_{\Lambda(n)}\Big(\frac{M(\Lambda(n))}{|\Lambda(n)|} \geq \alpha\Big) - \mu^+_{\Lambda(n)}\Big(\frac{M(\Lambda(n))}{|\Lambda(n)|} \geq \alpha + \varepsilon\Big)\,.$$

The map $J$ is convex non–negative and vanishes only on $[0, m^*]$, thus it is increasing on $[m^*, 1]$ and $J(\alpha + \varepsilon) > J(\alpha)$. By the identity defining $J$ in theorem 8.12, the second term on the right–hand side is negligible compared to the first, whence

$$\liminf_{n\to\infty} \frac{1}{n^d} \ln \mu^+_{\Lambda(n)}\Big(\Big|\frac{M(\Lambda(n))}{|\Lambda(n)|} - \alpha\Big| < \varepsilon\Big) \geq -J(\alpha) = -I(\alpha)\,.$$

Now suppose that $m^* > 0$. Let $\alpha \in [0, m^*[$ and let $\varepsilon, \eta > 0$. By corollary 8.13, there exists an integer $m \geq 1$ such that

$$\forall n \geq m \qquad \mu^+_{\Lambda(n),T}\Big(\Big|\frac{M(\Lambda(n))}{|\Lambda(n)|} - m^*\Big| > \varepsilon\Big) < \eta\,.$$

Let $n$ be an integer larger than $m$. Let $s = \lfloor (1-\alpha/m^*)^{1/d} n/2 \rfloor$, $q_1 = \lfloor n/2m \rfloor$, $q_2 = \lfloor s/(2m) \rfloor + 1$. We partition $\Lambda(2mq_1) \setminus \Lambda(2mq_2)$ into $N = 2^d q_1^d - 2^d q_2^d$ disjoint translates of $\Lambda(m)$, denoted by $\Lambda_i$, $1 \leq i \leq N$. Let $E, F$ be the events

$$E = \Big\{ \text{all the sites of } \big(\Lambda(n) \setminus \Lambda(2mq_1)\big) \cup \bigcup_{1\leq i\leq N} \partial^{in} \Lambda_i \text{ are set to } + \Big\}\,,$$

$$F = \Big\{ \text{all the sites of } \partial^{in} \Lambda(2mq_2) \text{ are set to } - \Big\}\,.$$

On the event

$$E \cap F \cap \Big\{ \Big|\frac{M(\Lambda(2mq_2))}{|\Lambda(2mq_2)|} + m^*\Big| < \varepsilon \Big\} \cap \bigcap_{1\leq i\leq N} \Big\{ \Big|\frac{M(\Lambda_i)}{|\Lambda_i|} - m^*\Big| < \varepsilon \Big\}$$

we have $\big|M(\Lambda(n))/|\Lambda(n)| - \alpha\big| \leq 2\varepsilon$, whence, using the Markov property (see section 2.4),

$$\begin{aligned}
\mu^+_{\Lambda(n)}\Big(\Big|\frac{M(\Lambda(n))}{|\Lambda(n)|} - \alpha\Big| < 2\varepsilon\Big) &\geq \mu^+_{\Lambda(n)}(E \cap F) \times \\
\mu^+_{\Lambda(n)}\Big(\Big|\frac{M(\Lambda(2mq_2))}{|\Lambda(2mq_2)|} + m^*\Big| < \varepsilon, \Big|&\frac{M(\Lambda_i)}{|\Lambda_i|} - m^*\Big| < \varepsilon, 1 \leq i \leq N \,\Big|\, E \cap F\Big) \\
&\geq \mu^+_{\Lambda(n)}(E \cap F)\, \mu^-_{\Lambda(2mq_2)}\Big(\Big|\frac{M(\Lambda(2mq_2))}{|\Lambda(2mq_2)|} + m^*\Big| < \varepsilon\Big) \\
&\qquad \times \mu^+_{\Lambda(m)}\Big(\Big|\frac{M(\Lambda(m))}{|\Lambda(m)|} - m^*\Big| < \varepsilon\Big)^N\,.
\end{aligned}$$

The number of sites involved in the definition of the events $E, F$ is negligible compared to $|\Lambda(n)|$. Taking logarithms and sending $n$ to infinity, we get

$$\liminf_{n\to\infty} \frac{1}{|\Lambda(n)|} \ln \mu^+_{\Lambda(n)}\Big(\Big|\frac{M(\Lambda(n))}{|\Lambda(n)|} - \alpha\Big| < 2\varepsilon\Big) \geq \frac{\alpha}{m^*|\Lambda(m)|} \ln(1-\eta)\,.$$

Sending $\eta$ to 0 yields the local large deviation estimate we need. □

Large deviations of volume order for the Gibbs measures have been investigated for a long time. An important issue was to provide a rigorous derivation of the famous Gibbs variational principle, which characterizes the Gibbs states as the solutions to a certain variational problem. The relevant object is the empirical measure, defined by

$$\forall \sigma \in \{-,+\}^{\mathbb{Z}^d} \qquad T_\Lambda(\sigma) = \frac{1}{|\Lambda|}\sum_{x\in\Lambda} \delta_{T_x(\sigma)}$$

where $T_x$ is the translation operator on $\{-,+\}^{\mathbb{Z}^d}$ defined by

$$\forall x, y \in \mathbb{Z}^d \qquad T_x(\sigma)(y) = \sigma(x+y)\,,$$

and $\delta_{T_x(\sigma)}$ is the Dirac mass on the configuration $T_x(\sigma)$. Hence $T_\Lambda(\sigma)$ is a random measure on the set $\{-,+\}^{\mathbb{Z}^d}$. If $\sigma$ is a sample from a translation invariant Gibbs measure $\mu$, then $T_{\Lambda(n)}(\sigma)$ satisfies a large deviation principle in the set $\mathcal{M}_1(\{-,+\}^{\mathbb{Z}^d})$ of the probability measures on $\{-,+\}^{\mathbb{Z}^d}$ equipped with a suitable weak topology, with speed $n^d$ and with rate function $s$ defined as follows: for any translation invariant probability measure $\nu$ on $\{-,+\}^{\mathbb{Z}^d}$,

$$s(\nu|\mu) = \lim_{n\to\infty} \frac{1}{|\Lambda(n)|}\mathcal{H}_{\Lambda(n)}(\nu|\mu)$$

and $s(\nu|\mu) = +\infty$ otherwise; for a box $\Lambda$, $\mathcal{H}_\Lambda(\nu|\mu)$ is the relative entropy given by

$$\mathcal{H}_\Lambda(\nu|\mu) = \sum_{\eta\in\{-,+\}^\Lambda} \nu\big(\forall x\in\Lambda \quad \sigma(x)=\eta(x)\big) \ln \frac{\nu\big(\forall x\in\Lambda \quad \sigma(x)=\eta(x)\big)}{\mu\big(\forall x\in\Lambda \quad \sigma(x)=\eta(x)\big)}\,.$$

This rate function $s$ is the limit of finite volume entropies. This result has been proven in several works [47, 70, 71, 108]. The most recent and efficient presentation can be found in [110]. For a related result for the FK model, see [121]. By considering the map

$$\nu \in \mathcal{M}_1\big(\{-,+\}^{\mathbb{Z}^d}\big) \mapsto \nu(\sigma(0))$$

the large deviation principle stated in theorem 8.14 can be obtained from this large deviation principle by a routine application of the contraction principle (see theorem 6.8).

# Part V

# Fundamental probabilistic estimates

# 9

# Coarse graining

A successful strategy to understand the behavior of the percolation model or the Ising model on the macroscopic scale is to first understand its typical behavior on a scale intermediate between the microscopic and the macroscopic scales, sometimes called the mesoscopic scale. The coarse graining techniques aim at describing the typical behavior of the percolation configuration in a large box.

## 9.1 The good blocks

We first state the results for Bernoulli percolation; the extension to FK percolation is discussed in the next section. Let $\Lambda$ be a box. We define the inner vertex boundary $\partial^{in}\Lambda$ of $\Lambda$ as

$$\partial^{in}\Lambda = \{ x \in \Lambda : \exists y \in \mathbb{Z}^d \setminus \Lambda \quad |x-y| = 1 \} .$$

An open cluster within $\Lambda$ is called crossing for $\Lambda$ if it intersects each of the $2d$ faces of $\partial^{in}\Lambda$. For $m \in \mathbb{N}$ and $\varepsilon \in ]0,1[$, we define the following events:

$$\begin{aligned} U(\Lambda) &= \{ \text{there exists an open crossing cluster } C^* \text{ in } \Lambda \} , \\ R(\Lambda, m) &= U(\Lambda) \cap \{ \, C^* \text{ crosses every sub–box of } \Lambda \text{ with diameter } \geq m \} \\ &\qquad \cap \{ \text{ there exists a unique open cluster in } \Lambda \text{ with diameter } \geq m \} , \\ V(\Lambda, \varepsilon) &= U(\Lambda) \cap \{ \theta(1-\varepsilon) \, |\Lambda| \leq |C^*| \} , \\ W(\Lambda, \varepsilon) &= \{ \, |\{ x \in \Lambda : x \longleftrightarrow \partial^{in}\Lambda \}| \, \leq (1+\varepsilon) \, \theta \, |\Lambda| \, \} . \end{aligned}$$

We will next show that all these events are typical, i.e., their probabilities converge to 1 as the size of the box goes to $\infty$. We provide here estimates which are weaker and simpler than those of Pisztora, because we do not make appeal to renormalization at this stage, yet their essence and their form come from Pisztora's work [112].

**Lemma 9.1.** *For any $d \geq 2$, any $p > p_c$, any $\varepsilon > 0$, we have*

$$\lim_{n\to\infty} P\big(W(\Lambda(n),\varepsilon)\big) = 1\,.$$

*Proof.* This is a straightforward consequence of theorem 8.3 (or of theorem 8.8 for the FK percolation model). We even know that the convergence to 1 occurs at speed of order $\exp -cn^d$ for some positive constant $c$. □

We deal next with the other events. The mechanism of proof is rather different according to the dimension. In dimensions $d=2$, everything relies on the exponential decay of the connectivity function for the dual model. In dimensions $d \geq 3$, everything relies on the slab technology. We shall sketch the proofs in dimensions $d \geq 3$ and we refer to [51] in dimensions $d=2$.

Let $d \geq 3$. Grimmett and Marstrand [81] have proved that the critical point $p_c$ of Bernoulli percolation is the limit of the critical points associated to percolation in finite slabs $S(L) = \mathbb{Z}^{d-1} \times [1,L]$ as $L$ goes to $\infty$. This in turn implies that long range order occurs uniformly in the finite slab

$$S(n,L) = [-n,n]^{d-1} \times [1,L]$$

for $L$ large enough and $n \geq 1$. More precisely, there exist $L \in \mathbb{N}$ and $\alpha > 0$ such that

$$\forall n \geq 1 \quad \forall x,y \in S(n,L) \qquad P\big(x \longleftrightarrow y \text{ inside } S(n,L)\big) \geq \alpha\,.$$

In accordance with the literature, we keep here the notation $L$ for the height of the slab. This $L$ is distinct from the $L$ used later on when performing block arguments (in chapters 16, 18, 19). For a precise proof of this result, see [78], Lemma 7.78. This uniform estimate is the basic ingredient to build the coarse–graining estimates. It implies also the following lemma, which is a key to get the surface decoupling (see chapter 10 and proposition 10.3).

**Lemma 9.2.** *For any $d \geq 2$, any $p > p_c$, any $l \geq 1$,*

$$P\begin{pmatrix}\text{inside } S(n,l) \text{ there exist two disjoint clusters}\\ \text{joining } [-n,n]^{d-1}\times\{0\} \text{ to } [-n,n]^{d-1}\times\{l\}\end{pmatrix} \leq (2n+1)^{d-1}(1-\alpha)^{\lfloor l/L\rfloor}\,.$$

*Proof.* For $d=2$, this estimate is a consequence of the exponential decay for the dual connections. Indeed, if there exist two disjoint open clusters joining $[-n,n]^{d-1}\times\{0\}$ to $[-n,n]^{d-1}\times\{l\}$, then they are separated by a dual path of length at least $l$. We sketch now the proof in dimensions $d \geq 3$. Let $k \geq 0$. For $x \in [-n,n]^{d-1}\times\{0\}$, we define

$$V_k(x) = \big\{\, z \in [-n,n]^{d-1}\times\{kL\} : x \longleftrightarrow z \text{ in } [-n,n]^{d-1}\times[0,kL]\,\big\}\,.$$

For $x,y \in [-n,n]^{d-1}\times\{0\}$, we consider the event

$$E_k(x,y) = \big\{\, x,y \longleftrightarrow [-n,n]^{d-1}\times\{kL\},\ x \not\longleftrightarrow y \text{ in } [-n,n]^{d-1}\times[0,kL-1]\,\big\}\,.$$

Let $x, y \in [-n, n]^{d-1} \times \{0\}$. For $k \geq 1$, we have

$$\begin{aligned}
P(E_k(x,y)) &= P(E_k(x,y) \cap E_{k-1}(x,y)) \\
&= \sum_{A,B} P\big(E_k(x,y) \cap E_{k-1}(x,y),\, V_{k-1}(x) = A,\, V_{k-1}(y) = B\big) \\
&= \sum_{A,B} P\big(E_k(x,y) \,\big|\, E_{k-1}(x,y),\, V_{k-1}(x) = A,\, V_{k-1}(y) = B\big) \\
&\qquad\qquad \times P\big(E_{k-1}(x,y),\, V_{k-1}(x) = A,\, V_{k-1}(y) = B\big)\,.
\end{aligned}$$

If $E_k(x,y)$ occurs and $V_{k-1}(x) = A$, $V_{k-1}(y) = B$, then no vertex of $A$ is connected to a vertex of $B$ inside the set $[-n, n]^{d-1} \times [(k-1)L, kL-1]$, which is isometric to the slab $S(n, L)$. Therefore

$$P\big(E_k(x,y) \,\big|\, E_{k-1}(x,y),\, V_{k-1}(x) = A,\, V_{k-1}(y) = B\big) \;\leq\; 1 - \alpha$$

and $P(E_k(x,y)) \leq (1-\alpha)P(E_{k-1}(x,y))$. Iterating this inequality, we conclude that

$$P(E_k(x,y)) \;\leq\; (1-\alpha)^k\,.$$

The inequality stated in the lemma follows then easily. □

**Lemma 9.3.** *For any $d \geq 2$, any $p > p_c$, any $\varepsilon > 0$, we have*

$$\lim_{n \to \infty} P\big(V(\Lambda(n), \varepsilon)\big) \;=\; 1\,.$$

*Proof.* We provide here a sketch of the proof for $d \geq 3$. It is directly drawn from the proof for Bernoulli percolation presented in [78], chapter 7, Theorem 7.61. The original idea comes from Kesten and Zhang [93], which was also an important source of inspiration to [112]. For $m \in \mathbb{N}$ and $x, y \in \Lambda(m)$, we denote by $A_k(x,y)$ the event

$$A_k(x,y) \;=\; \big\{\, x, y \longleftrightarrow \partial^{in}\Lambda(m+2kL),\; x \not\longleftrightarrow y \text{ in } \Lambda(m+2kL) \,\big\}\,.$$

The shell $\Lambda(m+2kL) \setminus \Lambda(m+2(k-1)L)$ of width $L$ around $\Lambda(m+2(k-1)L)$ can be written as a finite union of $2d$ slabs which are obtained as translations or rotations of the slab $S(m+2kL, L)$. The long range order in each of these slabs implies that there is also long range order in the shell, i.e., there exists $\beta > 0$ such that

$$\begin{aligned}
&\forall k \geq 1 \quad \forall u, v \in \Lambda(m+2kL) \setminus \Lambda(m+2(k-1)L) \\
&\qquad P\big(u \longleftrightarrow v \text{ in } \Lambda(m+2kL) \setminus \Lambda(m+2(k-1)L)\big) \;\geq\; \beta\,.
\end{aligned}$$

If the event $A_k(x,y)$ occurs, then there exist two disjoint open clusters joining $\partial^{in}\Lambda(m+2(k-1)L)$ to $\partial^{in}\Lambda(m+2kL)$ and these open clusters are not connected together inside $\Lambda(m+2kL)$. By conditioning on the configuration in $\partial^{in}\Lambda(m+2(k-1)L)$, one gets

$$\forall k \geq 1 \qquad P\big(A_k(x,y)\,|\,A_{k-1}(x,y)\big) \,\leq\, 1-\beta\,.$$

By iterated conditioning, we have for any $x,y \in \Lambda(m)$,

$$P\big(\,x,y \longleftrightarrow \partial^{in}\Lambda(n),\, x \not\longleftrightarrow y \text{ in } \Lambda(n)\,\big) \,\leq\, \prod_{k=1}^{\lfloor\frac{n-m}{2L}\rfloor} P\big(A_k(x,y)\,|\,A_{k-1}(x,y)\big)\,.$$

These inequalities imply that

$$P\begin{pmatrix}\text{all the vertices belonging to } \Lambda(m)\\ \text{which are connected to } \partial^{in}\Lambda(n)\\ \text{are connected together inside } \Lambda(n)\end{pmatrix} \,\geq\, 1-|\Lambda(m)|^2(1-\beta)^{\left\lfloor\frac{n-m}{2L}\right\rfloor}\,.$$

We apply this inequality with $m = n - c\ln n$ where $c$ is a sufficiently large constant, so that the right–hand side goes to 1 as $n$ goes to $\infty$. By proposition 8.1,

$$\lim_{n\to\infty} P\Big(\frac{1}{|\Lambda(n)|}\,\big|\{\,x\in\Lambda(n) : x \longleftrightarrow \partial^{in}\Lambda(n)\,\}\big| \,>\, \theta(1-\frac{\varepsilon}{2})\Big) \,=\, 1\,,$$

whence

$$\lim_{n\to\infty} P\big(\partial^{in}\Lambda(n-c\ln n) \longleftrightarrow \partial^{in}\Lambda(n)\big) \,=\, 1\,.$$

Indeed, the shell $\Lambda(n)\setminus\Lambda(n-c\ln n)$ has a negligible fractional volume in $\Lambda(n)$ and we know that $\theta > 0$. Let $F_i$, $1\leq i\leq 2d$, be the faces of $\partial^{in}\Lambda(n)$. By the FKG inequality and the lattice symmetries, we have

$$P\big(\partial^{in}\Lambda(n-c\ln n) \not\longleftrightarrow \partial^{in}\Lambda(n)\big) \,=\, P\Big(\partial^{in}\Lambda(n-c\ln n) \not\longleftrightarrow \bigcup_{1\leq i\leq 2d} F_i\Big)$$

$$\geq \prod_{1\leq i\leq 2d} P\big(\partial^{in}\Lambda(n-c\ln n) \not\longleftrightarrow F_i\big) \,=\, P\big(\partial^{in}\Lambda(n-c\ln n) \not\longleftrightarrow F_1\big)^{2d}\,,$$

therefore $\lim_{n\to\infty} P\big(\partial^{in}\Lambda(n-c\ln n) \longleftrightarrow F_1\big) \,=\, 1$ and this holds for all the faces $F_i$, $1\leq i\leq 2d$. Suppose that $n$ is large enough so that

$$\big|\Lambda(n)\setminus\Lambda(n-c\ln n)\big| \,<\, \theta\frac{\varepsilon}{2}\,|\Lambda(n)|\,.$$

We claim that the event $V(\Lambda(n),\varepsilon)$ occurs if:

- $\Lambda(n-c\ln n)$ is connected to all the faces of $\partial^{in}\Lambda(n)$,
- the number of vertices of $\Lambda(n)$ connected to $\partial^{in}\Lambda(n)$ is larger than $\theta(1-\varepsilon/2)|\Lambda(n)|$,
- all the vertices of $\Lambda(n-c\ln n)$ which are connected to $\partial^{in}\Lambda(n)$ are connected together inside $\Lambda(n)$.

Indeed, if these events occur simultaneously, then the number of vertices of $\Lambda(n-c\ln n)$ connected to $\partial^{in}\Lambda(n)$ is larger than $\theta(1-\varepsilon)|\Lambda(n)|$ and all these vertices belong to a single cluster which is connected to all the faces of $\Lambda(n)$. Yet the events listed above have a probability going to 1 as $n$ goes to $\infty$. □

**Lemma 9.4.** *For any $d \geq 2$, any $p > p_c$, there exists a constant $\kappa$ such that*

$$\lim_{n\to\infty} P\big(R(\Lambda(n), \kappa \ln n)\big) = 1 .$$

*Remark 9.5.* Let $f : \mathbb{N} \to \mathbb{N}$ be a function such that $\lim_{n\to\infty} f(n)/\ln n = +\infty$ and $f(n) \leq n$. Since we have the inclusion $R(\Lambda(n), \kappa \ln n) \subset R(\Lambda(n), f(n))$ for $n$ large enough, the result holds also for $R(\Lambda(n), f(n))$.

*Proof.* We provide a sketch of proof for $d \geq 3$. From lemma 9.3 we already know that

$$\lim_{n\to\infty} P\big(U(\Lambda(n))\big) = 1 .$$

Suppose that $U(\Lambda(n))$ occurs but not $R(\Lambda(n), \kappa \ln n)$. This implies that inside the box $\Lambda(n)$ there exists a crossing cluster $C^*$ and there is either another disjoint cluster $D$ or a disjoint sub–box $\Lambda'$, with $D$ or $\Lambda'$ having a diameter larger than or equal to $\kappa \ln n$. Let us assume for instance that the diameter of $D$ or of $\Lambda'$ is $\kappa \ln n$ along the vertical axis. We partition $\Lambda(n)$ with vertical translates of the slab $S(n, L)$. There exist at least $\lfloor \kappa \ln n / L \rfloor$ consecutive slabs which are crossed vertically by two mutually disconnected open clusters or by a cluster disconnected from $\Lambda'$. By repeated conditioning on the configuration in each of these slabs (the procedure is similar to the one employed in lemma 9.2), we obtain that

$$P\big(U(\Lambda(n)) \setminus R(\Lambda(n), \kappa \ln n)\big) \leq dn^{2d+1}(1-\alpha)^{\lfloor \kappa \ln n / L \rfloor} .$$

The factor $d$ accounts for the choice of the direction, the factor $n^{2d+1}$ for the choices of the starting slab and the initial points of the open paths. See [78], Lemma 7.104 for a precise clean derivation of this inequality. Since the right–hand side goes to 0 as $n$ goes to $\infty$, we are done. □

The results stated in lemmas 9.3 and 9.4 are much weaker than the results of [112]. Indeed, with the help of a general and powerful renormalization procedure, Pisztora provides quantitative estimates on the probability of the events $R(\Lambda, m)$ and especially $V(\Lambda, \varepsilon)$.

We finally combine the previous events as follows. Let $n \in \mathbb{N}$, let $\Lambda \subset \Lambda'$ be two boxes of diameter larger than $n$ and let $\varepsilon > 0$. We define the event $RVW(\Lambda, \Lambda', n, \varepsilon)$ as the intersection

$$RVW(\Lambda, \Lambda', n, \varepsilon) = R(\Lambda', n) \cap V(\Lambda, \varepsilon) \cap W(\Lambda, \varepsilon) .$$

The event $RVW(\Lambda, \Lambda', n, \varepsilon)$ occurs if

- Inside $\Lambda'$ there is exactly one crossing cluster $C'$, it is the only cluster in $\Lambda'$ having diameter larger than or equal to $n$ and it crosses every sub–box of $\Lambda$ with diameter larger than or equal to $n$.
- Inside $\Lambda$ there is a crossing cluster $C^*$ and it satisfies

$$|C^*| \geq (1-\varepsilon)\,\theta\,|\Lambda|\,.$$

- Moreover
$$\Big|\{x\in\Lambda : x \longleftrightarrow \partial^{in}\Lambda\}\Big| \leq (1+\varepsilon)\,\theta\,|\Lambda|\,.$$

Because the diameter of $\Lambda$ is larger than $n$, the first condition implies that $C^* \subset C'$. Whenever the distance from $\Lambda$ to $\mathbb{R}^d \setminus \Lambda'$ is larger than $n$, these conditions imply that any cluster $C$ intersecting both $\Lambda$ and $\mathbb{Z}^d \setminus \Lambda'$ satisfies

$$(1-\varepsilon)\,\theta\,|\Lambda| \leq |C\cap\Lambda| \leq (1+\varepsilon)\,\theta\,|\Lambda|\,.$$

Lemmas 9.1, 9.3, 9.4 yield the following corollary.

**Corollary 9.6.** *For any $d\geq 2$, any $p > p_c$, any $\varepsilon > 0$, we have*

$$\lim_{n\to\infty} P\big(RVW(\Lambda(n),\Lambda(3n),n,\varepsilon)\big) = 1\,.$$

## 9.2 Extension to FK measures

If $\omega$ is a percolation configuration on $\mathbb{Z}^d$ and $A$ is a subset of $\mathbb{Z}^d$, we denote by $\omega_A$ the configuration restricted to the edges having at least one endpoint in $A$. Here are the crucial properties needed to prove the coarse graining results stated in lemmas 9.1, 9.3 and 9.4:

- $P$ is invariant under the isometries which leave the lattice $\mathbb{Z}^d$ invariant.
- $P$ satisfies the FKG inequality and the Markov property (see section 4.4).
- For any $\varepsilon > 0$, any percolation configuration $\eta$ on $\mathbb{Z}^d$,

$$\lim_{n\to\infty} P\Big(\Big|\frac{\big|\{x\in\Lambda(n) : x \longleftrightarrow \partial^{in}\Lambda(n)\,\}\big|}{|\Lambda(n)|} - \theta\Big| \geq \varepsilon \mid \omega_{\mathbb{Z}^d\setminus\Lambda(n)} = \eta_{\mathbb{Z}^d\setminus\Lambda(n)}\Big) = 0.$$

- There exist $L\in\mathbb{N}$ and $\alpha > 0$ such that

$$\forall n\geq 1 \quad \forall x,y\in S(n,L) \qquad P\big(x \longleftrightarrow y \,|\, \omega_{\mathbb{Z}^d\setminus S(n,L)} = 0\big) \geq \alpha\,.$$

Whenever $p > \widehat{p}_c(d)$, $p \notin \mathcal{U}(q)$, the infinite volume FK measure $\Phi^{p,q}_\infty$ corresponding to the parameters $p,q$ satisfies these conditions [112]. For finite volume FK measures, we provide in the next proposition estimates which are uniform with respect to the boundary conditions.

**Proposition 9.7.** *Let $d\geq 2$, $q\geq 1$, $p > \widehat{p}_c(d)$, $p\notin\mathcal{U}(q)$. The coarse graining results stated in lemmas 9.1, 9.3, 9.4 and corollary 9.6 hold for the finite volume FK measures corresponding to the parameters $p,q$ and they are uniform over the boundary conditions:*

$$\forall\varepsilon > 0 \qquad \lim_{n\to\infty} \sup_{\Phi\in\mathcal{FK}(p,q,\Lambda(n))} \Phi(\,*\,) = 1$$

*where $*$ stands for one of the events $R(\Lambda(n),\kappa\ln n)$, $V(\Lambda(n),\varepsilon)$, $W(\Lambda(n),\varepsilon)$, $RVW(\Lambda(n),\Lambda(3n),n,\varepsilon)$.*

*Proof.* The proofs are mere rewritings in the FK context of the proofs of lemmas 9.1, 9.3 and 9.4. Since we were already very sketchy when explaining these proofs, there is no sense in giving here a detailed account of the proofs! The additional difficulties are linked with the conditioning and the boundary conditions. Yet the last two properties in the above list provide estimates with the worst possible boundary conditions, so that the chain of inequalities in the conditioning argument still holds. □

Theorem 3.1 in [112] implies the following more refined estimates:
For $d \geq 3$, $q \geq 1$, $\delta > 0$ and $p > \widehat{p}_c$, $p \notin \mathcal{U}(q)$, there exist positive constants $b, c$ depending on $p, q, d, \delta$ such that for any $n \geq 1$, any box $\Lambda$ with side-lengths between $n$ and $2n$, and any finite volume FK measure $\Phi$ in $\Lambda$ corresponding to the parameters $p, q$,

$$\Phi\left(R(\Lambda, n/2)^c\right) + \Phi\left(V(\Lambda, \delta)^c\right) \leq b\exp(-cn)\,.$$

We will also need a coarse graining estimate involving the FK–Ising coupling (see section 4.3). Let $\Lambda$ be a box. We denote by $\mathcal{CP}(\Lambda)$ the set of all the couplings $\mathbb{P}$ between a FK measure $\Phi_\Lambda$ in $\Lambda$ with arbitrary boundary conditions and a Gibbs measure $\mu_\Lambda$ with the property that the colours of the clusters not touching the boundary are i.i.d. equal to $-$ or $+$ with probability $1/2$ each. For $\varepsilon > 0$, let $T(\Lambda, \varepsilon)$ be the event

$$T(\Lambda,\varepsilon) = \Big\{ \Big| \sum_{\substack{x\in\Lambda \\ x \not\leftrightarrow \partial^{in}\Lambda}} \sigma(x) \Big| \leq \varepsilon|\Lambda| \Big\}\,.$$

**Lemma 9.8.** *For any $\varepsilon > 0$, we have*

$$\lim_{n\to\infty} \sup_{\mathbb{P}\in\mathcal{CP}(\Lambda(n))} \mathbb{P}\big(T(\Lambda(n),\varepsilon)\big) = 1\,.$$

*Proof.* A stronger version of this result is proved in theorem 1.1 of [112]. Let $\mathbb{P}$ belong to $\mathcal{CP}(\Lambda(n))$. Let $\kappa$ be as in lemma 9.4. We condition on the edge configuration $\omega$ observed on the FK level and we assume that $\omega$ belongs to the event $R(\Lambda(n), \kappa\ln n)$. Let $\mathcal{C}$ be the collection of the open clusters in $\Lambda(n)$ which do not intersect $\partial^{in}\Lambda(n)$. On the event $R(\Lambda(n), \kappa \ln n)$, the clusters of $\mathcal{C}$ have a diameter less than $\kappa\ln n$ and a cardinality bounded by $(\kappa \ln n)^d$. By Chebyshev's inequality,

$$\begin{aligned} \mathbb{P}\big(T(\Lambda(n),\varepsilon)^c \big| \omega\big) &\leq \mathbb{P}\Big( \Big| \frac{1}{|\Lambda(n)|} \sum_{C\in\mathcal{C}} \sigma(C)|C| \Big| \geq \varepsilon \Big| \omega \Big) \\ &\leq \frac{1}{\varepsilon^2|\Lambda(n)|^2} \sum_{C\in\mathcal{C}} |C|^2 \leq \frac{(\kappa\ln n)^{2d}}{\varepsilon^2 n^d}\,. \end{aligned}$$

We integrate this estimate over $R(\Lambda(n), \kappa\ln n)$ and we conclude with the help of lemma 9.4 and proposition 9.7. □

## 9.3 The rescaled lattice

To put into action the coarse graining estimates, we will rescale the lattice. Let $K$ be a fixed positive integer. We divide $\mathbb{Z}^d$ into small boxes called blocks of size $K$ in the following way. For $\underline{x} \in \mathbb{Z}^d$, we define the block indexed by $\underline{x}$ as

$$B(\underline{x}) = K\underline{x} + \Lambda(K).$$

Note that the blocks partition $\mathbb{R}^d$. Let $A$ be a region in $\mathbb{R}^d$. We define the rescaled region $\underline{A}$ as

$$\underline{A} = \{\underline{x} \in \mathbb{Z}^d : B(\underline{x}) \cap A \neq \emptyset\}.$$

In general, we use underline in the notation to emphasize that we are dealing with rescaled objects. For instance, we denote by $\underline{\Lambda}(n)$ the box $\Lambda(n)$ rescaled by a factor $K$.

**Block events:** With a block we associate events which can be observed in the block or in a neighbourhood of the block. For $\underline{x} \in \underline{\mathbb{Z}}^d$, we introduce a larger block $B'(\underline{x})$ around $K\underline{x}$, called the event–block, by setting

$$B'(\underline{x}) = \bigcup_{\substack{\underline{z} \in \underline{\mathbb{Z}}^d \\ |\underline{z}-\underline{x}|_\infty \leq 1}} B(\underline{z})$$

where $|\cdot|_\infty$ is the usual supremum norm (see paragraph below).

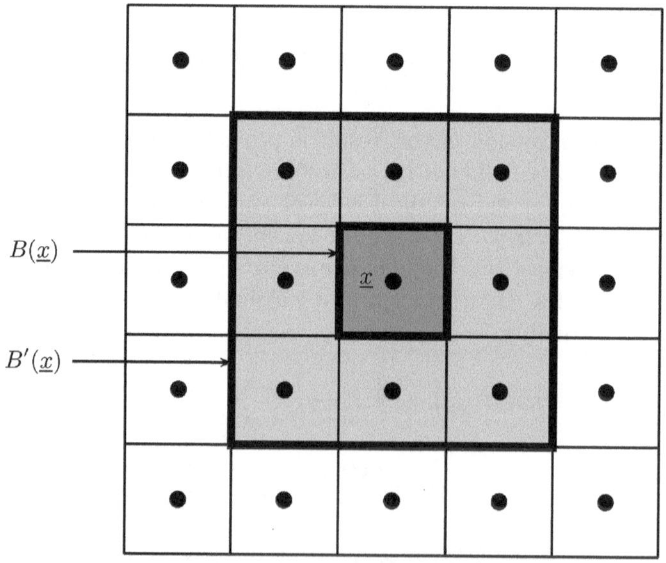


A block and the associated event–block

For convenience, we use the following notation: for $n \geq 1$,

$$\forall \underline{x} \in \mathbb{Z}^d \qquad B_n(\underline{x}) = \frac{1}{n} B(\underline{x}) .$$

**Block variables:** In the course of the proofs we will often use coarse graining in $\Lambda(n)$ by looking at a block process $(X(\underline{x}), \underline{x} \in \underline{\Lambda}(n))$ indicating the occurrence of a typical event in the event-blocks. By controlling the coarse grained process $X$ we can extract useful information about the underlying percolation process; this is a basic tool to analyze the microscopic behavior of the model.

The rescaled lattice is isomorphic to $\mathbb{Z}^d$ and we equip it with the graph structures corresponding to $\mathbb{L}^d$ or $\mathbb{L}^{d,\infty}$. Since we will use the $\mathbb{L}^{d,\infty}$ graph structure on the rescaled lattice, we describe it more precisely in the next paragraph.

**The lattice $\mathbb{L}^{d,\infty}$.** We introduce another graph structure on $\mathbb{Z}^d$. We denote by $\mathrm{d}_\infty$ the metric associated with the $\infty$-norm, i.e., $\mathrm{d}_\infty(x,y) = |x-y|_\infty$ for any $x, y$ in $\mathbb{R}^d$, where

$$\forall (x_1, \dots, x_d) \in \mathbb{R}^d \qquad |(x_1, \dots, x_d)|_\infty = \max_{1 \leq i \leq d} |x_i| .$$

The distance between two subsets $A$ and $B$ of $\mathbb{R}^d$ is

$$\mathrm{d}_\infty(A, B) = \inf \{ |x-y|_\infty : x \in A,\, y \in B \} .$$

The neighbourhood of $A \subset \mathbb{R}^d$ of radius $r$ is the set

$$\mathcal{V}_\infty(A, r) = \{ x \in \mathbb{R}^d : \mathrm{d}_\infty(x, A) < r \} .$$

The diameter of a subset $A$ of $\mathbb{R}^d$ is

$$\mathrm{diam}_\infty A = \sup \{ |x-y|_\infty : x, y \in A \} .$$

We define the edge set

$$\mathbb{E}^{d,\infty} = \{ \{x, y\} : x, y \in \mathbb{Z}^d,\, |x-y|_\infty = 1 \} .$$

The lattice $\mathbb{L}^{d,\infty}$ is the graph $(\mathbb{Z}^d, \mathbb{E}^{d,\infty})$. The relevance of this lattice stems from the fact that the exterior boundary $\partial^{\,out,ext} A$ of any $\mathbb{L}^d$–connected finite set $A$ in $\mathbb{Z}^d$ is itself connected when regarded as a subgraph of $\mathbb{L}^{d,\infty}$ (but not of $\mathbb{L}^d$). Let us be more precise. Let $A$ be a subset of $\mathbb{Z}^d$. We define its edge boundary

$$\partial^{edge} A = \{ \{x, y\} \in \mathbb{E}^d : x \in A, y \in \mathbb{Z}^d \setminus A \} ,$$

its inner vertex boundaries

$$\partial^{in} A = \{ x \in A : \exists y \in \mathbb{Z}^d \setminus A \quad \{x, y\} \in \mathbb{E}^d \} ,$$
$$\partial^{in}_\infty A = \{ x \in A : \exists y \in \mathbb{Z}^d \setminus A \quad \{x, y\} \in \mathbb{E}^{d,\infty} \} ,$$

its outer vertex boundaries

$$\partial^{out} A = \left\{ x \in \mathbb{Z}^d \setminus A : \exists y \in A \quad \{x,y\} \in \mathbb{E}^d \right\},$$
$$\partial^{out}_\infty A = \left\{ x \in \mathbb{Z}^d \setminus A : \exists y \in A \quad \{x,y\} \in \mathbb{E}^{d,\infty} \right\}.$$

A subset $A$ of $\mathbb{Z}^d$ is $\mathbb{L}^d$-connected (respectively $\mathbb{L}^{d,\infty}$–connected) if the graph $(A, \mathbb{E}^d(A))$ (resp. $(A, \mathbb{E}^{d,\infty}(A))$) is connected, where $\mathbb{E}^d(A)$ (resp. $\mathbb{E}^{d,\infty}(A)$) is the set of the edges of $\mathbb{E}^d$ (resp. $\mathbb{E}^{d,\infty}$) whose both endpoints belong to $A$. Note that connectedness in the usual $\mathbb{L}^d$ sense implies $\mathbb{L}^{d,\infty}$-connectedness. We define the $\mathbb{L}^{d,\infty}$ residual components of a subset $A$ of $\mathbb{Z}^d$ as the $\mathbb{L}^{d,\infty}$ connected components of $\mathbb{Z}^d \setminus A$. Let $A$ be a $\mathbb{L}^d$-connected subset of $\mathbb{Z}^d$. If $R$ is a $\mathbb{L}^{d,\infty}$ residual component of $A$, then $\partial^{out}_\infty R$ is $\mathbb{L}^d$-connected while $\partial^{in} R$, $\partial^{out} R$, $\partial^{in}_\infty R$ are $\mathbb{L}^{d,\infty}$-connected (see [58]). Suppose in addition that $A$ is finite. Then exactly one of its residual components, say $R_\infty$, is infinite. The external outer vertex boundaries of $A$ are

$$\partial^{out,ext} A = \partial^{in} R_\infty, \qquad \partial^{out,ext}_\infty A = \partial^{in}_\infty R_\infty.$$

## 9.4 Two rough estimates

In order to use the block estimates of section 9.1, we shall work on the rescaled lattice. A typical block process looks like supercritical site percolation with parameter close to 1. We shall rely on rather rough estimates for such processes which we present next. Let us fix $\delta \in ]0,1[$. Throughout the section, we consider a block process $(X(\underline{x}),\ \underline{x} \in \mathbb{Z}^d)$ with values in $\{0,1\}$ satisfying

$$\forall \underline{x} \in \mathbb{Z}^d \qquad P\Big( X(\underline{x}) = 0 \,\big|\, X(\underline{z}),\ |\underline{x} - \underline{z}|_\infty \geq 3 \Big) \leq \delta.$$

We introduce an equivalence relation on $\mathbb{Z}^d$: $z \approx y$ if and only if 3 divides each component of $z - y$. There are $3^d$ distinct classes in $\mathbb{Z}^d$. Let $x^* \in \mathbb{Z}^d$. The above condition implies that the field $X(\underline{x})$, $\underline{x} \in \mathbb{Z}^d$, $\underline{x} \approx \underline{x}^*$, stochastically dominates a Bernoulli product field with parameter $1 - \delta$ (see [78], p. 179).

**Lemma 9.9.** *For any subset $\underline{A}$ of $\mathbb{Z}^d$,*

$$P(\forall\, \underline{x} \in \underline{A} \quad X(\underline{x}) = 0) \leq \exp\Big( 3^{-d} |\underline{A}| \ln \delta \Big).$$

*Proof.* Since there are $3^d$ distinct classes in $\mathbb{Z}^d$ for the relation $\approx$, certainly there exists $\underline{x}^* \in \mathbb{Z}^d$ such that the intersection of $\underline{A}$ and the equivalence class of $\underline{x}^*$ has cardinality at least $3^{-d}|\underline{A}|$. Thus

$$\begin{aligned} P(\forall\, \underline{x} \in \underline{A} \quad X(\underline{x}) = 0) &\leq P(\forall\, \underline{x} \in \{\, \underline{y} \in \underline{A} : \underline{y} \approx \underline{x}^* \,\} \quad X(\underline{x}) = 0) \\ &\leq \exp\big( 3^{-d} |\underline{A}| \ln \delta \big) \end{aligned}$$

as claimed. □

The block process can be viewed as a dependent site percolation process where a site $\underline{x}$ is occupied if and only if $X(\underline{x}) = 0$. The occupied $\mathbb{L}^{d,\infty}$ cluster of the site $\underline{x}$, i.e., the $\mathbb{L}^{d,\infty}$ connected component of the occupied sites containing $\underline{x}$, is then denoted by $\underline{C}(\underline{x})$.

**Lemma 9.10.** *There exists a dimension dependent constant $b(d) > 0$ such that, for any bounded open subset $O$ of $\mathbb{R}^d$, any $s,t > 0$, any $K,n \in \mathbb{N}$ with $n \geq K$,*

$$P\Big(\big|\{\,\underline{x}\in\mathbb{Z}^d : B(\underline{x})\cap O\neq\emptyset\,,\ |\underline{C}(\underline{x})|\geq t\,\}\big| \geq s\Big) \leq 2\sum_{j\geq s}\exp j\,\Big(\frac{1}{t}\ln\mathcal{L}^d\big(\mathcal{V}(O,d)\big)+\ln b+\frac{1}{3^d}\ln\delta\Big)$$

*where* $\mathcal{V}(O,d) = \{\,x\in\mathbb{R}^d : \mathrm{dist}(x,O)\leq d\,\}$.

*Proof.* The proof is based on a standard counting Peierls argument. By decomposing the event appearing in the statement of the lemma, we can bound its probability by

$$\sum_{j\geq s}\ \sum_{1\leq i\leq j/t}\ \sum_{\substack{m_1,m_2,\dots,m_i\geq t\\ m_1+m_2+\dots+m_i=j}}\ \sum_{\underline{A}_1,\underline{A}_2,\dots,\underline{A}_i} P\big(\forall\underline{x}\in\underline{A}_1\cup\underline{A}_2\cup\ldots\cup\underline{A}_i\quad X(\underline{x})=0\big)$$

The ultimate summation extends over the pairwise disjoint sets $\underline{A}_1,\dots,\underline{A}_i$ such that for $1\leq l\leq i$, $\underline{A}_l$ is $\mathbb{L}^{d,\infty}$-connected, $|\underline{A}_l| = m_l$ and $\overline{\mathcal{V}_\infty(K\underline{A}_l, K/2)}$ intersects $O$; this implies that $K\underline{A}_l$ intersects $K\mathbb{Z}^d\cap\mathcal{V}(O,dK/2)$. The cardinality of this set is bounded by $K^{-d}\mathcal{L}^d\big(\mathcal{V}(O,dK)\big)$. By lemma 9.9, the probability appearing in the summation is less than $\exp(3^{-d}j\ln\delta)$. For fixed $j$ and $i$, there are at most $2^j$ ways to choose the values $m_1,\dots,m_i$. Moreover there exists a constant $b = b(d) > 0$ such that the number of $\mathbb{L}^{d,\infty}$-connected sets of size $m$ containing the origin is bounded by $(b/2)^m$. For $1\leq l\leq i$, the number of possibilities for choosing the set $\underline{A}_l$ is bounded by

$$\frac{1}{K^d}\mathcal{L}^d\big(\mathcal{V}(O,dK)\big)\Big(\frac{b}{2}\Big)^{m_l}.$$

Since $\mathcal{L}^d\big(\mathcal{V}(O/K,d)\big)\geq 2$, the number of terms involved in the last three summations is less than

$$\sum_{1\leq i\leq j/t}\mathcal{L}^d\big(\mathcal{V}(O/K,d)\big)^i b^j \leq 2b^j\mathcal{L}^d\big(\mathcal{V}(O/K,d)\big)^{j/t}.$$

Putting together the previous bounds, we obtain the desired inequality. □

For $m\geq K$, we denote by $\underline{\Lambda}(m)$ the box $\Lambda(m)$ rescaled by a factor $K$, defined by

$$\underline{\Lambda}(m) = \{\,\underline{x}\in\mathbb{Z}^d : B(\underline{x})\cap\Lambda(m)\neq\emptyset\,\}.$$

**Lemma 9.11.** *For any $m, K, \varepsilon, \delta$ satisfying $m \geq 6K$, $0 < \delta < \varepsilon$, we have*

$$P\Big(\frac{1}{|\underline{\Lambda}(m)|}\sum_{\underline{x}\in\underline{\Lambda}(m)} 1_{X(\underline{x})=0} \geq \varepsilon\Big) \leq 3^d \exp\Big(-\Lambda^*(\varepsilon,\delta)\left\lfloor\frac{m}{6K}\right\rfloor^d\Big)$$

*where*

$$\Lambda^*(\varepsilon,\delta) = \varepsilon \ln\frac{\varepsilon}{\delta} + (1-\varepsilon)\ln\frac{1-\varepsilon}{1-\delta}$$

*is the Cramér transform of a Bernoulli variable with parameter $\delta$.*

*Proof.* We partition $\underline{\Lambda}(m)$ into the equivalence classes $V_1, \ldots, V_{3^d}$ associated to the relation $\approx$ defined at the beginning of the section. By the classical exponential Chebyshev inequality (see for instance [59]),

$$\begin{aligned} P\Big(\frac{1}{|\underline{\Lambda}(m)|}\sum_{\underline{x}\in\underline{\Lambda}(m)} 1_{X(\underline{x})=0} \geq \varepsilon\Big) &\leq \sum_{i=1,\ldots,3^d} P\Big(\frac{1}{|V_i|}\sum_{\underline{x}\in V_i} 1_{X(\underline{x})=0} \geq \varepsilon\Big) \\ &\leq 3^d \exp\Big(-\Lambda^*(\varepsilon,\delta)\min_{i=1,\ldots,3^d}|V_i|\Big) \end{aligned}$$

and the claim follows from the fact that $|V_i| \geq \lfloor m/6K\rfloor^d$ for $i = 1, \ldots, 3^d$. □

# 10

# Decoupling

A fundamental difficulty for the study of the FK model is the control of the stochastic dependence. We state here a weak decoupling result which allows to decouple rare events in distant regions at the level of the large deviations of surface order.

## 10.1 Half–space clusters

We denote by $\mathbb{H}^d$ the half–space

$$\mathbb{H}^d = \{ (x_1, \dots, x_d) \in \mathbb{Z}^d : x_d \geq 0 \}$$

and by $\mathbb{D}^d$ its boundary

$$\mathbb{D}^d = \{ (x_1, \dots, x_d) \in \mathbb{Z}^d : x_d = 0 \}.$$

Let $A \subset \mathbb{D}^d$ and let $h > 0$. Given an edge configuration $\omega$ in $\mathbb{Z}^d$ or in a box $\Lambda$ containing the cylinder $A \times [0, h]$, we define $K(A, h)$ as the number of clusters in the configuration $\omega$ restricted to the slab $\{ (x_1, \dots, x_d) \in \mathbb{Z}^d : 0 \leq x_d \leq h \}$ which intersect $A$. We define also $K^w(A, h)$ as the number of clusters in the configuration $\omega$ restricted to the same slab which intersect $A$ and which don't intersect the inner vertex boundary of the cylinder $A \times [0, h]$ outside $A$. Notice that $K(A, h)$ and $K^w(A, h)$ are decreasing random variables. Moreover, the map $K$ is a non–increasing function of the height $h$, while $K^w$ is a non–decreasing function of $h$. Finally the map $K$ is subadditive with respect to $A$, while the map $K^w$ is superadditive: if $A, B$ are two disjoint subsets of $\mathbb{D}^d$, then for any $h > 0$,

$$K(A \cup B, h) \leq K(A, h) + K(B, h),$$
$$K^w(A \cup B, h) \geq K^w(A, h) + K^w(B, h).$$

For $n \geq 1$, we denote by $D(n)$ the hypersquare $D(n) = \mathbb{D}^d \cap \Lambda(n)$. For $h > 0$, we denote by $C(n, h)$ the cylinder

$$C(n, h) = D(n) \times [-h, h].$$

**Lemma 10.1.** *Let $q \geq 1$ and let $p \in ]0,1[$. Let $f, g, h$ be three maps from $\mathbb{N}$ to $\mathbb{R}^+$ such that $f(n) \geq n$, $g(n) \geq h(n)$ for $n \geq 1$ and $\lim_{n\to\infty} h(n) = \infty$. The following limits exist:*

$$\kappa^+ = \lim_{n\to\infty} \frac{1}{n^{d-1}} \Phi^{f,p,q}_{C(f(n),g(n))}\big(K(D(n), h(n))\big),$$
$$\kappa^{w,+} = \lim_{n\to\infty} \frac{1}{n^{d-1}} \Phi^{w,p,q}_{C(f(n),g(n))}\big(K^w(D(n), h(n))\big).$$

*The limiting values $\kappa^+$ and $\kappa^{w,+}$ are independent of the choice of $f, g, h$. Moreover $\kappa^+ \geq \kappa^{w,+}$.*

*Proof.* Let $f, g, h, f', g', h'$ be six functions from $\mathbb{N}$ to $\mathbb{R}^+$ such that $f, g, h$ and $f', g', h'$ satisfy the hypothesis of the lemma. Let $m \leq n$ be two integers. We suppose that $n$ is large enough so that $f(n) \geq f'(m)$, $g(n) \geq g'(m)$, $h(n) \geq h'(m)$. Let $n - f'(m) = mq + r$ be the Euclidean division of $n - f'(m)$ by $m$. We partition $D(n - f'(m))$ into $q^{d-1}$ disjoint translates of $D(m)$, denoted by $D_i$, $1 \leq i \leq q^{d-1}$, which cover $D(qm)$, and a remaining region $D(n) \setminus D(qm)$. By the subadditivity of $K$, we have

$$K(D(n), h(n)) \leq \sum_{1\leq i\leq q^{d-1}} K(D_i, h(n)) + K(D(n) \setminus D(qm), h(n))$$
$$\leq \sum_{1\leq i\leq q^{d-1}} K(D_i, h'(m)) + n^{d-1} - (qm)^{d-1}.$$

Taking expectation with respect to $\Phi^{f,p,q}_{C(f(n),g(n))}$, we obtain

$$\Phi^{f,p,q}_{C(f(n),g(n))}\big(K(D(n), h(n))\big)$$
$$\leq \sum_{1\leq i\leq q^{d-1}} \Phi^{f,p,q}_{C(f(n),g(n))}\big(K(D_i, h'(m))\big) + n^{d-1} - (qm)^{d-1}$$
$$\leq q^{d-1}\, \Phi^{f,p,q}_{C(f'(m),g'(m))}\big(K(D(m), h'(m))\big) + n^{d-1} - (qm)^{d-1}.$$

The last inequality is a consequence of the FKG inequality and the Markov property (see section 4.4). Dividing by $n^{d-1}$ and sending $n$ to $\infty$, we see that

$$\limsup_{n\to\infty} \frac{1}{n^{d-1}} \Phi^{f,p,q}_{C(f(n),g(n))}\big(K(D(n), h(n))\big)$$
$$\leq \frac{1}{m^{d-1}} \Phi^{f,p,q}_{C(f'(m),g'(m))}\big(K(D(m), h'(m))\big).$$

Sending now $m$ to $\infty$, we conclude that the limit defining $\kappa^+$ exists. The argument for the limit defining $\kappa^{w,+}$ is similar. Starting with integers $n, m$ as above and considering the same partition of $D(n)$, we have

$$K^w(D(n), h(n)) \geq \sum_{1\leq i\leq q^{d-1}} K^w(D_i, h'(m)).$$

We then take the expectation with respect to $\Phi^{w,p,q}_{C(f(n),g(n))}$ and we conclude as above. □

**Proposition 10.2.** *Let $q \geq 1$ and let $p \in ]0,1[$. Let $f, g, h$ be three maps from $\mathbb{N}$ to $\mathbb{R}^+$ such that $f(n) \geq n$, $g(n) \geq h(n)$ for $n \geq 1$ and $\lim_{n\to\infty} h(n) = \infty$. We have*

$$\forall \alpha > \kappa^+ \qquad \limsup_{n\to\infty} \frac{1}{n^{d-1}} \ln \Phi^{f,p,q}_{C(f(n),g(n))}\Big(\frac{1}{n^{d-1}} K(D(n),h(n)) \geq \alpha\Big) < 0\,,$$

$$\forall \alpha < \kappa^{w,+} \quad \limsup_{n\to\infty} \frac{1}{n^{d-1}} \ln \Phi^{w,p,q}_{C(f(n),g(n))}\Big(\frac{1}{n^{d-1}} K^w(D(n),h(n)) \leq \alpha\Big) < 0\,.$$

*Proof.* Let $f, g, h$ be three maps from $\mathbb{N}$ to $\mathbb{R}^+$ satisfying the above hypothesis. Let $\alpha > \kappa^+$ and let $m \leq n$ be two integers. Let $n = mq + r$ be the Euclidean division of $n$ by $m$. We partition $D(n)$ into $q^{d-1}$ disjoint translates of $D(m)$, denoted by $D_i$, $1 \leq i \leq q^{d-1}$, which cover $D(qm)$, and a remaining region $D(n) \setminus D(qm)$. Let $E$ be the event

$$\Big\{ \text{all the edges having an endpoint in } \bigcup_{1\leq i\leq q^{d-1}} \partial^{in}\big(D_i \times [0,h(m)]\big) \text{ are closed} \Big\}.$$

By the FKG inequality, we have

$$\Phi^{f,p,q}_{C(f(n),g(n))}\Big(\frac{K(D(n),h(n))}{n^{d-1}} \geq \alpha\Big) \leq \Phi^{f,p,q}_{C(f(n),g(n))}\Big(\frac{K(D(n),h(n))}{n^{d-1}} \geq \alpha \Big| E\Big)\,.$$

For $i \in \{1 \cdots q^{d-1}\}$, we set $D'_i = D_i \setminus \partial^{in} D_i$ (here $\partial^{in} D_i$ is defined as if $D_i$ was a subset of $\mathbb{Z}^{d-1}$). On the event $E$, we have

$$\begin{aligned} K(D(n),h(n)) &\leq \sum_{1\leq i\leq q^{d-1}} \big(K(D'_i,h(n)) + |\partial^{in} D_i|\big) + K(D(n)\setminus D(qm),h(n)) \\ &\leq \sum_{1\leq i\leq q^{d-1}} K(D'_i,h(m)) + q^{d-1}|\partial^{in} D(m)| + |D(n)\setminus D(qm)|\,. \end{aligned}$$

By lemma 10.1, for $m$ large enough, we have

$$\Phi^{f,p,q}_{C(f(m),g(m))}\Big(\frac{1}{|D'(m)|} K(D'(m),h(m))\Big) \leq \frac{\alpha+\kappa^+}{2}\,.$$

We suppose also that $m, n$ are large enough so that

$$\alpha - \frac{1}{n^{d-1}}\big(q^{d-1}|\partial^{in} D(m)| + |D(n)\setminus D(qm)|\big) > \frac{\alpha+\kappa^+}{2}\,.$$

Combining the previous inequalities, we have then

$$\Phi^{f,p,q}_{C(f(n),g(n))}\Big(\frac{K(D(n),h(n))}{n^{d-1}} \geq \alpha\Big) \leq$$
$$\Phi^{f,p,q}_{C(f(n),g(n))}\Big(\frac{1}{q^{d-1}} \sum_{1\leq i\leq q^{d-1}} \frac{K(D'_i,h(m))}{|D'_i|} > \frac{\alpha+\kappa^+}{2} \,\Big|\, E\Big)\,.$$

By the Markov property (see section 4.4), conditionally on $E$, the random variables $K(D'_i,h(m))/|D'_i|$, $1 \leq i \leq q^{d-1}$, are i.i.d., their common law being the one of $K(D'(m),h(m))/|D'(m)|$, whose expectation is less than or equal to $(\alpha+\kappa^+)/2$. By the classical Cramér theorem in $\mathbb{R}$ (see theorems 1.3 and 1.5), there exists $c>0$ such that the right–hand quantity of the previous inequality is bounded by $\exp(-cq^{d-1})$, whence

$$\limsup_{n\to\infty}\frac{1}{n^{d-1}}\ln\Phi^{f,p,q}_{C(f(n),g(n))}\Big(\frac{1}{n^{d-1}}K(D(n),h(n))\geq\alpha\Big)\leq-\frac{c}{m^{d-1}}<0\,.$$

The argument to prove the second estimate is very similar, only we open the edges in the event $E$ instead of closing them. □

**Proposition 10.3.** *Let $d\geq 2$, $q\geq 1$, $p>\widehat{p}_c(d)$, $p\notin\mathcal{U}(q)$. Then $\kappa^+=\kappa^{w,+}$.*

*Proof.* Let $h:\mathbb{N}\to\mathbb{N}$ be a map such that $\lim_{+\infty}h=+\infty$. Let $p,q$ be as in the statement of the proposition. For such parameters, there exists a unique infinite volume FK measure $\Phi^{p,q}_\infty$ (see theorem 4.2), so that for any $n\geq 1$,

$$\begin{aligned}\lim_{N\to\infty}\Phi^{f,p,q}_{\Lambda(N)}\big(K(D(n),h(n))\big)&=\Phi^{p,q}_\infty\big(K(D(n),h(n))\big)\,,\\ \lim_{N\to\infty}\Phi^{w,p,q}_{\Lambda(N)}\big(K^w(D(n),h(n))\big)&=\Phi^{p,q}_\infty\big(K^w(D(n),h(n))\big)\,.\end{aligned}$$

Thus there exists a map $f:\mathbb{N}\to\mathbb{N}$ such that for any $n\geq 1$, $f(n)\geq n$ and

$$\begin{aligned}\Big|\Phi^{f,p,q}_{\Lambda(f(n))}\big(K(D(n),h(n))\big)-\Phi^{p,q}_\infty\big(K(D(n),h(n))\big)\Big|&\leq 1\,,\\ \Big|\Phi^{w,p,q}_{\Lambda(f(n))}\big(K^w(D(n),h(n))\big)-\Phi^{p,q}_\infty\big(K^w(D(n),h(n))\big)\Big|&\leq 1\,.\end{aligned}$$

Lemma 10.1 implies then that

$$\begin{aligned}\kappa^+&=\lim_{n\to\infty}\frac{1}{n^{d-1}}\Phi^{p,q}_\infty\big(K(D(n),h(n))\big)\,,\\ \kappa^{w,+}&=\lim_{n\to\infty}\frac{1}{n^{d-1}}\Phi^{p,q}_\infty\big(K^w(D(n),h(n))\big)\,.\end{aligned}$$

Let us fix $n\geq 1$ and let $N$ be the number of clusters in the slab $[-n,n]^{d-1}\times[0,h(n)]$ joining $[-n,n]^{d-1}\times\{0\}$ to $[-n,n]^{d-1}\times\{h(n)\}$. The number of clusters intersecting $D(n)$ and which touch the inner vertex boundary of the cylinder $D(n)\times[0,h(n)]$ outside $D(n)$ is bounded by $2dn^{d-2}h(n)+N$, whence

$$K(D(n),h(n))-K^w(D(n),h(n))\leq 2dn^{d-2}h(n)+N\,.$$

By lemma 9.2,

$$\Phi^{p,q}_\infty(N)\leq 1+\big((2n)^{d-1}\big)^2(1-\alpha)^{h(n)/L}\,.$$

We choose $h(n)=c\ln n$, where $c$ is a sufficiently large constant. Taking the expectation with respect to $\Phi^{p,q}_\infty$, we conclude from the previous inequalities that

$$\lim_{n\to\infty}\frac{1}{n^{d-1}}\Phi^{p,q}_\infty\big(K(D(n),h(n))\big)=\lim_{n\to\infty}\frac{1}{n^{d-1}}\Phi^{p,q}_\infty\big(K^w(D(n),h(n))\big)\,,$$

so that $\kappa^+=\kappa^{w,+}$. □

We next state a weak law of large numbers for $K(D(n), h(n))$.

**Corollary 10.4.** *Let $d \geq 2$, $q \geq 1$, $p > \widehat{p}_c(d)$, $p \notin \mathcal{U}(q)$. Let $f, g, h$ be three maps from $\mathbb{N}$ to $\mathbb{R}^+$ such that $f(n) \geq n$, $g(n) \geq h(n)$ for $n \geq 1$ and $\lim_{n\to\infty} h(n) = \infty$. For each $n \in \mathbb{N}$, let $\pi(n)$ be a partially wired boundary condition on $C(f(n), g(n))$. We have*

$$\forall \varepsilon > 0 \quad \limsup_{n\to\infty} \frac{1}{n^{d-1}} \ln \Phi^{\pi(n),p,q}_{C(f(n),g(n))}\Big(\Big|\frac{1}{n^{d-1}} K(D(n), h(n)) - \kappa^+\Big| > \varepsilon\Big) < 0.$$

*Remark 10.5.* The multiparameter subadditive ergodic theorem (see [1], theorem 2.4 or [122], theorem 1.1) implies that, under $\Phi^{p,q}_\infty$, with probability 1,

$$\lim_{n\to\infty} \frac{1}{n^{d-1}} K(D(n), h(n)) = \kappa^+ .$$

*Proof.* We denote by $C(n)$ the cylinder $C(f(n), g(n))$. For any $\varepsilon > 0$ and any partially wired boundary condition $\pi(n)$ on $C(n)$, we have

$$\Phi^{\pi(n),p,q}_{C(n)}\Big(\frac{K(D(n), h(n))}{n^{d-1}} > \kappa^+ + \varepsilon\Big) \leq \Phi^{f,p,q}_{C(n)}\Big(\frac{K(D(n), h(n))}{n^{d-1}} \geq \kappa^+ + \varepsilon\Big),$$
$$\Phi^{\pi(n),p,q}_{C(n)}\Big(\frac{K(D(n), h(n))}{n^{d-1}} < \kappa^+ - \varepsilon\Big) \leq \Phi^{w,p,q}_{C(n)}\Big(\frac{K^w(D(n), h(n))}{n^{d-1}} \leq \kappa^+ - \varepsilon\Big).$$

Now by proposition 10.3, we have $\kappa^+ = \kappa^{w,+}$. The desired claim follows from the above inequalities and the estimates of proposition 10.2. □

## 10.2 Decoupling lemma

We quote here the decoupling lemma of [43]. This lemma is one of the key results which allow to extend to FK percolation the large deviation results for Bernoulli percolation. We provide a proof which avoids the use of renormalization.

Let $\Theta \subset \mathbb{R}^d$ be a box building, i.e., the union of finitely many $d$-dimensional boxes with non empty interior. Let $\phi : \mathbb{N} \to \mathbb{N}$ be a function satisfying $\lim_{n\to\infty} \phi(n) = \infty$ and $\lim_{n\to\infty} \phi(n)/n = 0$. We will consider the $\phi(n)$-interior of the building $n\Theta$, defined by

$$\mathrm{int}(n\Theta, \phi(n)) = \big\{ x \in n\Theta : \mathrm{d}_\infty(x, \partial^{in} n\Theta) > \phi(n) \big\} .$$

**Proposition 10.6.** *Let $d \geq 2$, $q \geq 1$, $p > \widehat{p}_c$, $p \notin \mathcal{U}(q)$. For $n \in \mathbb{N}$, let $\pi(n)$ be a partially wired boundary condition on $n\Theta$ and let $S_n$ be an event depending only on the edges in $\mathrm{int}(n\Theta, \phi(n))$. Then*

$$\limsup_{n\to\infty} \frac{1}{n^{d-1}} \ln \Phi^{\pi(n),p,q}_{n\Theta}(S_n) = \limsup_{n\to\infty} \frac{1}{n^{d-1}} \ln \Phi^{f,p,q}_{n\Theta}(S_n) .$$

*The same equality is valid when* lim sup *is replaced by* lim inf.

*Remark 10.7.* We will also work in the box $\Lambda(n)$ with wired boundary conditions, hence we will need the following slight generalization of proposition 10.6, whose proof is similar. Let $d \geq 2,\ q \geq 1,\ p > \widehat{p}_c,\ p \notin \mathcal{U}(q)$. Let $\Theta$ be a box building included in $\Lambda(1)$. We define the $\phi(n)$-interior of the building $n\Theta$ relative to $\Lambda(n)$ by

$$\mathrm{int}_{\Lambda(n)}(n\Theta, \phi(n)) = \big\{\, x \in n\Theta : \mathrm{d}_\infty(x, \partial^{in} n\Theta \setminus \partial^{in}\Lambda(n)) > \phi(n) \,\big\}\,.$$

For $n \in \mathbb{N}$, let $\pi(n)$ be a partially wired boundary condition on $n\Theta$ such that the sites in $\partial^{in} n\Theta \cap \partial^{in}\Lambda(n)$ are wired together and let $S_n$ be an event depending only on the edges in $\mathrm{int}_{\Lambda(n)}(n\Theta, \phi(n))$. We have

$$\limsup_{n\to\infty} \frac{1}{n^{d-1}} \ln \Phi^{\pi(n),p,q}_{n\Theta}\big(S_n\big) \;=\; \limsup_{n\to\infty} \frac{1}{n^{d-1}} \ln \Phi^{w,p,q}_{\Lambda(n)}\big(S_n\big)\,.$$

*Remark 10.8.* A direct consequence of proposition 10.6 is that

$$\begin{aligned}
\limsup_{n\to\infty} \frac{1}{n^{d-1}} \ln \Phi^{f,p,q}_{n\Theta}\big(S_n\big) \;&=\; \limsup_{n\to\infty} \frac{1}{n^{d-1}} \ln \max_{\Phi \in\, c\mathcal{FK}(p,q,n\Theta)} \Phi\,\big(S_n\big) \\
&=\; \limsup_{n\to\infty} \frac{1}{n^{d-1}} \ln \min_{\Phi \in\, c\mathcal{FK}(p,q,n\Theta)} \Phi\,\big(S_n\big)\,.
\end{aligned}$$

We will give the proof of proposition 10.6 for the choice $\Theta = \Lambda(1)$, so that $n\Theta = \Lambda(n)$. The generalization to arbitrary buildings is straightforward. We begin with the description of the idea behind the proof. Let $\pi$ be a partially wired boundary condition on $\Lambda(n)$. We build a monotone coupling of the two FK measures $\Phi^f_{\Lambda(n)}$ and $\Phi^\pi_{\Lambda(n)}$. We will show that for any $\varepsilon > 0$, with exceedingly high probability, i.e., up to large deviations of supersurface order, we are able to find a random centered box $B$, only a little smaller than $\Lambda(n)$, such that the boundary conditions $f \cdot W_B^{\Lambda(n)}$ and $\pi \cdot W_B^{\Lambda(n)}$ on $B$, induced by the configurations in $\Lambda(n) \setminus B$, satisfy

$$0 \leq |f \cdot W_B^{\Lambda(n)}| - |\pi \cdot W_B^{\Lambda(n)}| \leq \varepsilon n^{d-1}\,.$$

Whenever two boundary conditions $\pi, \pi'$ on a region are not too different, more precisely when $\pi'$ is finer than $\pi$ and the difference between their numbers of equivalence classes is bounded by some number $C$, then for any event $S$ depending on the edges in that region, we have $q^{-C} \leq \Phi^{\pi'}(S)/\Phi^{\pi}(S) \leq q^C$. Thus, for any event $S$ which depends on the edges in a region not too close to the boundary of $\Lambda(n)$, we have

$$q^{-\varepsilon n^{d-1}} \;\leq\; \frac{\Phi^\pi_{\Lambda(n)}\big(S\big)}{\Phi^f_{\Lambda(n)}\big(S\big)} \;\leq\; q^{\varepsilon n^{d-1}}\,.$$

Since $\varepsilon$ is arbitrarily small, the result follows easily. A key point for this argument is the next lemma, which allows to control monotone perturbations of boundary conditions.

**Lemma 10.9.** *Let $q \geq 1$ and let $V$ be a finite subset of $\mathbb{Z}^d$. Let $\pi'$, $\pi$ be two boundary conditions on $V$ such that $\pi'$ is finer than $\pi$ and $|\pi'|-|\pi| \leq C$. For any event $S$ depending only on the edges in $V$, we have*

$$q^{-C} \leq \frac{\Phi_V^{\pi'}(S)}{\Phi_V^{\pi}(S)} \leq q^C .$$

*Proof.* Since $\pi'$ is finer than $\pi$ and $|\pi'| - |\pi| \leq C$, then

$$\forall \omega \in \Omega_V \qquad 0 \leq cl^{\pi'}(\omega) - cl^{\pi}(\omega) \leq C .$$

For any event $A$, we define the partition sum $Z^{\pi}[A]$ by

$$Z^{\pi}[A] = \sum_{\omega \in A} \Big( \prod_{e \in \mathbb{E}^d(V)} p^{\omega(e)}(1-p)^{1-\omega(e)} \Big) q^{cl^{\pi}(\omega)} .$$

We have the inequalities: $Z^{\pi}[A] \leq Z^{\pi'}[A] \leq q^C Z^{\pi}[A]$. The first inequality is obvious and the second follows from

$$Z^{\pi'}[A] = \sum_{\omega \in A} \Big( \prod_{e \in \mathbb{E}^d(V)} p^{\omega(e)}(1-p)^{1-\omega(e)} \Big) q^{cl^{\pi}(\omega)} q^{cl^{\pi'}(\omega)-cl^{\pi}(\omega)} \leq q^C Z^{\pi}[A] .$$

The desired inequalities are then obtained as follows:

$$\Phi_V^{\pi}(S) = \frac{Z^{\pi}[S]}{Z^{\pi}[\Omega_V]} \leq \frac{Z^{\pi'}[S]}{q^{-C} Z^{\pi'}[\Omega_V]} = q^C \Phi_V^{\pi'}(S) ,$$

$$\Phi_V^{\pi'}(S) = \frac{Z^{\pi'}[S]}{Z^{\pi'}[\Omega_V]} \leq \frac{q^C Z^{\pi}[S]}{Z^{\pi}[\Omega_V]} = q^C \Phi_V^{\pi}(S) . \qquad \square$$

*Proof of proposition 10.6.* Let $\pi = \pi(n)$ be a boundary condition on $\Lambda(n)$. We begin the proof with the description of a monotone coupling $\mathbb{P}_{\Lambda(n)}$ of the measures $\Phi^f_{\Lambda(n)}$ and $\Phi^{\pi}_{\Lambda(n)}$ governing two layers of configurations $\omega = (\omega^f, \omega^{\pi})$ with

$$\mathbb{P}_{\Lambda(n)}\big(\omega^f \leq \omega^{\pi}\big) = 1 .$$

We first choose an arbitrary inward spiral ordering of all the bonds $b_1, b_2, \ldots$ in the box $\Lambda(n)$ beginning with some edge on the boundary (such an edge links two sites in $\partial^{in}\Lambda(n)$) and we assign i.i.d. random variables $X_i$ to the bonds $b_i$ in $\Lambda(n)$ which are uniformly distributed on [0,1]. By an inward spiral ordering we simply mean that for $m = 1, \ldots, n$, each bond between sites in $\partial^{in}\Lambda(m)$ has a smaller index than the bonds linking $\partial^{in}\Lambda(m)$ to $\partial^{in}\Lambda(m-1)$ which themselves have smaller indices than bonds in $\partial^{in}\Lambda(m-1)$. The coupling is constructed in an algorithmic way as follows. For $* = f, \pi$, we declare the first bond $b_1$ on the $*$-layer to be open if $X_1 \leq \Phi^*_{\Lambda(n)}(b_1 \text{ is open})$, otherwise $b_1$ is closed. Thanks to the FKG inequality, we have $\omega^f(b_1) \leq \omega^{\pi}(b_1)$. The second bond $b_2$ on the $*$-layer will be open or closed according to whether

$$X_2 \leq \Phi^*_{\Lambda(n)}\big(b_2 \text{ is open} \;\big|\; \text{the status of } b_1 \text{ is given by } \omega^*(b_1)\big)$$

or not. The fact that $\omega^f(b_1) \leq \omega^\pi(b_1)$ and the strong FKG inequality guarantee the monotonicity of the coupling. In general, the $k$-th bond $b_k$ on the $*$-layer is open if and only if

$$X_k \leq \Phi^*_{\Lambda(n)}\left(b_k \text{ is open} \,\middle|\, \begin{array}{l}\text{the status of } b_1, ..., b_{k-1} \text{ is} \\ \text{given by } \omega^*(b_1), \dots \omega^*(b_{k-1})\end{array}\right).$$

We proceed in this way until all the bonds have been assigned their status. An important property of this coupling is that, for any $k \geq 1$ and any double configuration $\eta = (\eta^f, \eta^\pi) \in \{0,1\}^{\{b_1,\dots,b_k\}} \times \{0,1\}^{\{b_1,\dots,b_k\}}$, the conditional measure

$$\mathbb{P}_{\Lambda(n)}\Big(\,\cdot\,\Big|\,\omega^f(b_i) = \eta^f(b_i),\, \omega^\pi(b_i) = \eta^\pi(b_i),\, 1 \leq i \leq k\Big)$$

restricted to the $*$-layer agrees with

$$\Phi^*_{\Lambda(n)}\big(\,\cdot\,\big|\,\omega^*(b_i) = \eta^*(b_i),\, 1 \leq i \leq k\big).$$

In particular, if $\eta$ is a double configuration defined on $\Lambda(n) \setminus \Lambda(m)$ for some $m \in \{1 \cdots n\}$, the same statement is true thanks to the particular choice of the ordering of the edges.

For $\varepsilon > 0$ and for $m < n$, we set

$$\mathcal{E}_m = \Big\{(\omega^f, \omega^\pi) \in \big(\Omega_{\Lambda(n)\setminus\Lambda(m)}\big)^2 : |f\cdot W^{\Lambda(n)}_{\Lambda(m)}(\omega^f)| - |\pi\cdot W^{\Lambda(n)}_{\Lambda(m)}(\omega^\pi)| < \varepsilon n^{d-1}\Big\}.$$

Let us set $L = \sqrt{\phi(n)/2}$. Let $i \in \{1 \cdots L\}$. Let $F_j$, $1 \leq j \leq 2d$, be the $2d$ faces of the box $\Lambda(n - 2iL)$. For $j \in \{1 \cdots 2d\}$, let $\nu_j$ be the unit outward normal vector to $F_j$ and let $P_j$ be the cylinder

$$P_j = \{x + t\nu_j : x \in F_j,\, t \geq 0\} \cap \Lambda(n - 2(i-1)L).$$

For $\omega$ a configuration in $\Lambda(n)$, we denote by $K(F_j, L)(\omega)$ the number of clusters in $\omega$ restricted to $P_j$ which intersect the face $F_j$ and by $K^w(F_j, L)(\omega)$ the number of clusters in $\omega$ restricted to $P_j$ which intersect the face $F_j$ and which don't intersect the inner vertex boundary of $P_j$ outside $F_j$. We have

$$|f \cdot W^{\Lambda(n)}_{\Lambda(n-2iL)}(\omega^f)| \leq |f \cdot W^{\Lambda(n-2(i-1)L)}_{\Lambda(n-2iL)}(\omega^f)| \leq \sum_{1\leq j\leq 2d} K(F_j, L)(\omega^f),$$

$$\sum_{1\leq j\leq 2d} K^w(F_j, L)(\omega^\pi) \leq |w \cdot W^{\Lambda(n-2(i-1)L)}_{\Lambda(n-2iL)}(\omega^\pi)| \leq |\pi \cdot W^{\Lambda(n)}_{\Lambda(n-2iL)}(\omega^\pi)|.$$

Let $A_i$ and $A_i^w$ be the events

$$A_i = \Big\{ \sum_{1\leq j\leq 2d} \frac{K(F_j, L)(\omega^f)}{2d|F_j|} \geq \kappa^+ + \frac{\varepsilon}{4d} \Big\},$$

$$A_i^w = \Big\{ \sum_{1\leq j\leq 2d} \frac{K^w(F_j, L)(\omega^\pi)}{2d|F_j|} \leq \kappa^+ - \frac{\varepsilon}{4d} \Big\}.$$

By the previous inequalities, if $\omega^f \notin A_i$ and $\omega^\pi \notin A_i^w$, then $(\omega^f, \omega^\pi) \in \mathcal{E}_{n-2iL}$. Thus

$$\Big(\bigcup_{n-\phi(n)\le m<n} \mathcal{E}_m\Big)^c \subset \bigcap_{1\le i\le L} (\mathcal{E}_{n-2iL})^c \subset \bigcap_{1\le i\le L} (A_i \cup A_i^w),$$

whence, by iterated conditioning,

$$\mathbb{P}_{\Lambda(n)}\Big(\Big(\bigcup_{n-\phi(n)\le m<n} \mathcal{E}_m\Big)^c\Big) \le \prod_{1\le i\le L} \mathbb{P}_{\Lambda(n)}\Big(A_i \cup A_i^w \,\Big|\, \bigcap_{1\le j<i} (A_j \cup A_j^w)\Big).$$

By corollary 10.4 and the Markov property (see section 4.4), there exist $b, c > 0$ such that,

$$\forall n \ge 1 \quad \forall i \in \{1 \cdots L\}$$
$$\mathbb{P}_{\Lambda(n)}\Big(A_i \cup A_i^w \,\Big|\, \bigcap_{1\le j<i} (A_j \cup A_j^w)\Big) \le b\exp\big(-c(n-2iL)^{d-1}\big).$$

This implies that

$$\limsup_{n\to\infty} \frac{1}{n^{d-1}} \ln \mathbb{P}_{\Lambda(n)}\Big(\Big(\bigcup_{n-\phi(n)\le m<n} \mathcal{E}_m\Big)^c\Big) = -\infty.$$

Let $S_n$ be an event which depends only on the edges in the box $\Lambda(n-2\phi(n))$. For $* = f$ or $\pi$, we denote by $S_n^*$ the event that $S_n$ occurs on the $*$-layer. For $\eta = (\eta^f, \eta^\pi)$ a configuration in $\Lambda(n) \setminus \Lambda(m)$, we have

$$\begin{aligned}
\mathbb{P}_{\Lambda(n)}\Big(S_n^* \,\Big|\, \forall e \in \mathbb{E}^d(\Lambda(n)) \setminus \mathbb{E}^d(\Lambda(m)) \quad \omega^f(e) = \eta^f(e),\, \omega^\pi(e) = \eta^\pi(e)\Big)& \\
= \Phi^*_{\Lambda(n)}\Big(S_n \,\Big|\, \forall e \in \mathbb{E}^d(\Lambda(n)) \setminus \mathbb{E}^d(\Lambda(m)) \quad \omega^*(e) = \eta^*(e)\Big)& \\
= \Phi_{\Lambda(m)}^{*\cdot W_{\Lambda(m)}^{\Lambda(n)}(\eta^*)}\big(S_n\big)&.
\end{aligned}$$

The first equality is a consequence of the specific choice of the coupling and the second one follows from the Markov property (see section 4.4). If we assume that $\eta \in \mathcal{E}_m$, we have by lemma 10.9

$$\Phi_{\Lambda(m)}^{f\cdot W_{\Lambda(m)}^{\Lambda(n)}(\eta^f)}\big(S_n\big) \le q^{\varepsilon n^{d-1}} \Phi_{\Lambda(m)}^{\pi\cdot W_{\Lambda(m)}^{\Lambda(n)}(\eta^\pi)}\big(S_n\big).$$

Hence

$$\begin{aligned}
&\mathbb{P}_{\Lambda(n)}\big(S_n^f \cap \mathcal{E}_m\big) \\
&\quad = \sum_{\eta\in\mathcal{E}_m} \mathbb{P}_{\Lambda(n)}\big(S_n^f \,\big|\, \omega = \eta \text{ in } \Lambda(n)\setminus\Lambda(m)\big)\, \mathbb{P}_{\Lambda(n)}\big(\omega = \eta \text{ in } \Lambda(n)\setminus\Lambda(m)\big) \\
&\quad = \sum_{\eta\in\mathcal{E}_m} \Phi_{\Lambda(m)}^{f\cdot W_{\Lambda(m)}^{\Lambda(n)}(\eta^f)}\big(S_n\big)\, \mathbb{P}_{\Lambda(n)}\big(\omega = \eta \text{ in } \Lambda(n)\setminus\Lambda(m)\big)
\end{aligned}$$

$$\leq \sum_{\eta\in\mathcal{E}_m} q^{\varepsilon n^{d-1}} \Phi_{\Lambda(m)}^{\pi\cdot W_{\Lambda(m)}^{\Lambda(n)}(\eta^\pi)}(S_n)\,\mathbb{P}_{\Lambda(n)}(\omega=\eta \text{ in } \Lambda(n)\setminus\Lambda(m))$$

$$= \sum_{\eta\in\mathcal{E}_m} q^{\varepsilon n^{d-1}} \mathbb{P}_{\Lambda(n)}(S_n^\pi\,|\,\omega=\eta \text{ in } \Lambda(n)\setminus\Lambda(m))\,\mathbb{P}_{\Lambda(n)}(\omega=\eta \text{ in } \Lambda(n)\setminus\Lambda(m))$$

$$= q^{\varepsilon n^{d-1}}\mathbb{P}_{\Lambda(n)}(S_n^\pi\cap\mathcal{E}_m) \leq q^{\varepsilon n^{d-1}}\mathbb{P}_{\Lambda(n)}(S_n^\pi)\,.$$

Therefore, using the previous inequalities, we obtain

$$\limsup_{n\to\infty}\frac{1}{n^{d-1}}\ln\Phi_{\Lambda(n)}^f(S_n) = \limsup_{n\to\infty}\frac{1}{n^{d-1}}\ln\mathbb{P}_{\Lambda(n)}\Big(S_n^f\cap\Big(\bigcup_{n-\phi(n)\leq m<n}\mathcal{E}_m\Big)\Big)$$

$$\leq \limsup_{n\to\infty}\frac{1}{n^{d-1}}\ln\Big(\phi(n)\max_{n-\phi(n)\leq m<n}\mathbb{P}_{\Lambda(n)}(S_n^f\cap\mathcal{E}_m)\Big)$$

$$\leq \limsup_{n\to\infty}\frac{1}{n^{d-1}}\ln\Big(q^{\varepsilon n^{d-1}}\mathbb{P}_{\Lambda(n)}(S_n^\pi)\Big)$$

$$\leq \varepsilon\ln q + \limsup_{n\to\infty}\frac{1}{n^{d-1}}\ln\Phi_{\Lambda(n)}^{\pi(n)}(S_n)\,.$$

Since $\varepsilon>0$ can be chosen arbitrarily small, we have

$$\limsup_{n\to\infty}\frac{1}{n^{d-1}}\ln\Phi_{\Lambda(n)}^f(S_n) \leq \limsup_{n\to\infty}\frac{1}{n^{d-1}}\ln\Phi_{\Lambda(n)}^{\pi(n)}(S_n)\,.$$

By interchanging the roles of $f$ and $\pi(n)$ and using the first inequality given by lemma 10.9, we derive similarly the opposite inequality and we obtain the claim of the proposition for the supremum limits. The derivation of the equality involving the infimum limits is analogous. □

**Lemma 10.10 (Decoupling lemma).** *Let $d\geq 2$, $q\geq 1$, $p>\widehat{p}_c$, $p\notin\mathcal{U}(q)$. Let $D_i$, $i\in I$, be a finite collection of disjoint compact subsets of $[-1/2,1/2]^d$ having non-empty connected interiors. For $i\in I$ and $n\in\mathbb{N}$, let $S_n^i$ be an event depending only on the edges in $nD_i\cap\Lambda(n)$. Then*

$$\limsup_{n\to\infty}\frac{1}{n^{d-1}}\ln\Phi_{\Lambda(n)}^w\Big(\bigcap_{i\in I}S_n^i\Big) \leq \sum_{i\in I}\limsup_{n\to\infty}\frac{1}{n^{d-1}}\ln\Phi_{\Lambda(n)}^w(S_n^i)\,.$$

*Remark 10.11.* The same result is valid for lim inf with the opposite inequality. Again, $w$ could be replaced by $f$, or by any boundary condition. We can even take the maximum over $\Phi$ in $c\mathcal{FK}(\Lambda(n))$. The result is also valid for the infinite volume measure $\Phi_\infty$.

*Proof.* It is sufficient to prove the statement for two sets $D_1$ and $D_2$ only. We suppose first that $D_1$ and $D_2$ are closed subsets of the interior of $[-1/2,1/2]^d$. Let $\Theta_1$ be a box building such that $D_1\subset\overset{\circ}{\Theta}_1$, $D_2\cap\overline{\Theta}_1=\emptyset$. Let $\phi:\mathbb{N}\to\mathbb{N}$ be an increasing function such that $\lim_{n\to\infty}\phi(n)=\infty$ and $\lim_{n\to\infty}\phi(n)/n=0$. For $n$ large enough, the event $S_n^1$ depends only on the edges in $\operatorname{int}(n\Theta_1,\phi(n))$, while $S_n^2$ depends only on the edges of $\mathbb{E}^d(\Lambda(n))\setminus\mathbb{E}^d(n\Theta_1)$. Therefore

$$\begin{aligned}\Phi^w_{\Lambda(n)}\big(S^1_n\cap S^2_n\big) &= \Phi^w_{\Lambda(n)}\Big(1_{S^2_n}(\omega)\,\Phi^w_{\Lambda(n)}\big(\,S^1_n\,\big|\,\omega(e),\, e\in\mathbb{E}^d(\Lambda(n))\setminus\mathbb{E}^d(n\Theta_1)\big)\Big)\\ &\leq \Big(\max_{\Phi\in\, c\mathcal{FK}(n\Theta_1)}\Phi\big(S^1_n\big)\Big)\,\Phi^w_{\Lambda(n)}\big(S^2_n\big)\,.\end{aligned}$$

We conclude by taking the supremum limits, with the help of proposition 10.6 and remark 10.8. For sets intersecting the boundary of $[-1/2,1/2]^d$, the argument is analogous and it relies on remark 10.7. □

# 11

# Surface tension

In this chapter, we define the surface tension in the percolation setting and we study its basic properties. Throughout the chapter, $P$ denotes a probability measure on the edge configurations $\{0,1\}^{\mathbb{E}^d}$ which is translation invariant, satisfies the FKG inequality and

$$\forall e \in \mathbb{E}^d \qquad P\big(e \text{ is closed}\big) \geq 1-p\,.$$

We suppose also that $P$ is invariant under the isometries which leave $\mathbb{Z}^d$ globally invariant. Most of the arguments are very robust because they rely essentially on the FKG inequality. The proof of the positivity of the surface tension requires in addition the coarse–graining estimates of chapter 9. Only the final estimate given in lemma 11.13 requires in addition the decoupling property (see lemma 10.10). In particular, the results are valid for the Bernoulli percolation model with parameter $p > p_c$ and more generally for the FK model with $q \geq 1$ in the uniqueness region $p \in ]\widehat{p}_c, 1] \setminus \mathcal{U}(q)$, in any dimensions $d \geq 2$.

## 11.1 Existence

Let $x = (x_1, \cdots, x_d)$ be a point of $\mathbb{R}^d$ and let $w$ be a vector in the unit sphere $S^{d-1}$. The hyperplane containing $x$ with normal vector $w$ is

$$\mathrm{hyp}\,(x,w) \;=\; \{\, y \in \mathbb{R}^d : (y-x)\cdot w = 0\,\}\,.$$

Let $A$ be a subset of $\mathbb{R}^d$ of linear dimension $d-1$, that is $A$ spans a hyperplane of $\mathbb{R}^d$, which we denote by $\mathrm{hyp}\,A$. We call such a set a hyperset. Let $r > 0$. The cylinder of basis $A$ and height $2r$ is the set

$$\mathrm{cyl}\,(A,r) \;=\; \big\{\, x + t\,\nu : t \in \mathbb{R},\ |t| \leq r,\ x \in A \,\big\}$$

where $\nu$ is one of the two unit vectors orthogonal to $\mathrm{hyp}\,A$. We set also

$$\mathrm{cyl}\,A \;=\; \mathrm{cyl}\,(A, +\infty)\,.$$

A hyperrectangle is a hyperset which, up to an orthonormal change of coordinates, is a $d-1$ dimensional parallelepiped with sides parallel to the axis. By disc $(x,r,w)$ we denote the closed disc centered at $x$ of radius $r$ and normal vector $w$.

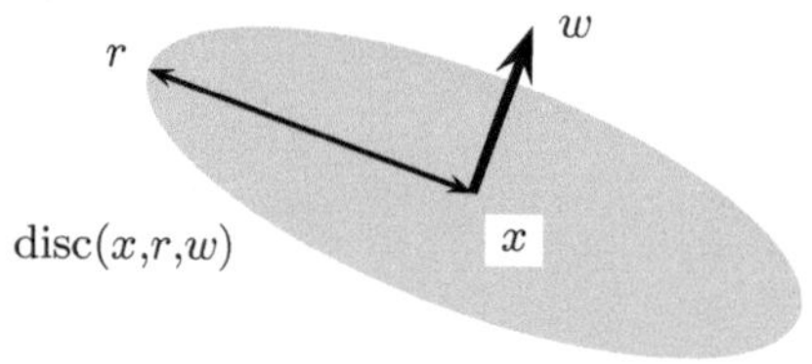


For $r>0$, the $r$–neighbourhood $\mathcal{V}(A,r)$ of a subset $A$ of $\mathbb{R}^d$ is

$$\mathcal{V}(A,r) = \{x \in \mathbb{R}^d : \inf_{y\in A} |x-y| < r\}.$$

We will have to work with enlargements of continuous subsets of $\mathbb{R}^d$ so that they have a significant trace on the discrete lattice $\mathbb{Z}^d$. We fix a real number $\zeta > 2d$ and we enlarge a subset $A$ of $\mathbb{R}^d$ by considering its $\zeta$–neighbourhood $\mathcal{V}(A,\zeta)$. A minimal requirement to choose $\zeta$ is that, whenever $A$ is an arcwise connected subset of $\mathbb{R}^d$, the graph having for vertex set $\mathbb{Z}^d \cap \mathcal{V}(A,\zeta)$ and for edge set $\mathbb{E}^d(\mathcal{V}(A,\zeta))$ is also connected; here $\mathbb{E}^d(\mathcal{V}(A,\zeta))$ is the set of the edges of $\mathbb{E}^d$ included in $\mathcal{V}(A,\zeta)$. Some of the constants appearing in the statements and the proofs depend on $\zeta$. However the direction dependent surface tension and the rough probabilistic estimates are independent of the particular choice of $\zeta > 2d$.

**Definition 11.1.** *Let $D$ be a subset of $\mathbb{R}^d$. A set of edges $E$ of $\mathbb{E}^d$ is said to separate $\infty$ in $D$ if the graph $(\mathbb{Z}^d \cap D, \mathbb{E}^d(D)\setminus E)$ has at least two unbounded connected components ($\mathbb{E}^d(D)$ is the set of the edges of $\mathbb{E}^d$ included in $D$).*

Let $A$ be a closed hyperrectangle and let $s$ be positive, possibly equal to $+\infty$. We denote by $W(\partial A, s, \zeta)$ the event that there exists a finite set of closed edges $E$ inside $\mathcal{V}(\operatorname{hyp} A, s)$ such that

- $E$ separates $\infty$ in $\operatorname{cyl} A$.
- $E$ contains no edge in $\mathcal{V}(\operatorname{cyl}\partial A,\zeta)\setminus\mathcal{V}(\operatorname{hyp} A,\zeta)$.

Loosely speaking, the second condition means that the "boundary" of $E$ is "pinned down" at $\partial A$ within a distance $\zeta$. Note that the event $W(\partial A, s, \zeta)$ is decreasing and it depends only on the edges inside $\operatorname{cyl} A \cap \mathcal{V}(\operatorname{hyp} A, s)$.

We denote by $\mathcal{H}^{d-1}(A)$ the $d-1$ dimensional Hausdorff measure of $A$ (see section 13.1).

**Proposition 11.2.** *Let $p \in ]0,1[$. Let $A$ be a hyperrectangle and let $\nu$ be one of the two unit vectors orthogonal to* $\operatorname{hyp} A$. *Let $\phi(n)$ be a function from $\mathbb{N}$ to $\mathbb{R}^+ \cup \{\infty\}$ such that $\lim_{n\to\infty} \phi(n) = \infty$. The limit*

$$\lim_{n\to\infty} -\frac{1}{\mathcal{H}^{d-1}(nA)} \ln P\big(W(\partial nA, \phi(n), \zeta)\big)$$

*exists in $[0,\infty]$ and it depends only on the vector $\nu$ normal to $A$. We denote it by $\tau(\nu)$ and we call it the surface tension in the direction $\nu$.*

*Proof.* This result is proved with the help of the same subadditivity argument used in [3], proposition 2.4. The only additional problem is that we work with curves whose position with respect to the discrete lattice $\mathbb{Z}^d$ is arbitrary. Let $w$ be a unit vector of $\mathbb{R}^d$ and let $A, A'$ be two hyperrectangles which are both orthogonal to $w$. Let $\phi(n), \phi'(n)$ be two functions from $\mathbb{N}$ to $\mathbb{R}^+ \cup \{\infty\}$ such that $\lim_{n\to\infty} \phi(n) = \infty$, $\lim_{n\to\infty} \phi'(n) = \infty$. Let $\zeta, \zeta'$ be two real numbers larger than $2d$. Let $n, m$ in $\mathbb{N}$ be such that

$$n \operatorname{diam} A > m \operatorname{diam} A' > \zeta + \zeta' + d \,.$$

Because we deal with hyperrectangles, certainly there exists a finite collection of sets $(T(i), i \in I)$ such that:

- each set $T(i)$ is a translate of $mA'$ intersecting the set

$$D(m,n) = \{x \in nA : \operatorname{dist}(x, n\partial A) > 2m \operatorname{diam} A'\}\,,$$

- the sets $(T(i), i \in I)$ have pairwise disjoint interiors,
- their union $\bigcup_{i\in I} T(i)$ contains the set $D(m,n)$.

Since $A$ is a hyperrectangle, then

$$\mathcal{H}^{d-1}(nA) - 2m(\operatorname{diam} A')\mathcal{H}^{d-2}(n\partial A) \le \mathcal{H}^{d-1}(D(m,n)) \le |I|\mathcal{H}^{d-1}(mA') \le \mathcal{H}^{d-1}(nA)\,.$$

For each $i$ in $I$, let $t(i)$ be a vector in $\mathbb{R}^d$ such that $|t(i)|_\infty \le 1$ and $t(i) + T(i)$ is the image of $mA'$ by an integer translation (a translation that leaves $\mathbb{Z}^d$ globally invariant). We set $T'(i) = t(i) + T(i)$ for $i \in I$. Suppose that all the events $W(\partial T'(i), \phi(n), \zeta')$, $i \in I$, occur and let $E(i)$, $i \in I$, be finite sets of closed edges realizing these events. Let $c(d,\zeta')$ be a positive constant and let $E_0$ be the set of the edges included in

$$\Big(\operatorname{cyl}(nA \setminus D(m,n)) \cap \mathcal{V}(\operatorname{hyp} nA, \zeta)\Big) \cup \bigcup_{i\in I} \Big(\mathcal{V}(\operatorname{cyl} \partial T'(i), c(d,\zeta')) \cap \mathcal{V}(\operatorname{hyp} nA, c(d,\zeta'))\Big)\,.$$

Let $E = E_0 \cup \bigcup_{i\in I} E(i)$. Clearly $E$ is finite. The constant $c(d,\zeta')$ can be chosen large enough (depending only on $d, \zeta'$) to guarantee that the edges of

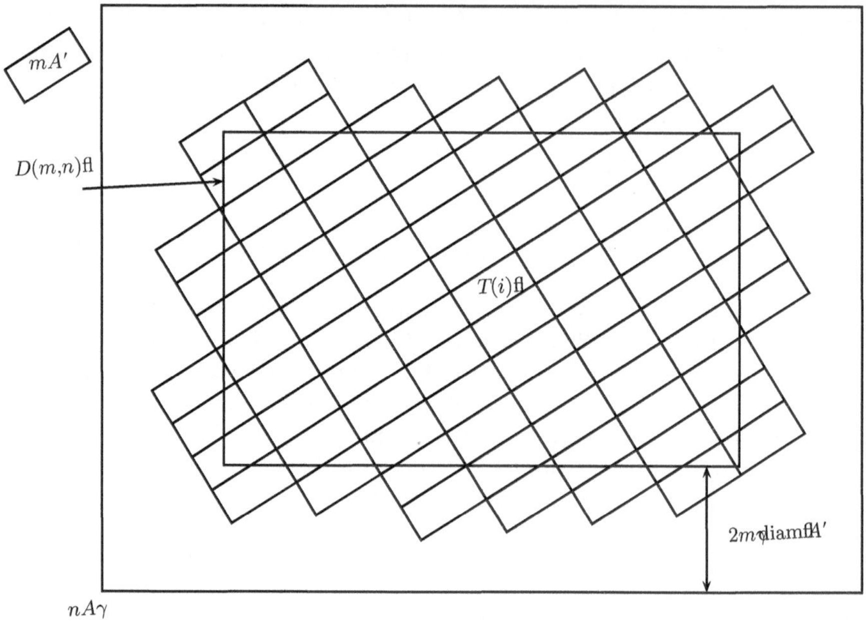


$E$ separate $\infty$ inside $\mathrm{cyl}\, nA$. For $m$ such that $m \operatorname{diam} A' > \zeta + d + c(d, \zeta')$, the set of edges $E$ realizes the event $W(\partial nA, \phi(n), \zeta)$. An attempt at a proof is done in [38], proposition 5.2. Therefore

$$\Big\{\omega : \forall e \in E_0 \quad \omega(e) = 0\Big\} \cap \Big(\bigcap_{i \in I} W(\partial T'(i), \phi(n), \zeta')\Big) \subset W(\partial nA, \phi(n), \zeta).$$

Since all these events are decreasing, by the FKG inequality,

$$P\big(W(\partial nA, \phi(n), \zeta)\big) \geq (1-p)^{|E_0|} \prod_{i \in I} P\big(W(\partial T'(i), \phi(n), \zeta')\big).$$

Since the model is invariant under the integer translations, for any $i$ in $I$,

$$P\big(W(\partial T'(i), \phi(n), \zeta')\big) = P\big(W(\partial mA', \phi(n), \zeta')\big).$$

Because $\phi(n)$ goes to $\infty$ as $n$ goes to $\infty$,

$$\lim_{n \to \infty} P\big(W(\partial mA', \phi(n), \zeta')\big) = P\big(W(\partial mA', \infty, \zeta')\big)$$

whence, for $n$ sufficiently large,

$$P\big(W(\partial mA', \phi(n), \zeta')\big) \geq (1/2) P\big(W(\partial mA', \infty, \zeta')\big).$$

For such integers $n$, combining the previous inequalities and passing to the logarithm,

$$\ln P\big(W(\partial nA,\phi(n),\zeta)\big) \geq \\ |I|\ln P\big(W(\partial mA',\phi'(m),\zeta')\big) + |E_0|\ln(1-p) - |I|\ln 2\,.$$

There exists a further constant $c(d,\zeta,\zeta',A,A')$ such that

$$|E_0| \leq c(d,\zeta,\zeta',A,A')(n^{d-2}m + n^{d-1}/m + 1)\,.$$

Using the previous inequalities, we obtain

$$\frac{\ln P\big(W(\partial nA,\phi(n),\zeta)\big)}{\mathcal{H}^{d-1}(nA)} \geq \frac{\ln P\big(W(\partial mA',\phi'(m),\zeta')\big)}{\mathcal{H}^{d-1}(mA')} - \frac{\ln 2}{\mathcal{H}^{d-1}(mA')} \\ + c(d,\zeta,\zeta',A,A')\mathcal{H}^{d-1}(A)^{-1}(m/n + 1/m + 1/n^{d-1})\ln(1-p)\,.$$

Sending successively $n$ to $\infty$ and then $m$ to $\infty$ yields

$$\liminf_{n\to\infty} \frac{\ln P\big(W(\partial nA,\phi(n),\zeta)\big)}{\mathcal{H}^{d-1}(nA)} \geq \limsup_{m\to\infty} \frac{\ln P\big(W(\partial mA',\phi'(m),\zeta')\big)}{\mathcal{H}^{d-1}(mA')}$$

which implies the result of the proposition. □

## 11.2 Finite volume definition

For the Bernoulli percolation model, it is very natural to define the surface tension with the help of the infinite volume product measure $P$. The analogue in the FK percolation setting is the infinite volume FK measure $\Phi_\infty$. However, the definition of this measure requires already to take a limit of finite volume FK measures. In addition, for the FK and Ising models, we are interested in understanding the large deviation behavior of the system in a finite box, rather than in infinite volume. Therefore it seems judicious to provide a definition of the surface tension as a limit of finite volume quantities. Since the event $W(\partial A,s,\zeta)$ depends only on the edges inside $\text{cyl}\,(A,s)$, we need only to put adequate boundary conditions on this set to obtain an operational formula.

**Definition 11.3.** *Let $D, A_1, A_2$ be three subsets of $\mathbb{R}^d$. We denote by*

$$\{\, A_1 \not\longleftrightarrow A_2 \text{ in } D \,\}$$

*the event that there is no open path in the graph $(D\cap\mathbb{Z}^d,\mathbb{E}^d(D))$ from $A_1$ to $A_2$ (where $\mathbb{E}^d(D)$ is the set of the edges of $\mathbb{E}^d$ included in $D$).*

Let $A$ be a closed hyperrectangle and let $\phi(n)$ be a function from $\mathbb{N}$ to $\mathbb{R}^+$ such that $\lim_{n\to\infty}\phi(n)=\infty$. Let $D_n = \text{cyl}\,(nA,\phi(n))$. The set $D_n\setminus \text{hyp}\,nA$ has two connected components, which we denote by $C_1^n$ and $C_2^n$. For $i=1,2$

and $n \in \mathbb{N}$, let $A_i^n$ be the set of the points of $C_i^n \cap \mathbb{Z}^d$ which have a nearest neighbour in $\mathbb{Z}^d \setminus D_n$:

$$A_i^n = \{ x \in C_i^n \cap \mathbb{Z}^d : \exists y \in \mathbb{Z}^d \setminus D_n \quad |x-y| = 1 \}.$$

See figure in section 7.1.

**Proposition 11.4.** *Let $q \geq 1$ and $p \in ]0,1[$. The limit*

$$\lim_{n\to\infty} -\frac{1}{\mathcal{H}^{d-1}(nA)} \ln \Phi_{D_n}^{w,p,q}(\, A_1^n \not\leftrightarrow A_2^n \text{ in } D_n \,)$$

*exists in $[0,\infty]$. In the uniqueness region $p \in ]\widehat{p}_c, 1] \setminus \mathcal{U}(q)$, this limit is equal to the surface tension defined in proposition 11.2 (with $\Phi_\infty^{p,q}$ in place of $P$).*

*Proof.* The existence of the limit involving $\Phi_{D_n}^{w,p,q}$ can be proved for $p \in ]0,1[$ with a subadditive argument which is a variant of the one used in the proof of proposition 11.2. From now onwards, we suppose that $p$ is in $]\widehat{p}_c, 1] \setminus \mathcal{U}(q)$. Since $\Phi_\infty^{p,q}$ is translation invariant, has the finite energy property and satisfies the FKG inequality, the limit appearing in proposition 11.2 with $\Phi_\infty^{p,q}$ instead of $P$ exists, we denote it by $\tau(\nu)$. Our goal is to prove that the limit involving $\Phi_{D_n}^{w,p,q}$ is also equal to $\tau(\nu)$. Let $\varepsilon > 0$ and let $A^{-\varepsilon}$, $A^\varepsilon$ be two hyperrectangles in $\operatorname{hyp} A$ such that

$$A^{-\varepsilon} \subset A \subset A^\varepsilon, \qquad A^\varepsilon \setminus A^{-\varepsilon} \subset \mathcal{V}(\partial A, 4\varepsilon),$$
$$\operatorname{dist}(A^{-\varepsilon}, \partial A) > 2\varepsilon, \qquad \operatorname{dist}(A, \partial A^\varepsilon) > 2\varepsilon.$$

Let $E_+$ be the set of the edges included in

$$\operatorname{cyl}\big(\operatorname{hyp} nA \cap \mathcal{V}(\partial nA, 5\varepsilon n)\big) \cap \mathcal{V}(\operatorname{hyp} nA, \zeta).$$

There exists a constant $c = c(d,\zeta)$ such that $|E_+| \leq c\varepsilon n^{d-1}$. For $n$ large enough, so that $\varepsilon n > \zeta$, if the events

$$\{\, A_1^n \not\leftrightarrow A_2^n \text{ in } D_n \,\}, \quad \{\text{all the edges of } E_+ \text{ are closed}\}$$

occur simultaneously, then the event $W(\partial nA^\varepsilon, \phi(n), \zeta)$ occurs as well. Thus, by the FKG inequality,

$$\begin{aligned}
&\Phi_\infty^{p,q}(W(\partial nA^\varepsilon, \phi(n), \zeta))\\
&\quad\geq \Phi_\infty^{p,q}(\text{all the edges of } E_+ \text{ are closed})\, \Phi_\infty^{p,q}(\, A_1^n \not\leftrightarrow A_2^n \text{ in } D_n \,)\\
&\quad\geq (1-p)^{|E_+|}\, \Phi_{D_n}^{w,p,q}(\, A_1^n \not\leftrightarrow A_2^n \text{ in } D_n \,).
\end{aligned}$$

Taking logarithms, and sending successively $n$ to $\infty$ and $\varepsilon$ to 0, we obtain

$$\tau(\nu) \leq \lim_{n\to\infty} -\frac{1}{\mathcal{H}^{d-1}(nA)} \ln \Phi_{D_n}^{w,p,q}(\, A_1^n \not\leftrightarrow A_2^n \text{ in } D_n \,).$$

Similarly, letting $E_-$ to be the set of the edges included in

$$\big(\operatorname{cyl}\big(nA\cap\mathcal{V}(\partial nA,5\varepsilon n)\big)\cap\mathcal{V}(\operatorname{hyp}nA,\zeta)\big)\cup\mathcal{V}(\partial nA^{-\varepsilon},d\zeta)\,,$$

we have $|E_-|\leq c\varepsilon n^{d-1}$ for some constant $c=c(d,\zeta)$ and if the events

$$\{\text{ all the edges of } E_- \text{ are closed}\}\,,\quad W(\partial nA^{-\varepsilon},\phi(n),\zeta)$$

both occur, then $\{\,A_1^n \not\leftrightarrow A_2^n \text{ in } D_n\,\}$ occurs as well. Therefore

$$\Phi_{D_n}^{w,p,q}\big(\,A_1^n \not\leftrightarrow A_2^n \text{ in } D_n\,\big)\,\geq\,(1-p)^{|E_-|}\,\Phi_{D_n}^{w,p,q}\big(W(\partial nA^{-\varepsilon},\phi(n),\zeta)\big)\,.$$

Taking logarithms, sending $n$ to $\infty$, we get with the help of proposition 10.6,

$$\lim_{n\to\infty}\,-\frac{1}{n^{d-1}}\ln\Phi_{D_n}^{w,p,q}\big(\,A_1^n \not\leftrightarrow A_2^n \text{ in } D_n\,\big)\,\leq\,\tau(\nu)\mathcal{H}^{d-1}(A^{-\varepsilon})-c\varepsilon\ln(1-p)\,.$$

Sending now $\varepsilon$ to 0, we see that the limits appearing in propositions 11.2 and 11.4 coincide. □

We finally reformulate the definition of the surface tension in the spin setting. We consider a closed hyperrectangle $A$. To simplify the formula, we suppose that $\mathcal{H}^{d-1}(A)=1$. We define the sets $D_n$, $A_1^n$, $A_2^n$ as above. The Ising Hamiltonian in $D_n$ is

$$\forall\sigma\in\{-,+\}^{D_n}\qquad H_n(\sigma)\,=\,-\frac{1}{2}\sum_{\substack{x,y\in D_n\\|x-y|=1}}\sigma(x)\sigma(y)\,.$$

Let $\mathcal{E}_n$ be the set of the spin configurations $\sigma$ inside $D_n$ such that $\sigma(x)=+$ for $x\in A_1^n\cup A_2^n$. Let $\mathcal{F}_n$ be the set of the spin configurations $\sigma$ inside $D_n$ such that $\sigma(x)=-$ for $x\in A_1^n$ and $\sigma(x)=+$ for $x\in A_2^n$. The partition functions $Z_n^+$, $Z_n^{-,+}$ corresponding to pure $+$ and mixed $-,+$ boundary conditions at temperature $T$ are

$$Z_n^+\,=\,\sum_{\sigma\in\mathcal{E}_n}\exp-\frac{H_n(\sigma)}{T}\,,\qquad Z_n^{-,+}\,=\,\sum_{\sigma\in\mathcal{F}_n}\exp-\frac{H_n(\sigma)}{T}\,.$$

**Proposition 11.5.** *Let $T>0$. The limit*

$$\lim_{n\to\infty}\,-\frac{1}{n^{d-1}}\ln\frac{Z_n^{-,+}}{Z_n^+}$$

*exists in $[0,\infty]$. It coincides with the limit appearing in proposition 11.4 with $q=2$ and $p=1-\exp(-2/T)$.*

*Proof.* We use the FK–Ising coupling measure $\mathbb{P}_n$ on edge–spin configurations in the domain $D_n$ with free boundary conditions to rewrite the partition functions $Z_n^+$ and $Z_n^{-,+}$ (as done in [23]). We say that a percolation configuration $\omega$

is compatible with a spin configuration $\sigma$, which we denote by $\omega \sim \sigma$, if the endpoints of each open edge in $\omega$ have the same sign in $\sigma$. Let $Z_n$ be the partition function of the Ising model in $D_n$ with free boundary conditions. From section 4.3, we know that, setting $p = 1 - \exp(-2/T)$, for any $\sigma \in \{-,+\}^{D_n}$,

$$\frac{1}{Z_n} \exp -\frac{H_n(\sigma)}{T} = \sum_{\omega \sim \sigma} \frac{1}{Z} \prod_{e \in \mathbb{E}^d(D_n)} p^{\omega(e)}(1-p)^{1-\omega(e)}.$$

Thus

$$\begin{aligned} Z_n^+ &= \frac{Z_n}{Z} \sum_{\sigma \in \mathcal{E}_n} \sum_{\omega \sim \sigma} \prod_{e \in \mathbb{E}^d(D_n)} p^{\omega(e)}(1-p)^{1-\omega(e)} \\ &= \frac{Z_n}{Z} \sum_{\omega \in \{0,1\}^{\mathbb{E}^d(D_n)}} \Big( \prod_{e \in \mathbb{E}^d(D_n)} p^{\omega(e)}(1-p)^{1-\omega(e)} \Big) q^{cl^w(\omega)}. \end{aligned}$$

Similarly, $Z_n^{-,+}$ is equal to an analogous expression where the sum is restricted to the configurations belonging to the event $\{ A_1^n \not\leftrightarrow A_2^n \text{ in } D_n \}$. Indeed, the percolation configurations compatible with a spin configuration of $\mathcal{F}_n$ are precisely the configurations where there is no connection between $A_1^n$ and $A_2^n$ in $D_n$. Therefore

$$\frac{Z_n^{-,+}}{Z_n^+} = \Phi_{D_n}^{w,p,q=2}\big( A_1^n \not\leftrightarrow A_2^n \text{ in } D_n \big).$$

The claim of the proposition is a direct consequence of this finite volume identity and proposition 11.4. □

## 11.3 Basic properties

We derive next some basic properties of the surface tension. In the context of lattice spin systems, where the definition of the surface tension is significantly different, the corresponding properties have been derived in [102].

The surface tension $\tau$ inherits automatically some symmetry properties from the model. For instance, if $f$ is a linear isometry of $\mathbb{R}^d$ such that $f(0) = 0$ and $f(\mathbb{Z}^d) = \mathbb{Z}^d$ then $\tau \circ f = \tau$. Besides, the surface tension $\tau$ satisfies another important inequality called the weak triangle inequality. For details and results concerning this kind of inequalities, see [64, 102, 106].

**Proposition 11.6 (weak triangle inequality).** *Let $(ABC)$ be a non degenerate triangle in $\mathbb{R}^d$ and let $\nu_A$, $\nu_B$, $\nu_C$ be the exterior normal unit vectors to the sides $[BC]$, $[AC]$, $[AB]$ in the plane spanned by $A, B, C$. Then*

$$\mathcal{H}^1([BC])\tau(\nu_A) \le \mathcal{H}^1([AC])\tau(\nu_B) + \mathcal{H}^1([AB])\tau(\nu_C).$$

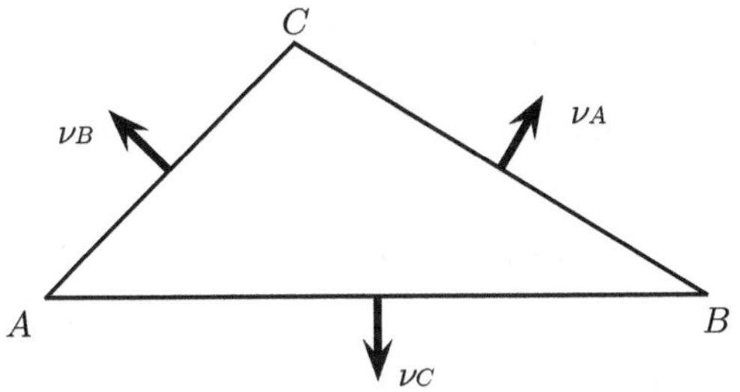


*Proof.* We consider first the case where the triangle $(ABC)$ is such that $BA \cdot BC \geq 0$ and $CA \cdot CB \geq 0$. Let $\varepsilon, h$ be such that $0 < \varepsilon \leq 1 \leq h$. Let $(e_1, \cdots, e_d)$ be an orthonormal basis of $\mathbb{R}^d$ such that $e_1, e_2$ belong to the two dimensional space spanned by $A, B, C$. Let $K$ be the compact convex set defined by

$$K = \left\{ x + \sum_{3 \leq i \leq d} u_i e_i \,:\, x \in (ABC),\, (u_3, \cdots, u_d) \in [0, h]^{d-2} \right\}.$$

The boundary of $K$ consists of the three hyperrectangles $R_A$, $R_B$, $R_C$ defined by

$$R_A = \left\{ x + \sum_{3 \leq i \leq d} u_i e_i \,:\, x \in [BC],\, (u_3, \cdots, u_d) \in [0, h]^{d-2} \right\},$$
$$R_B = \left\{ x + \sum_{3 \leq i \leq d} u_i e_i \,:\, x \in [AC],\, (u_3, \cdots, u_d) \in [0, h]^{d-2} \right\},$$
$$R_C = \left\{ x + \sum_{3 \leq i \leq d} u_i e_i \,:\, x \in [AB],\, (u_3, \cdots, u_d) \in [0, h]^{d-2} \right\},$$

and the set

$$T = \bigcup_{3 \leq j \leq d} \left\{ x + \sum_{3 \leq i \leq d} u_i e_i : x \in (ABC),\, u_j \in \{0, h\}, \right.$$
$$\left. (u_3, ..., u_{j-1}, u_{j+1}, ..., u_d) \in [0, h]^{d-3} \right\}.$$

Notice that the set $T$ is connected in dimensions $d \geq 4$ and consists of two disjoint triangles in dimensions $d = 3$. The intersection of the hyperrectangles $R_A$ and $R_B$ is a $d-2$ dimensional rectangle and it is denoted by $R_{A,B}$. Similar notation is used for the other intersections. Let $E_0$ be the set of the edges included in

$$\Big( \mathrm{cyl} \Big( \mathrm{hyp}\, nR_A \cap \mathcal{V}(\partial nR_A, 4\varepsilon n) \Big) \cap \mathcal{V}( \mathrm{hyp}\, nR_A, \zeta) \Big)$$
$$\cup\, \mathcal{V}(nR_{B,C}, 2\zeta) \cup \mathcal{V}(nR_{A,C}, 2\zeta) \cup \mathcal{V}(nR_{A,B}, 2\zeta) \cup \mathcal{V}(nT, 2\zeta) .$$

There exists a constant $c = c(d, \zeta)$ such that

$$|E_0| \leq c\Big(\varepsilon n^{d-1}h^{d-2} + (nh)^{d-2} + 2(d-2)n^{d-1}h^{d-3}\Big)\,.$$

Let $R^\varepsilon$ be a hyperrectangle in $\operatorname{hyp} R_A$ such that $R_A \subset R^\varepsilon \subset \mathcal{V}(R_A, 4\varepsilon)$ and $\operatorname{dist}(\partial R^\varepsilon, R_A) > 2\varepsilon$. For $n$ large enough, so that $\varepsilon n > \zeta$, if the events

$$W(\partial n R_B, \varepsilon n, \zeta)\,,\quad W(\partial n R_C, \varepsilon n, \zeta)\,,\quad \{\text{ all the edges of } E_0 \text{ are closed }\}$$

occur simultaneously, then the event $W(\partial n R^\varepsilon, \infty, \zeta)$ occurs as well; by the assumptions $BA \cdot BC \geq 0$ and $CA \cdot CB \geq 0$, the set $\mathcal{V}(nR_B \cup nR_C, \varepsilon n)$ is included in $\mathcal{V}(\operatorname{cyl} nR_A, \varepsilon n)$ and does not intersect $\mathcal{V}(\operatorname{cyl} \partial n R^\varepsilon, \zeta)$, so that the separating sets will be correctly localized. By the FKG inequality, this inclusion implies

$$(1-p)^{|E_0|} P\big(W(\partial n R_B, \varepsilon n, \zeta)\big)\, P\big(W(\partial n R_C, \varepsilon n, \zeta)\big) \leq P\big(W(\partial n R^\varepsilon, \infty, \zeta)\big)\,.$$

The previous inequalities and proposition 11.2 yield

$$\begin{aligned}\mathcal{H}^{d-1}(R^\varepsilon)\tau(\nu_A) \leq \mathcal{H}^{d-1}(R_B)\tau(\nu_B) + \mathcal{H}^{d-1}(R_C)\tau(\nu_C)&\\ -c\big(\varepsilon h^{d-2} + 2(d-2)h^{d-3}\big)\ln(1-p)\,.&\end{aligned}$$

We observe that

$$\begin{gathered}\mathcal{H}^{d-1}(R_B) = h^{d-2}\mathcal{H}^1([AC]),\ \mathcal{H}^{d-1}(R_C) = h^{d-2}\mathcal{H}^1([AB]),\\ h^{d-2}\mathcal{H}^1([BC]) \leq \mathcal{H}^{d-1}(R^\varepsilon)\,.\end{gathered}$$

Combining the last two inequalities and dividing by $h^{d-2}$, we get

$$\mathcal{H}^1([BC])\tau(\nu_A) \leq \mathcal{H}^1([AC])\tau(\nu_B) + \mathcal{H}^1([AB])\tau(\nu_C) - c\,(\varepsilon + \frac{2}{h}(d-2))\ln(1-p)\,.$$

By letting $h$ go to $\infty$ and $\varepsilon$ go to 0, we obtain the weak triangle inequality for the triangle $(ABC)$. Let now $A, B, C$ be three points such that $BA \cdot BC < 0$,

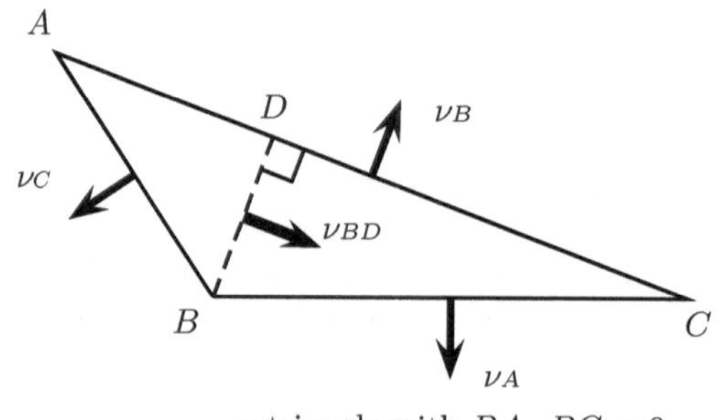


a triangle with $BA \cdot BC < 0$

$CA \cdot CB \geq 0$. Let $D$ be the orthogonal projection of $B$ on $[AC]$. Then

$$BC \cdot BD > 0\,, \quad DB \cdot DA = 0\,, \quad BA \cdot BD > 0\,.$$

We apply the weak triangle inequality to the triangles $(BCD)$ and $(BDA)$:

$$\mathcal{H}^1([BC])\tau(\nu_A) \le \mathcal{H}^1([BD])\tau(\nu_{BD}) + \mathcal{H}^1([DC])\tau(\nu_B)\,,$$
$$\mathcal{H}^1([BD])\tau(\nu_{BD}) \le \mathcal{H}^1([AB])\tau(\nu_C) + \mathcal{H}^1([AD])\tau(\nu_B)\,,$$

where $\nu_{BD}$ is a unit vector orthogonal to $[BD]$. Combining the two inequalities, we get the weak triangle inequality for the triangle $(ABC)$. The case $BA \cdot BC \ge 0$, $CA \cdot CB < 0$ is similar. □

The weak triangle inequality implies several nice properties for the surface tension.

**Corollary 11.7.** *The homogeneous extension $\tau_0$ of $\tau$ to $\mathbb{R}^d$ defined by*

$$\tau_0(0) = 0\,, \qquad \forall w \in \mathbb{R}^d \setminus \{0\} \quad \tau_0(w) \;=\; |w|_2\tau(w/|w|_2)$$

*is a convex function.*

*Proof.* Let $u, v$ be two linearly independent vectors in $\mathbb{R}^d$ and let $w = u + v$. Let $T$ be the rotation of angle $\pi/2$ in the plane spanned by $u$ and $v$. Applying the weak triangle inequality to the triangle having for sides $T(u)$, $T(v)$ and $T(w)$, we obtain

$$|w|_2\tau(w/|w|_2) \;\le\; |u|_2\tau(u/|u|_2) + |v|_2\tau(v/|v|_2)\,,$$

so that $\tau_0(w) \le \tau_0(u) + \tau_0(v)$. Since $\tau_0$ is also homogeneous, it is convex. □

The convexity of $\tau_0$ is in fact equivalent to the weak triangle inequality. The next corollary is a consequence of [64], theorem 3.1: the weak triangle inequality automatically implies the weak simplex inequality.

**Corollary 11.8 (weak simplex inequality).** *Let $A_0, \cdots, A_d$ be $d+1$ points in generic position in $\mathbb{R}^d$. For $i$ in $\{0 \cdots d\}$, let $\Delta(i)$ be the hypersimplex defined by the points $\{A_0, \cdots, A_d\} \setminus \{A_i\}$. Modulo signs, there exists a unique family of unit vectors $\nu_0, \cdots, \nu_d$ such that for $i$ in $\{0 \cdots d\}$, the vector $\nu_i$ is orthogonal to the vector space spanned by the hypersimplex $\Delta(i)$. Then*

$$\mathcal{H}^{d-1}(\Delta(0))\tau(\nu_0) \le \mathcal{H}^{d-1}(\Delta(1))\tau(\nu_1) + \cdots + \mathcal{H}^{d-1}(\Delta(d))\tau(\nu_d)\,.$$

**Proposition 11.9.** *The surface tension $\tau : S^{d-1} \to \mathbb{R}^+$ is bounded and continuous. In the Bernoulli percolation model, and in the FK percolation model with parameter $q = 2$, it is equal to $0$ for $p \le p_c$ and it is strictly positive for $p > p_c$. In the FK percolation model with parameter $q \ge 1$, the surface tension is strictly positive for $p \in ]\widehat{p}_c, 1] \setminus \mathcal{U}(q)$.*

*Proof.* Let $w \in S^{d-1}$ and let $A$ be a hyperrectangle orthogonal to $w$ such that $\mathcal{H}^{d-1}(A) = 1$. Let $E(n)$ be the set of the edges included in $\mathrm{cyl}\,(nA, 2d)$. Then $|E(n)| \le c(d)n^{d-1}$ and

$$\Phi_\infty^{p,q}\Big(W(\partial nA,\infty,2d)\Big)\geq\Phi_\infty^{p,q}\Big(\text{the edges of } E(n) \text{ are closed}\Big)\geq(1-p)^{|E(n)|}.$$

Passing to the limit, we get $\tau(w)\leq -c(d)\ln(1-p)$. Since $\tau_0$ is homogeneous and bounded on $S^{d-1}$, it is finite everywhere. The map $\tau_0$ being also convex (by corollary 11.7), it follows from a standard result of convex analysis ([113], corollary 10.1.1) that $\tau_0$ is continuous, as well as $\tau$.

We consider next the FK percolation model with parameter $q\geq 1$, and we prove that the surface tension is positive whenever $p\in]\widehat{p}_c,1]\setminus\mathcal{U}(q)$; this includes the cases of Bernoulli percolation with $p>p_c$ and of the FK percolation model with $q=2, p>p_c(2)$, since $\widehat{p}_c(1)=p_c$ by [81] and $]\widehat{p}_c(2),1]\setminus\mathcal{U}(2)=]p_c(2),1]$ by [26, 27]. A direct proof of the positivity of the surface tension for the subcritical Ising model can be found in [97]; in fact, this is a key input to prove that $\widehat{p}_c(2)=p_c(2)$ [26].

Let $\nu$ belong to the unit sphere $S^{d-1}$. We will show that $\tau(\nu)>0$. Let $A$ be a unit hypersquare orthogonal to $\nu$. We work with the lattice rescaled by a factor $K$. For $\underline{x}\in\mathbb{Z}^d$, the block variable $X(\underline{x})$ is the indicator function of the event $R(B'(\underline{x}),K)$, i.e.,

- Inside the event block $B'(\underline{x})$ there is exactly one crossing cluster $C(\underline{x})$, it is the only cluster in $B'(\underline{x})$ having diameter larger than or equal to $K$ and it crosses every sub–box of $B'(\underline{x})$ with diameter larger than or equal to $K$.

Suppose that the event $W(n\partial A,n,\zeta)$ occurs. Then there is no $\mathbb{L}^d$ connected path $\underline{x}_1,\dots,\underline{x}_r$ in the renormalized set

$$\big\{\,\underline{x}\in\mathbb{Z}^d : B(\underline{x})\subset \operatorname{cyl}(nA,n+2dK)\,\big\}$$

such that

$$B(\underline{x}_1)\cap(nA-(n+dK)\nu)\neq\emptyset,\quad B(\underline{x}_r)\cap(nA+(n+dK)\nu)\neq\emptyset,$$

and $X(\underline{x}_1)=\cdots=X(\underline{x}_r)=1$. Indeed, the existence of such a path of good blocks would entail the existence of an open path inside $\operatorname{cyl}(nA,n)$ connecting $nA-n\nu$ and $nA+n\nu$ and this would contradict the occurrence of the event $W(n\partial A,n,\zeta)$. This implies furthermore that there exists a $\mathbb{L}^{d,\infty}$ connected subset $\underline{E}$ of bad blocks which disconnects $nA-(n+dK)\nu$ and $nA+(n+dK)\nu$ in $\big\{\,\underline{x}\in\mathbb{Z}^d : B(\underline{x})\subset\operatorname{cyl}(nA,n+2dK)\,\big\}$. A rigorous proof of this intuitive fact is not that simple. A possible strategy is to look at the external vertex boundary of the set of the vertices which are connected to $nA-(n+dK)\nu$ by $\mathbb{L}^d$ connected paths of good blocks. This external vertex boundary is $\mathbb{L}^{d,\infty}$ connected and consists of bad blocks. However we refrain from entering into the details of the proof here. Necessarily, the cardinality of $\underline{E}$ is larger than $(n/2K)^{d-1}$. Now, there exists a positive constant $b$ depending on the dimension only such that the number of $\mathbb{L}^{d,\infty}$-connected sets of size $m$ containing a fixed point is bounded by $b^m$. Let $\delta>0$. By lemma 9.4 and proposition 9.7, for $K$ large enough, the block process $(X(\underline{x}),\ \underline{x}\in\mathbb{Z}^d)$ satisfies the condition stated at the beginning of section 9.4. Using lemma 9.9, we can bound the probability of $W(n\partial A, n, \zeta)$ by

$$\Phi_\infty^{p,q}(W(n\partial A, n, \zeta)) \leq 4n \sum_{i \geq (n/2K)^{d-1}} b^i \exp(3^{-d} i \ln \delta) .$$

For $\delta$ sufficiently small, we conclude that

$$\tau(\nu) = \lim_{n\to\infty} -\frac{1}{n^{d-1}} \ln \Phi_\infty^{p,q}(W(\partial nA, n, \zeta)) \geq -\frac{1}{(2K)^{d-1}}\big(\ln b + 3^{-d} \ln \delta\big)$$

and therefore $\tau(\nu) > 0$. Thus the surface tension $\tau$ does not vanish on $S^{d-1}$. Since $\tau$ is continuous on $S^{d-1}$, it is bounded away from 0 on $S^{d-1}$. □

The previous properties of $\tau$ can equivalently be described through its Wulff crystal

$$\mathcal{W}_\tau = \Big\{ x \in \mathbb{R}^d \,:\, x \cdot w \leq \tau(w) \text{ for all } w \text{ in } S^{d-1} \Big\} .$$

**Corollary 11.10.** *Let $q \geq 1$ and let $p \in ]\widehat{p}_c(q), 1] \setminus \mathcal{U}(q)$. The Wulff crystal $\mathcal{W}_\tau$ associated with $\tau$ is bounded, closed, convex and contains 0 in its interior. If $f$ is a linear isometry of $\mathbb{R}^d$ such that $f(0) = 0$ and $f(\mathbb{Z}^d) = \mathbb{Z}^d$ then $f(\mathcal{W}_\tau) = \mathcal{W}_\tau$. The surface tension $\tau$ is the support function of its Wulff crystal, i.e.,*

$$\forall \nu \in S^{d-1} \qquad \tau(\nu) = \sup \big\{ x \cdot \nu : x \in \mathcal{W}_\tau \big\} .$$

These properties are equivalent to the symmetry properties of $\tau$ and the facts stated in corollary 11.7, proposition 11.9. The function $\tau$ is the support function of $\mathcal{W}_\tau$ because $\tau_0$ is convex and coincides with its bipolar, see for instance [113], corollary 13.2.1, [72], proposition 3.5, or [64], theorem 2.1, corollary 3.6. A proof of these facts is provided in proposition 14.1.

## 11.4 Separating sets

With the help of the surface tension, we next estimate the probability of the occurrence of a separating set of closed edges near a hyperplane. Let $A$ be a hyperset in $\mathbb{R}^d$ and let $r$ be positive or infinite. We denote by $S(A,r)$ the event that there exists a finite set of closed edges in $\operatorname{cyl} A \cap \mathcal{V}(\operatorname{hyp} A, r)$ which separates $\infty$ in $\operatorname{cyl} A$, that is,

$$S(A,r) = \Big\{ \exists\, E \subset \mathbb{E}^d\big(\operatorname{cyl} A \cap \mathcal{V}(\operatorname{hyp} A, r)\big),\ |E| < \infty,$$
$$\forall e \in E \quad \omega(e) = 0,\ E \text{ separates } \infty \text{ in } \operatorname{cyl} A \Big\} .$$

From now on, we work with a fixed value of $\zeta$ larger than $2d$ and we drop $\zeta$ in the notation $W(\partial A, s, \zeta)$, thus writing simply $W(\partial A, s)$.

**Lemma 11.11.** *Let $O$ be an open hyperset in $\mathbb{R}^d$, let $\nu$ be one of the two unit vectors orthogonal to $O$ and let $\phi(n)$ be a function from $\mathbb{N}$ to $\mathbb{R}^+ \cup \{\infty\}$ such that* $\lim_{n\to\infty} \phi(n) = \infty$. *We have*

$$\liminf_{n\to\infty} \frac{1}{n^{d-1}} \ln P\big(S(nO, \phi(n))\big) \;\geq\; -\mathcal{H}^{d-1}(O)\,\tau(\nu)\,.$$

*Proof.* Let $(A_i, i \in I)$ be a finite family of hyperrectangles in $\operatorname{hyp} O$ having disjoint relative interiors and covering $O$. Let $c = c(d, \zeta)$ be a large constant. Let $E_0$ be the set of the edges included in the union

$$\bigcup_{i\in I} \mathcal{V}(\,\operatorname{cyl}\partial nA_i, c) \cap \mathcal{V}(\,\operatorname{hyp} nO, c)\,.$$

There exists a further constant $c' = c'(d, \zeta, c)$ such that $|E_0| \leq c'|I|n^{d-2}$. If all the events $W(\partial nA_i, \phi(n))$, $i \in I$, occur and all the edges of $E_0$ are closed, then $S(nO, \phi(n))$ occurs as well, provided the constant $c$ is large enough. By the FKG inequality,

$$P\big(S(nO, \phi(n))\big) \;\geq\; (1-p)^{|E_0|} \prod_{i\in I} P\big(W(\partial nA_i, \phi(n))\big)\,,$$

whence, by proposition 11.2 and the bound on $|E_0|$,

$$\liminf_{n\to\infty} \frac{1}{n^{d-1}} \ln P\big(S(nO, \phi(n))\big) \;\geq\; -\tau(\nu) \sum_{i\in I} \mathcal{H}^{d-1}(A_i)\,.$$

By taking the supremum of the right–hand side over all possible coverings of $O$, we obtain the claim of the lemma. □

**Lemma 11.12.** *Let $w \in S^{d-1}$. There exists a positive constant $c = c(p, d, \zeta)$ such that, for any hyperrectangle $A$ orthogonal to $w$, for any $r$ positive,*

$$\limsup_{n\to\infty} \frac{1}{n^{d-1}} \ln P\big(S(nA, nr)\big) \;\leq\; -\tau(w)\mathcal{H}^{d-1}(A) + cr\mathcal{H}^{d-2}(\partial A)\,.$$

*Proof.* Let $\varepsilon > 0$. Let $A^\varepsilon$ be a hyperrectangle such that

$$\mathcal{V}(A, 2\varepsilon) \cap \operatorname{hyp} A \;\subset\; A^\varepsilon \;\subset\; \mathcal{V}(A, 3\varepsilon) \cap \operatorname{hyp} A\,.$$

Let $E_0$ be the set of the edges included in

$$\Big(\mathcal{V}(\,\operatorname{cyl} n\partial A, \zeta) \cap \mathcal{V}(\,\operatorname{hyp} nA, nr)\Big) \cup \Big(\operatorname{cyl}\,(nA^\varepsilon \setminus nA) \cap \mathcal{V}(\,\operatorname{hyp} nA, \zeta)\Big)\,.$$

Suppose that the event $S(nA, nr)$ occurs, and let $E_S$ be a set of closed edges realizing it. Suppose also that all the edges of $E_0$ are closed. Then the set of closed edges $E_0 \cup E_S$ realizes the event $W(\partial nA^\varepsilon, nr)$. Therefore

$$\Big\{\omega : \forall e \in E_0 \quad \omega(e) = 0\Big\} \cap S(nA, nr) \;\subset\; W(\partial nA^\varepsilon, nr)\,.$$

Since all these events are decreasing, by the FKG inequality,

$$(1-p)^{|E_0|}\,P\big(S(nA,nr)\big)\;\le\;P\big(W(\partial nA^\varepsilon,nr)\big)\,.$$

There exists a constant $c=c(d,\zeta)$ such that

$$|E_0|\le cn^{d-1}\big(r\mathcal{H}^{d-2}(\partial A)+\mathcal{H}^{d-1}(A^\varepsilon\setminus A)\big)$$

whence, passing to the logarithm,

$$\ln P\big(S(nA,nr)\big)\;\le\;\\ \ln P\big(W(\partial nA^\varepsilon,nr)\big)-cn^{d-1}\big(r\mathcal{H}^{d-2}(\partial A)+\mathcal{H}^{d-1}(A^\varepsilon\setminus A)\big)\ln(1-p)\,.$$

Letting $n$ go to $\infty$, applying proposition 11.2, and sending $\varepsilon$ to 0, we get the desired inequality. □

**Lemma 11.13.** *There exists a positive constant $c=c(p,d,\zeta)$ such that: for any open hyperset $O$ in $\mathbb{R}^d$, for any finite family $(A_i,i\in I)$ of disjoint closed hyperrectangles included in $O$, for any positive $r$,*

$$\limsup_{n\to\infty}\frac{1}{n^{d-1}}\ln P\big(S(nO,nr)\big)\le-\tau(\nu)\sum_{i\in I}\mathcal{H}^{d-1}(A_i)+c\,r\sum_{i\in I}\mathcal{H}^{d-2}(\partial A_i)$$

*where $\nu$ is one of the two unit vectors orthogonal to $O$.*

*Remark 11.14.* This result requires the surface decoupling property stated in lemma 10.10, in addition to the hypothesis on $P$ stated at the beginning of the chapter.

*Proof.* The very definition of the event $S(nO,nr)$ implies that

$$S(nO,nr)\subset\bigcap_{i\in I}S(nA_i,nr)\,.$$

For $i$ in $I$, the event $S(nA_i,nr)$ depends only on the status of the edges inside $n\,\mathrm{cyl}\,(A_i,r)$. Since the hyperrectangles $(A_i,i\in I)$ are pairwise disjoint and compact, so are the sets $(\,\mathrm{cyl}\,(A_i,r),i\in I)$. Using independence in the case of Bernoulli percolation or the decoupling lemma 10.10 in the case of FK percolation with $q\ge1$, we get

$$\limsup_{n\to\infty}\frac{1}{n^{d-1}}\ln P\big(S(nO,nr)\big)\;\le\;\sum_{i\in I}\;\limsup_{n\to\infty}\frac{1}{n^{d-1}}\ln P\big(S(nA_i,nr)\big)\,.$$

The conclusion follows from lemma 11.12. □

By $\mathrm{disc}\,(x,r,w)$ we denote the closed disc centered at $x$ of radius $r$ with normal vector $w$; the $\mathcal{H}^{d-1}$ measure of a unit disc is denoted by $\alpha_{d-1}$.

**Corollary 11.15.** *There exists a positive constant $c = c(p, d, \zeta)$ such that, for any $x$ in $\mathbb{R}^d$, any positive $\varrho, \eta$ with $\eta \leq \varrho$, any $w$ in $S^{d-1}$,*

$$\limsup_{n \to \infty} \frac{1}{n^{d-1}} \ln P\big(S(n\,\mathrm{disc}\,(x, \varrho, w), n\eta)\big) \leq -\alpha_{d-1}\varrho^{d-1}\tau(w) + c\,\eta\varrho^{d-2}.$$

*Proof.* We apply lemma 11.13 with $O = \mathrm{disc}\,(x, \varrho, w)$, $r = \eta$. There exists a constant $c' = c'(d)$ such that, for any $\varepsilon$ positive, there exists a finite family $(A_i, i \in I)$ of disjoint closed hyperrectangles included in $O$ such that

$$\sum_{i \in I} \mathcal{H}^{d-1}(A_i) \geq \alpha_{d-1}\varrho^{d-1} - \varepsilon, \qquad \sum_{i \in I} \mathcal{H}^{d-2}(\partial A_i) \leq c'\varrho^{d-2}.$$

The result follows by taking the infimum over all possible families in the inequality stated in lemma 11.13. □

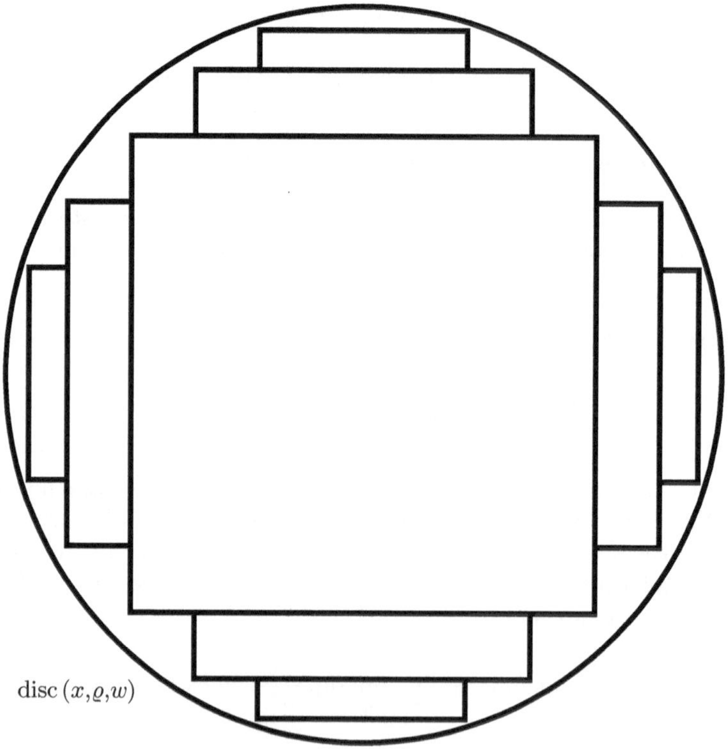

## 11.5 What do we know about the surface tension?

In dimensions $d = 2$, it is possible to obtain an exact formula for the surface tension of the Ising model.

**Theorem 11.16.** *Let $T < T_c$ and let $\varphi \in [0, \pi/4]$. The surface tension of the two dimensional Ising model at inverse temperature $\beta = 1/T$ is given by*

$$\tau(\beta, \varphi) = \operatorname{asinh}(s \cos\varphi)\cos\varphi + \operatorname{asinh}(s \sin\varphi)\sin\varphi$$

*where*

$$s = \frac{2}{m|\cos(2\varphi)|}\sqrt{1 - \sqrt{(1-m^2)\sin^2(2\varphi) + m^2}}\,,$$

$$m = \frac{2}{\sinh(2\beta) + (\sinh(2\beta))^{-1}}\,.$$

This beautiful and tricky computation is performed thanks to the dimer representation of the Ising model. The details are presented in [103] and the corresponding background material can be found in [99].

In dimensions $d = 3$, it is possible to obtain a first order expansion of the surface tension of the FK percolation model with the help of combinatorial computations involving domino tiling [41]. Theorem 5.6 is a consequence of this expansion.

**Theorem 11.17.** *For any $q \geq 1$, uniformly over $\nu$ in $S^2$, as $p$ goes to 1,*

$$\tau(\nu, p) = |\nu|_1 \ln\left(\frac{1}{1-p}\right) - |\nu|_1 \mathrm{ent}(\nu) + o(1)$$

*where for any $\nu = (a, b, c)$ in the unit sphere $S^2$, we set $|\nu|_1 = |a| + |b| + |c|$,*

$$\mathrm{ent}(\nu) = \frac{1}{\pi}L\left(\pi\frac{|a|}{|\nu|_1}\right) + \frac{1}{\pi}L\left(\pi\frac{|b|}{|\nu|_1}\right) + \frac{1}{\pi}L\left(\pi\frac{|c|}{|\nu|_1}\right)$$

*and $L$ is the Lobachevsky function given by*

$$\forall x \in [0, \pi] \qquad L(x) = -\int_0^x \ln(2\sin t)\,dt\,.$$

## 11.5 What do we know about the surface tension?

# 12

# Interface estimate

This chapter is devoted to the proof of a lemma which is crucial for linking the surface tension to the desired large deviation upper bounds. The interface lemma gives a probabilistic estimate for the local presence of a collection of open clusters creating a small flat interface near a middle hyperplane of a ball. The estimate is uniform with respect to the location, the size and the direction of the interface.

## 12.1 Interface lemma

We state here an improved version of the interface lemma used in [43]. The proof is due to a young Italian mathematician, David Barbato [15], who considerably enhanced the original argument of [38]. In the original proof, a surgical procedure was designed in order to relate the event "Sep" defined below to a disconnection event, whose probability was then estimated with the help of the surface tension. The surgical procedure consisted in closing by force a certain set of edges. David Barbato has found a way to choose this set of edges so that their positions can be recovered even after the surgery has been performed. This leads to great simplifications. In particular, it is not any more necessary to go to an intermediate scale to perform the surgery. As a consequence, the range of validity of this crucial estimate is in principle extended (it does not rely any more on the coarse graining results of Pisztora).

We need some notation to proceed further. Let $x$ be a point of $\mathbb{R}^d$. The closed ball of center $x$ and Euclidean radius $r > 0$ is denoted by $B(x,r)$. We denote by $\alpha_d$ the volume of the $d$–dimensional unit ball. For $w$ in the unit sphere $S^{d-1}$, we define the half balls

$$B_-(x,r,w) = B(x,r) \cap \left\{ y \in \mathbb{R}^d : (y-x)\cdot w \leq 0 \right\},$$
$$B_+(x,r,w) = B(x,r) \cap \left\{ y \in \mathbb{R}^d : (y-x)\cdot w \geq 0 \right\}.$$

To prove the local large deviation upper bound in the Ising setting, we will

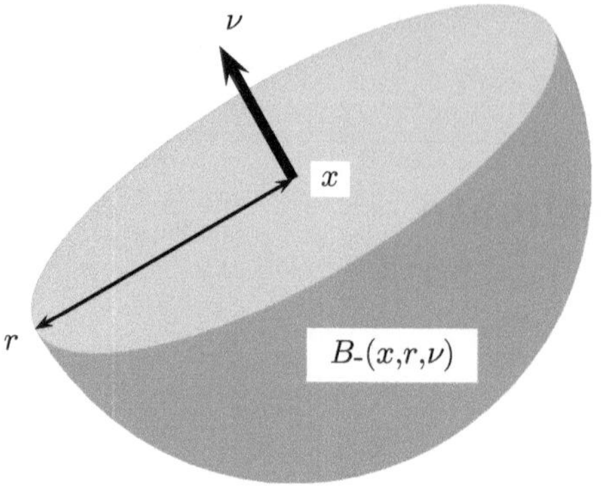


examine the empirical magnetization close to a flat macroscopic interface and we will translate the situation in the FK percolation model. This will typically result in the occurrence of the event that we define and estimate next. The open $B(nx, nr)$–clusters are the open clusters in the configuration restricted to the ball $B(nx, nr)$. Let $\mathrm{Sep}(n, x, r, w, \delta)$ be the following event: there exists a collection $\mathcal{C}$ of open $B(nx, nr)$–clusters such that

$$\sum_{C\in\mathcal{C}} \big|C \cap B_-(nx, nr, w)\big| \geq (1-\delta)\mathcal{L}^d\big(B_-(nx, nr, w)\big)\,,$$
$$\sum_{C\in\mathcal{C}} \big|C \cap B_+(nx, nr, w)\big| \leq \delta\mathcal{L}^d\big(B_+(nx, nr, w)\big)\,.$$

The event $\mathrm{Sep}(n, x, r, w, \delta)$ depends only on the status of the edges inside $B(nx, nr)$ and it is decreasing: it becomes easier to find a suitable collection $\mathcal{C}$ when additional edges are closed.

**Lemma 12.1.** *Let $p \in ]0,1[$. There exists a constant $c = c(p, d, \zeta)$ such that for any $x \in \mathbb{R}^d$, any $r \in ]0,1[$, any unit vector $w \in S^{d-1}$, and any $\delta \in ]0,1[$,*

$$\limsup_{n\to\infty} \frac{1}{n^{d-1}} \ln P\big(\mathrm{Sep}(n, x, r, w, \delta)\big) \leq -\alpha_{d-1} r^{d-1} \tau(w)\,(1 - c\delta^{1/2})\,.$$

For simplicity, we state the lemma for the Bernoulli percolation model. Yet the same result holds for the FK percolation model with $q \geq 1$ in the uniqueness region $p \in ]\widehat{p}_c(q), 1[\setminus \mathcal{U}(q)$: just replace $P$ by the infinite volume FK measure $\Phi_\infty$ in the statement and in the proof. The basic idea of the proof is as follows. We will show that whenever the event "Sep" occurs, it is always possible to perform a surgery by closing not too many edges so that in the modified configuration a separating set of closed edges appears near the middle hyperplane $\mathrm{hyp}\,(nx, w)$ of the ball $B(nx, nr)$. The occurrence of such a separating set of closed edges is directly related to our definition of the surface tension.

*Proof.* If $p \le p_c$ then $\tau(w) = 0$ and the statement is trivially true. Hence we need only to consider the case $p > p_c$ where $\tau(w) > 0$. Throughout the proof, we fix $x, r, w, \delta$ as in the statement of the interface lemma and we drop them from the notation. For instance, $B(n)$ stands for $B(nx, nr)$ and $\text{Sep}(n)$ stands for $\text{Sep}(n, x, r, w, \delta)$.

Let $\zeta > 2d$ be an arbitrary fixed constant like the one we used to define the surface tension (see proposition 11.2). Let $\eta$ be a positive real number (we shall later on choose $\eta$ small enough, depending on $r, \delta$). Let $\varrho$ be such that

$$0 < \eta < \varrho < r\,, \quad 0 < 2\eta < \sqrt{r^2 - \varrho^2}$$

($\varrho$ will be chosen later to be close to $r$). By disc $(x, \varrho, w)$ we denote the closed disc centered at $x$ of radius $\varrho$ and normal vector $w$. For $y \in \mathbb{R}^d$, $w$ in the unit sphere $S^{d-1}$, and $r_1, r_2$ in $\mathbb{R} \cup \{-\infty, +\infty\}$, we define

$$\text{slab}\,(y, w, r_1, r_2) \;=\; \big\{\, z \in \mathbb{R}^d : r_1 \le (z - y) \cdot w \le r_2 \,\big\}\,.$$

For $h > 0$, we define the following sets:

$$\begin{aligned}
D(h) &= \text{cyl}\,(n\,\text{disc}\,(x, \varrho, w)) \cap \text{slab}\,(nx, w, -h, h)\,,\\
\partial^+ D(h) &= \big\{\, y \in D(h) : \exists z \in \mathbb{Z}^d \quad (z - x) \cdot w > h,\, |z - y| = 1 \,\big\}\,,\\
\partial^- D(h) &= \big\{\, y \in D(h) : \exists z \in \mathbb{Z}^d \quad (z - x) \cdot w < -h,\, |z - y| = 1 \,\big\}\,.
\end{aligned}$$

The event $\text{Sep}(n)$ is defined with the help of sums on a volume. We shall define next a surface version of $\text{Sep}(n)$. For $h, \varepsilon > 0$, the event $\text{Sep}^*(n, h, \varepsilon)$ occurs if there exists a collection $\mathcal{C}(h)$ of open $D(h)$–clusters such that

$$\big|\partial^- D(h)\big| - \sum_{C \in \mathcal{C}(h)} \big|C \cap \partial^- D(h)\big| \le \varepsilon n^{d-1}\,,$$

$$\sum_{C \in \mathcal{C}(h)} \big|C \cap \partial^+ D(h)\big| \le \varepsilon n^{d-1}\,.$$

We first show that if $\text{Sep}(n)$ occurs, then $\text{Sep}^*(n, h, \varepsilon)$ occurs for a suitable choice of $h$ and $\varepsilon$. Indeed, suppose that the event $\text{Sep}(n)$ occurs and let $\mathcal{C}$ be a collection of open $B(n)$–clusters realizing it. We suppose that $n$ is large enough, so that $\eta n > 2$ and

$$|\mathbb{Z}^d \cap B_-(n)| - \mathcal{L}^d(B_-(n)) \le \frac{1}{2}\delta \alpha_d r^d n^d\,.$$

Let $h > 0$. The collection of the clusters of $\mathcal{C}$ intersected with $D(h)$ splits into a collection $\mathcal{C}(h)$ of $D(h)$–open clusters; in particular, for $C \in \mathcal{C}$, the set $C \cap D(h)$ is the union of some clusters of $\mathcal{C}(h)$. Thus

$$\begin{aligned}
\int_1^{\eta n} \sum_{C \in \mathcal{C}(h)} \big|C \cap \partial^+ D(h)\big|\, dh &= \sum_{C \in \mathcal{C}} \int_1^{\eta n} \big|C \cap \partial^+ D(h)\big|\, dh\\
&\le \sum_{C \in \mathcal{C}} \big|C \cap B_+(n)\big| \le \frac{1}{2}\delta \alpha_d r^d n^d\,,
\end{aligned}$$

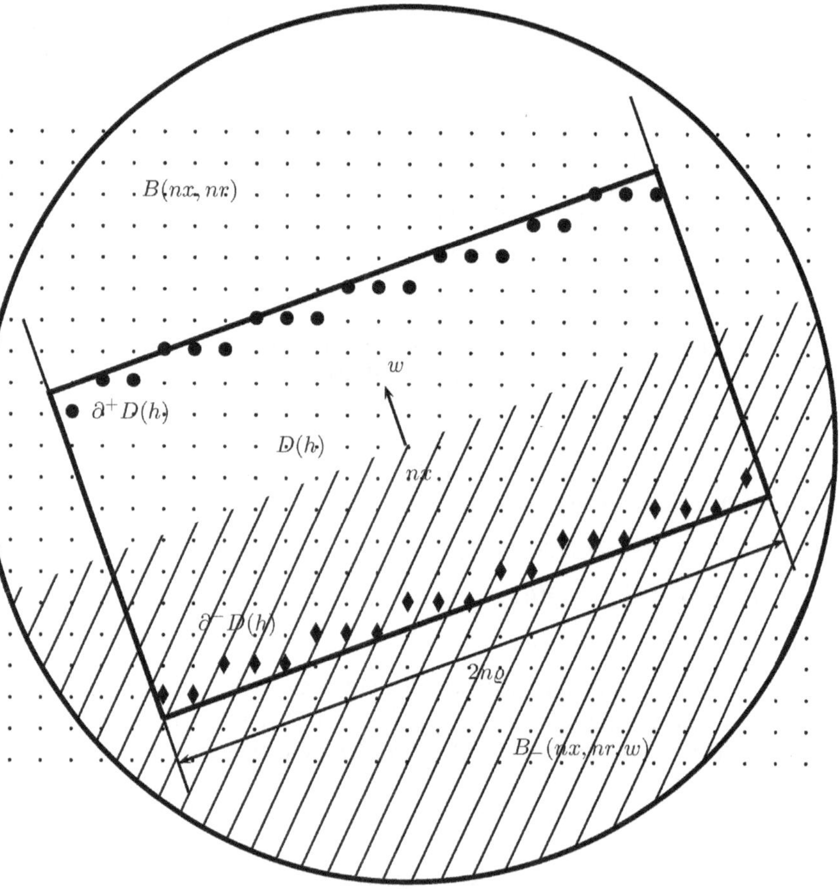


and similarly,

$$\int_1^{\eta n} \Big( |\partial^- D(h)| - \sum_{C \in \mathcal{C}(h)} |C \cap \partial^- D(h)| \Big)\, dh$$
$$\leq |\mathbb{Z}^d \cap B_-(n)| - \sum_{C \in \mathcal{C}} |C \cap B_-(n)| \leq \delta \alpha_d r^d n^d .$$

These inequalities imply that there exists $h \in [1, \eta n]$ such that

$$\sum_{C \in \mathcal{C}(h)} |C \cap \partial^+ D(h)| + \Big( |\partial^- D(h)| - \sum_{C \in \mathcal{C}(h)} |C \cap \partial^- D(h)| \Big) \leq 4 \frac{\delta}{\eta} \alpha_d r^d n^{d-1} .$$

Letting $\varepsilon = 4\delta\alpha_d r^d / \eta$, we conclude that

$$\mathrm{Sep}(n) \text{ occurs} \qquad \Longrightarrow \qquad \exists h \in [1, \eta n] \quad \mathrm{Sep}^*(n, h, \varepsilon) \text{ occurs} .$$

Suppose that $\mathrm{Sep}(n)$ occurs and let $h^*$ be the infimum in $[1, \eta n]$ of the real numbers $h$ such that $\mathrm{Sep}^*(n,h,\varepsilon)$ occurs. Since the set $D(h)$ is closed, then the event $\mathrm{Sep}^*(n,h^*,\varepsilon)$ occurs as well. Notice that $h^*$ is a random value which depends only on the configuration restricted to the edges included in $D(h^*)$. Let $E_-$ (respectively $E_+$) be the set of the edges of $\partial^{edge} D(h^*)$ having an endpoint in $\partial^- D(h^*)$ (respectively $\partial^+ D(h^*)$). Let $F_-$ be the subset of $E_-$ consisting of the edges which have an endpoint in

$$\partial^- D(h^*) \setminus \bigcup_{C \in \mathcal{C}(h^*)} C$$

and let $F_+$ be the subset of $E_+$ consisting of the edges which have an endpoint in

$$\partial^+ D(h^*) \cap \bigcup_{C \in \mathcal{C}(h^*)} C .$$

Since each vertex is the endpoint of at most $2d$ edges, then

$$|F_-| + |F_+| \le 8d\frac{\delta}{\eta}\alpha_d r^d n^{d-1} .$$

Let us set $D = D(n\eta + \zeta)$ and $\partial^- D = \partial^- D(n\eta+\zeta)$, $\partial^+ D = \partial^+ D(n\eta+\zeta)$. Let $\gamma$ be an open path inside $D$ joining the sets $\partial^- D$ and $\partial^+ D$. Certainly the path $\gamma$ has to cross $D(h^*)$ and it goes through an edge of $E_-$. Let $e$ be the last edge of $E_-$ visited by $\gamma$. After its passage through $e$, the path $\gamma$ lands in a vertex $y \in \partial^- D(h^*)$ and it exits from $D(h^*)$ through a vertex $z$ of $\partial^+ D(h^*)$ and an edge $f$ belonging to $E_+$. If $y$ does not belong to a cluster of $\mathcal{C}(h^*)$, then the edge $e$ is in $F_-$. If $y$ belongs to a cluster of $\mathcal{C}(h^*)$, then so does $z$ and the edge $f$ is in $F_+$. In conclusion, any open path in $D$ joining the sets $\partial^- D$ and $\partial^+ D$ has to go through an edge of $F_- \cup F_+$.

Let $\phi : \mathrm{Sep}(n) \to S(n\,\mathrm{disc}\,(x,\varrho,w), n\eta)$ be the map defined by

$$\forall \omega \in \mathrm{Sep}(n) \quad \forall e \in \mathbb{E}^d \qquad \phi(\omega)(e) = \begin{cases} 0 & \text{if } e \in F_- \cup F_+ \\ \omega(e) & \text{otherwise} \end{cases}$$

The event $S(n\,\mathrm{disc}\,(x,\varrho,w), n\eta)$ is defined before lemma 11.11 in section 11.4. The map $\phi$ closes by force the edges in $F_- \cup F_+$ and leaves the other edges unchanged. From the previous discussion, we see that this operation destroys every open connection in $D$ between $\partial^- D$ and $\partial^+ D$. Hence the map $\phi$ has values in the set of configurations $S(n\,\mathrm{disc}\,(x,\varrho,w),\, n\eta)$. Let

$$M = 8d\frac{\delta}{\eta}\alpha_d r^d n^{d-1} .$$

Since $|F_-| + |F_+| \le M$, then

$$\forall \omega \in \mathrm{Sep}(n) \qquad P(\omega) \le (1-p)^{-M} P(\phi(\omega)) .$$

Summing over $\omega$ in $\operatorname{Sep}(n)$, we obtain

$$P(\operatorname{Sep}(n)) \leq (1-p)^{-M} \sum_{\omega \in \operatorname{Sep}(n)} P(\phi(\omega)).$$

Moreover

$$\sum_{\omega \in \operatorname{Sep}(n)} P(\phi(\omega)) = \sum_{\omega \in \phi(\operatorname{Sep}(n))} P(\omega) \left|\phi^{-1}(\omega) \cap \operatorname{Sep}(n)\right|$$

and for any $\omega$ in $\phi(\operatorname{Sep}(n))$, the cardinality of $\phi^{-1}(\omega) \cap \operatorname{Sep}(n)$ is less than or equal to $2^M$. The crucial point is that the edges of $F_- \cup F_+$ are not included in $D(h^*)$, so that if $\omega \in \phi(\operatorname{Sep}(n))$, we can recover the value of $h^*$ by inspecting the configuration $\omega$ restricted to $D$, as well as the sets $F_-$, $F_+$. The map $\phi$ can alter only the states of the edges belonging to $F_-$ and $F_+$. Thus

$$P(\operatorname{Sep}(n)) \leq \Big(\frac{2}{1-p}\Big)^M P(\phi(\operatorname{Sep}(n))).$$

Since $\phi(\operatorname{Sep}(n))$ is included in $S(n \operatorname{disc}(x, \varrho, w), n\eta)$, we arrive at

$$P(\operatorname{Sep}(n)) \leq \Big(\frac{2}{1-p}\Big)^M P(S(n \operatorname{disc}(x, \varrho, w), n\eta)).$$

The previous estimates and corollary 11.15 together imply

$$\limsup_{n\to\infty} \frac{1}{n^{d-1}} \ln P\big(\operatorname{Sep}(n, x, r, w, \delta)\big) \leq$$
$$8d\frac{\delta}{\eta} r^d \alpha_d \ln \frac{2}{1-p} - \alpha_{d-1}\varrho^{d-1}\tau(w) + c\eta\varrho^{d-2}$$

where $c = c(p, d, \zeta)$ is the constant appearing in corollary 11.15. We choose now $\eta = \sqrt{\delta} r/3$, $\varrho = r\sqrt{1-\delta}$. Because $\tau$ is bounded away from 0 for $p > p_c$ (proposition 11.9), there exists a constant $c'' = c''(p, d, \zeta)$ such that

$$\limsup_{n\to\infty} \frac{1}{n^{d-1}} \ln P\big(\operatorname{Sep}(n, x, r, w, \delta)\big) \leq -\alpha_{d-1} r^{d-1}\tau(w)(1 - c''\delta^{1/2})$$

and this holds for any $x \in \mathbb{R}^d$, $r \in ]0,1[$, $\delta \in ]0,1[$ and $w \in S^{d-1}$. □

## 12.2 Near the boundary

In the FK percolation or Ising setting, we will work within a finite box $\Lambda(n)$, and we must take care of interfaces very close to the boundary. Let $Q = [-1/2, 1/2]^d$ be the $d$–dimensional unit cube. Let $B(x, r)$ be a ball such that $x$ belongs to $\partial^* Q$ and $B_-(x, r, \nu_Q(x))$ is included in $Q$; here $\nu_Q(x)$ is the exterior normal vector to $Q$ at $x$. Let $n \in \mathbb{N}$ and let $\delta > 0$. Let $\operatorname{Sep}^{\mathrm{bd}}(n, x, r, \delta)$ be the event: the collection $\mathcal{C}$ of the open $B_-(nx, nr, \nu_Q(x))$–clusters intersecting $\partial^{in}\Lambda(n)$ satisfies

$$\sum_{C\in\mathcal{C}} \big|C\cap B_-(nx,nr,\nu_Q(x))\big| \,\leq\, \delta\mathcal{L}^d\big(B_-(nx,nr,\nu_Q(x))\big)\,.$$

The event $\mathrm{Sep}^{\mathrm{bd}}(n,x,r,\delta)$ is decreasing. By the monotonicity of FK measures with respect to boundary conditions, we have

$$\Phi^w_{\Lambda(n)}\big(\mathrm{Sep}^{\mathrm{bd}}(n,x,r,\delta)\big) \,\leq\, \Phi_\infty\big(\mathrm{Sep}^{\mathrm{bd}}(n,x,r,\delta)\big)$$

and furthermore, for $n$ large enough,

$$\Phi_\infty\big(\mathrm{Sep}^{\mathrm{bd}}(n,x,r,\delta)\big) \,\leq\, \Phi_\infty\big(\mathrm{Sep}(n,x,r,\nu_Q(x),2\delta)\big)\,.$$

Indeed, suppose that $\mathrm{Sep}^{\mathrm{bd}}(n,x,r,\delta)$ occurs and let $\mathcal{D}$ be the collection of the open $B_-(nx,nr,\nu_Q(x))$–clusters which do not intersect $\partial^{in}\Lambda(n)$. Then

$$\begin{aligned}\sum_{C\in\mathcal{D}} &\big|C\cap B_-(nx,nr,\nu_Q(x))\big| \\ &\geq\, \big|\mathbb{Z}^d\cap B_-(nx,nr,\nu_Q(x))\big| - \delta\mathcal{L}^d\big(B_-(nx,nr,\nu_Q(x))\big) \\ &\geq\, (1-2\delta)\mathcal{L}^d\big(B_-(nx,nr,\nu_Q(x))\big)\,,\end{aligned}$$

where the last inequality holds for $n$ large enough. Moreover the clusters of $\mathcal{D}$ do not intersect $B_+(nx,nr,\nu_Q(x))$. Hence $\mathcal{D}$ realizes $\mathrm{Sep}(n,x,r,\nu_Q(x),2\delta)$. We can therefore bound $\Phi^w_{\Lambda(n)}\big(\mathrm{Sep}^{\mathrm{bd}}(n,x,r,\delta)\big)$ with the help of the previous inequalities and lemma 12.1.

## 12.3 Percolation setting

In the Bernoulli percolation setting, we are interested in the infinite cluster or in the large finite clusters. Thus we must rule out the small clusters, whose typical density is $1-\theta$. The event Sep is not directly relevant. We define and estimate next the adequate event to consider. For $A\subset\mathbb{R}^d$, we denote by $\partial^{in}A$ the inner vertex boundary of $A$, defined by

$$\partial^{in}A \,=\, \big\{\,x\in A\cap\mathbb{Z}^d : \exists\, y\in\mathbb{Z}^d\setminus A,\ |y-x|=1\,\big\}\,.$$

Let $\mathrm{Sep}_\theta(n,x,r,w,\delta)$ be the following event: there exists a collection $\mathcal{C}$ of open $B(nx,nr)$–clusters intersecting $\partial^{in}B(nx,nr)$ and such that

$$\sum_{C\in\mathcal{C}} \big|C\cap B_-(nx,nr,w)\big| \,\geq\, (\theta-\delta)\mathcal{L}^d\big(B_-(nx,nr,w)\big)\,,$$

$$\sum_{C\in\mathcal{C}} \big|C\cap B_+(nx,nr,w)\big| \,\leq\, \delta\mathcal{L}^d\big(B_+(nx,nr,w)\big)\,.$$

**Lemma 12.2.** *Let $p \in ]0,1[$. For any $x \in \mathbb{R}^d$, $r \in ]0,1[$, $w \in S^{d-1}$, $\delta \in ]0,1[$,*

$$\lim_{n\to\infty} \frac{1}{n^{d-1}} \ln P\big(\mathrm{Sep}_\theta(n,x,r,w,\delta) \setminus \mathrm{Sep}(n,x,r,w,2\delta)\big) = -\infty\,.$$

*Proof.* We fix $x, r, w, \delta$ as in the statement of the lemma and we note

$$B(n) = B(nx,nr)\,,\quad B_-(n) = B_-(nx,nr,w)\,,\quad B_+(n) = B_+(nx,nr,w)\,,$$
$$\mathrm{Sep}_\theta(n) = \mathrm{Sep}_\theta(n,x,r,w,\delta)\,.$$

Suppose that the event $\mathrm{Sep}_\theta(n)$ occurs and let $\mathcal{C}$ be a collection of open $B(n)$–clusters realizing it. Let $\mathcal{D}$ be the collection of all the open $B_-(n)$–clusters which do not intersect $\partial^{in} B_-(n)$. Notice that the clusters of $\mathcal{D}$ are also open $B(n)$–clusters and that the collections $\mathcal{C}$ and $\mathcal{D}$ are disjoint. Moreover, by the very definition of $\mathcal{D}$,

$$\sum_{C\in\mathcal{D}} |C| = \big|B_-(n)\cap\mathbb{Z}^d\big| - \sum_{x\in B_-(n)\cap\mathbb{Z}^d} 1_{\{\,x \longleftrightarrow \partial^{in} B_-(n)\,\}}\,.$$

Let $n$ be large enough so that

$$\big|\,B_-(n)\cap\mathbb{Z}^d\,\big| - \mathcal{L}^d(B_-(n)) \geq -\frac{\delta}{4}\mathcal{L}^d(B_-(n))\,.$$

Let $\Lambda_i$, $i \in I$, be a finite collection of disjoint boxes included in $\frac{1}{n}B_-(n)$ such that

$$\Big|\mathbb{Z}^d\cap B_-(n)\setminus \bigcup_{i\in I} n\Lambda_i\Big| \leq \frac{\delta}{2}\mathcal{L}^d(B_-(n))\,.$$

By subadditivity (see section 8.1), we have then

$$\sum_{x\in B_-(n)\cap\mathbb{Z}^d} 1_{\{x \longleftrightarrow \partial^{in} B_-(n)\}} \leq \sum_{i\in I}\sum_{x\in n\Lambda_i\cap\mathbb{Z}^d} 1_{\{\,x \longleftrightarrow \partial^{in} n\Lambda_i\,\}} + \frac{\delta}{2}\mathcal{L}^d(B_-(n)).$$

Let Fill be the event

$$\mathrm{Fill} = \Big\{\sum_{C\in\mathcal{D}} |C| > (1-\theta-\delta)\mathcal{L}^d(B_-(n))\Big\}.$$

The previous identities imply then that

$$\begin{aligned} P(\mathrm{Fill}^c) &\leq P\Big(\sum_{i\in I}\sum_{x\in n\Lambda_i\cap\mathbb{Z}^d} 1_{\{\,x \longleftrightarrow \partial^{in} n\Lambda_i\,\}} \geq \sum_{i\in I}\big(\theta+\frac{\delta}{4}\big)\mathcal{L}^d(n\Lambda_i)\Big)\\ &\leq \sum_{i\in I} P\Big(\sum_{x\in n\Lambda_i\cap\mathbb{Z}^d} 1_{\{\,x \longleftrightarrow \partial^{in} n\Lambda_i\,\}} \geq \big(\theta+\frac{\delta}{4}\big)\mathcal{L}^d(n\Lambda_i)\Big)\,. \end{aligned}$$

Let $A_i$ be the event appearing on the right–hand side. By theorem 8.3,

$$\forall i \in I \qquad \limsup_{n\to\infty} \frac{1}{n^d} \ln P(A_i) \,<\, 0\,,$$

therefore

$$\limsup_{n\to\infty} \frac{1}{n^{d-1}} \ln P(\mathrm{Fill}^c) \,=\, -\infty\,.$$

Let $\mathcal{E} = \mathcal{C} \cup \mathcal{D}$. On the event $\mathrm{Sep}_\theta(n) \cap \mathrm{Fill}$, we have

$$\sum_{C\in\mathcal{E}} \big|C \cap B_-(n)\big| \,\geq\, (1-2\delta)\mathcal{L}^d\big(B_-(n)\big)\,,$$

$$\sum_{C\in\mathcal{E}} \big|C \cap B_+(n)\big| \,\leq\, \delta\mathcal{L}^d\big(B_+(n)\big) \,\leq\, 2\delta\mathcal{L}^d\big(B_+(n)\big)\,,$$

so that the collection $\mathcal{E}$ realizes the event $\mathrm{Sep}(n,x,r,w,2\delta)$ and we conclude that $\mathrm{Sep}_\theta(n) \cap \mathrm{Fill} \subset \mathrm{Sep}(n,x,r,w,2\delta)$. □

## 12.4 Lower bound

It is possible to prove the corresponding lower bound

$$\liminf_{n\to\infty} \frac{1}{n^{d-1}} \ln\, P\big(\mathrm{Sep}(n,x,r,w,\delta)\big) \,\geq\, -\alpha_{d-1} r^{d-1} \tau(w)\,.$$

Since this lower bound is not useful for the proof of the large deviation principle, we do not give the detailed proof but merely a sketch. To prove it, one should use a block argument (see section 16.4) to ensure that the large clusters intersecting $\partial^{in} B(nx,nr)$ fill the lower half–ball $B_-(nx,nr,w)$ with the correct density, together with the following lower bound for the probability of the presence of a flat interface.

**Lemma 12.3.** *Let $F$ be a $d-1$ dimensional set such that $\mathcal{H}^{d-2}(\partial F) < \infty$ and let $\nu_F$ be a unit vector normal to $F$. We define* wall $(F,n)$ *as the event*

$$\begin{aligned}\mathrm{wall}\,(F,n) \,&=\, S(nF,\ln n) \cap \\ &\big\{ \textit{ all the edges in } \mathcal{V}(\,\mathrm{cyl}\,\partial nF, 2d) \cap \mathcal{V}(\,\mathrm{hyp}\, nF, \ln n) \textit{ are closed } \big\}\,,\end{aligned}$$

*where the event $S(nF,\ln n)$ is defined before lemma 11.11. Then*

$$\liminf_{n\to\infty} \frac{1}{n^{d-1}} \ln\, P\big(\mathrm{wall}\,(F,n)\big) \,\geq\, -\mathcal{H}^{d-1}(F)\,\tau(\nu_F)\,.$$

*Proof.* The number of edges in the set $\mathcal{V}(\,\mathrm{cyl}\,\partial nF, 2d) \cap \mathcal{V}(\,\mathrm{hyp}\, nF, \ln n)$ is less than $c(d)\mathcal{H}^{d-2}(\partial F)\, n^{d-2} \ln n$ for some positive constant $c(d)$ depending only on the dimension $d$. By the FKG inequality,

$$P\big(\mathrm{wall}\,(F,n)\big) \geq P\big(S(nF,\ln n)\big) \exp\Big(c(d)\mathcal{H}^{d-2}(\partial F)\, n^{d-2} \ln n\, \ln(1-p)\Big)\,.$$

The conclusion follows easily from lemma 11.11. □

# Part VI

## Basic geometric tools

# 13

# Sets of finite perimeter

This chapter is a rudimentary primer on geometric measure theory and sets of finite perimeter.

## 13.1 Basic definitions

For convenience, we recall here some basic definitions from measure theory. Let $(X, d)$ be a metric space. In geometric measure theory, one calls "measure" what probabilists would call "outer measure".

**Definition 13.1.** *A set function $\mu : \{ A : A \subset X \} \to [0, +\infty]$ is a measure if $\mu(\emptyset) = 0$ and*

$$\forall A \subset B \qquad \mu(A) \leq \mu(B)\,,$$
$$\forall (A_n)_{n \in \mathbb{N}} \in X^{\mathbb{N}} \qquad \mu\Big(\bigcup_{i \in \mathbb{N}} A_i\Big) \leq \sum_{i \in \mathbb{N}} \mu(A_i)\,.$$

Usually a measure is not defined on every subset of $X$, but rather on a $\sigma$–algebra included in $X$. We say that a subset $A$ of $X$ is $\mu$–measurable if

$$\forall E \subset X \qquad \mu(E) = \mu(E \cap A) + \mu(E \setminus A)\,.$$

The $\mu$–measurable sets form a $\sigma$–algebra.

Let $\mu$ be a measure on $X$. We say that

- $\mu$ is locally finite if any point admits a neighborhood having finite $\mu$–measure.
- $\mu$ is a Borel measure if all the Borel sets of $X$ are $\mu$–measurable.
- $\mu$ is Borel regular if it is a Borel measure and if:

$$\forall A \subset X \quad \exists B \text{ Borel subset of } X \quad A \subset B\,, \quad \mu(A) = \mu(B)\,.$$

- $\mu$ is a Radon measure if it is a Borel measure and if:
  i) $\forall K \subset X\,, \quad K \text{ compact}\,, \quad \mu(K) < \infty.$

ii) $\forall V \subset X\,,\quad V \text{ open}\,,\quad \mu(V) = \sup\{\,\mu(K) : K \subset V,\ K \text{ compact}\,\}.$
iii) $\forall A \subset X\,,\quad \mu(A) = \inf\{\,\mu(V) : A \subset V,\ V \text{ open}\,\}.$

In $\mathbb{R}^d$, a measure $\mu$ is a Radon measure if and only if it is a locally finite Borel regular measure. For $E$ a subset of $\mathbb{R}^d$, we define its Euclidean diameter as

$$\operatorname{diam} E = \sup\{\,|x-y| : x, y \in E\,\}$$

where $|\cdot|$ is the usual Euclidean norm. We define the distance $\operatorname{dist}(A,B)$ between two subsets $A, B$ of $\mathbb{R}^d$ by

$$\operatorname{dist}(A,B) \;=\; \inf\{\,|a-b| : x \in A,\ y \in B\,\}\,.$$

In the discrete setting, we will use a lot the $\infty$–diameter defined by

$$\operatorname{diam}_\infty E = \sup\{\,|x-y|_\infty : x, y \in E\,\}$$

where $|\cdot|_\infty$ is the usual supremum norm. Let $k$ be a non–negative integer. We denote by $\alpha_k$ the volume of the unit ball of $\mathbb{R}^k$. For any $A \subset \mathbb{R}^d$, the $k$–dimensional Hausdorff measure $\mathcal{H}^k(A)$ of $A$ is defined by

$$\mathcal{H}^k(A) \;=\; \sup_{\delta>0}\ \inf\ \Big\{\,\frac{\alpha_k}{2^k}\sum_{i\in I}(\operatorname{diam} E_i)^k \;:\; A \subset \bigcup_{i\in I} E_i,\ \sup_{i\in I}\ \operatorname{diam} E_i \le \delta\,\Big\}\,.$$

The Hausdorff measure $\mathcal{H}^k$ is a Borel regular measure. However, for $k < d$, the measure $\mathcal{H}^k$ is not a Radon measure in $\mathbb{R}^d$ because it is not locally finite.

## 13.2 Covering and differentiating

**Definition 13.2.** *Let $E$ be a Borel subset of $\mathbb{R}^d$. A collection of sets $\mathcal{U}$ is called a Vitali class for $E$ if for each $x \in E$ and $\delta > 0$, there exists a set $U \in \mathcal{U}$ containing $x$ such that* $0 < \operatorname{diam} U < \delta$.

We state next the Vitali covering theorem for $\mathcal{H}^{d-1}$ ([68], theorem 1.10).

**Theorem 13.3.** *Let $E$ be an $\mathcal{H}^{d-1}$–measurable subset of $\mathbb{R}^d$ and let $\mathcal{U}$ be a Vitali class of closed sets for $E$. Then we may select a (countable) disjoint sequence $(U_i)_{i\in I}$ from $\mathcal{U}$ such that*

$$\text{either}\ \sum_{i\in I}(\operatorname{diam} U_i)^{d-1} = \infty \quad \text{or}\quad \mathcal{H}^{d-1}\Big(E \setminus \bigcup_{i\in I} U_i\Big) = 0\,.$$

*If $\mathcal{H}^{d-1}(E) < \infty$ then, given $\varepsilon > 0$, we may also require that*

$$\mathcal{H}^{d-1}(E) \;\le\; \frac{\alpha_{d-1}}{2^{d-1}}\sum_{i\in I}(\operatorname{diam} U_i)^{d-1} + \varepsilon\,.$$

For a general version concerning Radon measures, see [100], theorem 2.8. We recall next the Besicovitch differentiation theorem in $\mathbb{R}^d$. For $x \in \mathbb{R}^d$ and $r > 0$, we denote by $B(x,r)$ the open ball of radius $r$ centered at $x$.

**Theorem 13.4.** *Let $\lambda$ be a finite positive Radon measure on $\mathbb{R}^d$. For any Borelian function $f \in L^1(\lambda)$, the quotient*

$$\frac{1}{\lambda\big(B(x,r)\big)} \int_{B(x,r)} f(y)\, d\lambda(y)$$

*converges $\lambda$ almost surely towards $f(x)$ as $r$ goes to 0.*

The version of the differentiation theorem for a general measure is due to Younovitch [132] and Besicovitch [19]. The classical textbooks on measure theory usually propose very general versions of this theorem (see for instance [84], Section 10.3, theorem 1 or [69], theorem 2.9.8). Here we will need only the version dealing with Radon measures and Euclidean balls in $\mathbb{R}^d$ ([133], theorem 1.3.8). Assouad and Quentin de Gromard [13] propose a proof of this result which is considerably simpler than the original proof of Besicovitch or its modern versions. We reproduce this proof below.

Let $m$ be the maximal number of points that can be put on the unit sphere $S^{d-1}$ of $\mathbb{R}^d$, but keeping their mutual distances strictly larger than 1.

**Lemma 13.5.** *For any $a, y_1, \ldots, y_{m+1} \in \mathbb{R}^d$, there exist indices $i \neq j$ such that*

$$|y_i - y_j| \leq \max\big(|a - y_i|, |a - y_j|\big).$$

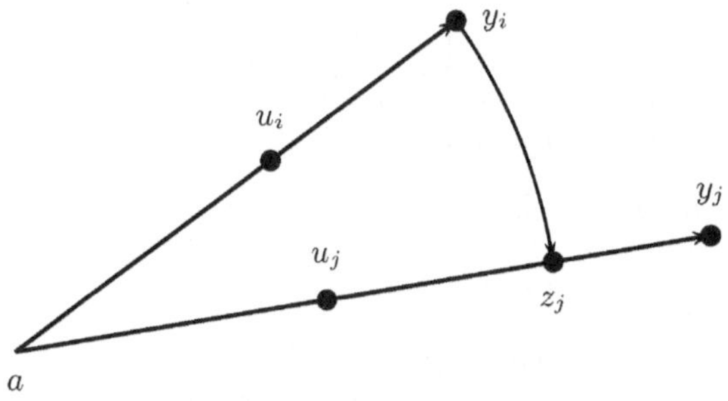


the points $y_i, y_j, z_j, u_i, u_j$

*Proof.* Let $a, y_1, \ldots, y_{m+1} \in \mathbb{R}^d$. If $y_i = a$ for some index $i$, we have equality in the above inequality for any $j$ distinct from $i$. Otherwise, let $i \neq j$ be two indices such that

$$|y_i - y_j| > \max\big(|a - y_i|, |a - y_j|\big).$$

We can assume for instance that $|a - y_i| \leq |a - y_j|$. Let $z_j$ be the point of the segment $[a, y_j]$ such that $|a - z_j| = |a - y_i|$. Then $|y_i - y_j| > |a - y_j|$ and by the triangle inequality

$$|y_i - z_j| \geq |y_i - y_j| - |y_j - z_j| > |a - y_j| - |y_j - z_j| = |a - y_i|.$$

For $k \in \{1 \cdots m+1\}$, let $u_k$ be the point of $[a, y_k]$ such that $|a - u_k| = 1$. The preceding inequalities imply that $|u_j - u_i| > 1$. Hence, if the claim of the lemma was wrong, we would get $m+1$ distinct points on the sphere $S^{d-1}$ whose mutual distances are strictly larger than one, thus contradicting the very definition of $m$. □

We prove next that the space $\mathbb{R}^d$ satisfies the following covering property.

**Weak covering property of degree** $m$. From any family of closed balls $B(y_i, r_i)$, $i \in I$, where the radii $r_i, i \in I$, belong to a fixed decreasing sequence, one can extract a subfamily $B(y_j, r_j)$, $j \in J$, where $J \subset I$, covering all the centers $y_i, i \in I$, and of order at most $m$, i.e.,

- $\forall i \in I \qquad y_i \in \bigcup_{j \in J} B(y_j, r_j)\,.$
- $\forall a \in \mathbb{R}^d \qquad \Big|\{\, j \in J : a \in B(y_j, r_j)\,\}\Big| \leq m\,.$

Since $\mathbb{R}^d$ is separable, such a family is finite or countable.

*Proof.* Let $B(y_i, r_i)$, $i \in I$, be a collection of balls whose radii belong to the decreasing sequence $(v_k)_{k \geq 1}$. We set $J_0 = \emptyset$ and we define inductively a sequence $(J_k)_{k \geq 1}$ of subsets of $I$ as follows. Suppose that $J_0, \ldots, J_{k-1}$ have been built. Let $J_k$ be a maximal subset of $I$ such that:

- $\forall i \in J_k, \quad r_i = v_k, \quad y_i \notin \bigcup_{1 \leq l \leq k-1} \bigcup_{j \in J_l} B(y_j, r_j)\,.$
- $\forall i, j \in J_k, \quad i \neq j, \quad |y_i - y_j| > v_k\,.$

Let $J = \bigcup_{n \in \mathbb{N}} J_n$. The balls $B(y_j, r_j)$, $j \in J$, cover the centers $y_i$, $i \in I$, and moreover

$$\forall i, j \in J, \quad i \neq j, \quad y_i \notin B(y_j, r_j)\,.$$

Let $a \in \mathbb{R}^d$ and suppose that $B(y_1, r_1), \ldots, B(y_k, r_k)$ are $k$ balls belonging to the collection $B(y_j, r_j)$, $j \in J$, and that each of these balls contains $a$. For any $i, j \in \{1 \cdots k\}$, we have then

$$|y_i - y_j| > \max(r_i, r_j) \geq \max\big(|a - y_i|, |a - y_j|\big)\,.$$

By lemma 13.5, we must have $k \leq m$, hence the collection of balls $B(y_j, r_j)$, $j \in J$, is of order at most $m$ as requested. □

The final ingredient we need to prove theorem 13.4 is the following maximal inequality.

**Maximal inequality**. For any finite positive Radon measures $\lambda, \nu$ on $\mathbb{R}^d$, any $\alpha > 0$, we have

$$\alpha\,\lambda^*\Big(\big\{\, y \in \mathbb{R}^d : \sup_{r>0} \frac{\nu(B(y,r))}{\lambda(B(y,r))} > \alpha \,\big\}\Big) \,\leq\, m\,\nu(\mathbb{R}^d)\,,$$

where $\lambda^*$ is the outer measure associated to $\lambda$. We make the convention that the quotient is 0 if $\nu(B(y,r)) = \lambda(B(y,r)) = 0$.

*Proof.* For $\alpha > 0$, let

$$Y(\alpha) \,=\, \Big\{\, y \in \mathbb{R}^d : \sup_{r>0} \frac{\nu(B(y,r))}{\lambda(B(y,r))} > \alpha \,\Big\}\,.$$

For $y \in Y(\alpha)$, let $r_y \in \mathbb{Q} \cap \mathbb{R}^+$ be such that $\alpha\lambda(B(y,r_y)) < \nu(B(y,r_y))$. Let $D_j, j \in \mathbb{N}$, be an increasing sequence of finite sets whose union is equal to $\mathbb{Q} \cap \mathbb{R}^+$. Let

$$Y(\alpha,j) \,=\, \big\{\, y \in Y(\alpha) : r_y \in D_j \,\big\}\,.$$

Let us fix $j \in \mathbb{N}$. By the weak covering property, we can extract from the family $B(y,r_y)$, $y \in Y(\alpha,j)$, a finite or countable family $B(y_i,r_i)$, $i \in I(\alpha,j)$, covering $Y(\alpha,j)$ and of order at most $m$. We have then

$$\begin{aligned} \alpha\,\lambda^*(Y(\alpha,j)) &\leq \alpha \sum_{i\in I(\alpha,j)} \lambda\big(B(y_i,r_i)\big) < \sum_{i\in I(\alpha,j)} \nu\big(B(y_i,r_i)\big) \\ &\leq m\,\nu\Big(\bigcup_{i\in I(\alpha,j)} B(y_i,r_i)\Big) \,\leq\, m\,\nu(\mathbb{R}^d)\,. \end{aligned}$$

Letting $j$ go to $\infty$, we get the maximal inequality we were seeking. □

We are now ready to prove the Besicovitch differentiation theorem in $\mathbb{R}^d$.

*Proof.* Let $\lambda, f$ be as in the hypothesis of theorem 13.4. We set for $r > 0$,

$$\forall x \in \mathbb{R}^d \qquad T(f,\lambda,r)(x) \,=\, \frac{1}{\lambda(B(x,r))} \int_{B(x,r)} f(y)\,d\lambda(y)\,,$$

with the convention that the quotient is 0 if $\lambda(B(x,r)) = 0$. Let $\varepsilon > 0$ and let $g$ be a continuous function such that $\int_{\mathbb{R}^d} |f-g|\,d\lambda < \varepsilon$. Setting $h = f - g$, we have

$$\big|f - T(f,\lambda,r)\big| \,\leq\, \big|g - T(g,\lambda,r)\big| \,+\, \big|h - T(h,\lambda,r)\big|\,.$$

Since $g$ is continuous,

$$\forall y \in \mathbb{R}^d \qquad \lim_{r\to 0} T(g,\lambda,r)(y) \,=\, g(y)\,,$$

whence

$$\forall y \in \mathbb{R}^d \qquad \limsup_{r \to 0} \big| f(y) - T(f, \lambda, r)(y) \big| \leq |h(y)| + \sup_{r>0} T(|h|, \lambda, r)(y)\,.$$

Let $\alpha > 0$ and let

$$C(\alpha) = \Big\{ y \in \mathbb{R}^d : \limsup_{r \to 0} \big| f(y) - T(f, \lambda, r)(y) \big| > \alpha \Big\}.$$

From the previous inequality, we deduce that

$$C(\alpha) \subset \Big\{ y \in \mathbb{R}^d : \sup_{r>0} T(|h|, \lambda, r)(y) > \frac{\alpha}{2} \Big\} \cup \Big\{ y \in \mathbb{R}^d : |h(y)| > \frac{\alpha}{2} \Big\}.$$

Applying the maximal inequality and Chebyshev's inequality, we get

$$\alpha\, \lambda^*(C(\alpha)) \leq 2m \int_{\mathbb{R}^d} |h|\, d\lambda + 2 \int_{\mathbb{R}^d} |h|\, d\lambda \leq 2(m+1)\varepsilon$$

where $\lambda^*$ is the outer measure associated to $\lambda$. Letting $\varepsilon$ go to 0, we conclude that $\lambda^*(C(\alpha)) = 0$. This being true for any $\alpha > 0$, the proof of the derivation theorem 13.4 is completed. □

## 13.3 Caccioppoli sets

Renato Caccioppoli is a genius mathematician belonging to the Italian school of the calculus of variations. He is even the main figure of a film, entitled "Morte da un matematico napoletano", directed by Mario Martone. This film is a daily account on the last week of his life: in fact, he committed suicide.

The original motivation for introducing "Caccioppoli sets" was to extend the classical geometric analysis of smooth sets to general (possibly irregular) sets. In particular, Caccioppoli wanted to answer the following question: Is it possible to assign an area to the boundary of general sets? He suggested the following approach. For $A$ a Borelian subset of $\mathbb{R}^d$, he defined its perimeter $\mathcal{P}(A)$ as

$$\mathcal{P}(A) = \inf \Big\{ \liminf_{n \to \infty} \mathcal{P}(A_n) : \\ (A_n)_{n \in \mathbb{N}} \text{ sequence of polyhedra },\ \lim_{n \to \infty} \mathcal{L}^d(A \Delta A_n) = 0 \Big\},$$

where $\mathcal{P}(A_n)$ is the area of the boundary of the polyhedron $A_n$. Somewhat later on, Ennio de Giorgi, another mythical Italian mathematician, was interested in extending the classical Gauss–Green theorem to non smooth sets. This theorem says that for any $C^1$ set $E$, any $C^1$ vector field $f : \mathbb{R}^d \to \mathbb{R}^d$ having compact support,

$$\int_E \text{div f}(x)\, d\mathcal{L}^d(x) = \int_{\partial E} f(x) \cdot \nu_E(x)\, d\mathcal{H}^{d-1}(x)$$

where $\nu_E(x)$ is the unit outward normal vector to $E$ at the point $x$. It turned out that the same concepts helped to answer these questions in a satisfactory way. For a complete account of the theory of Caccioppoli sets, see the references [54, 55, 56, 57, 67, 69, 77, 101, 133]. Next we give the modern distributional definition of the perimeter. The perimeter of a Borel set $E$ of $\mathbb{R}^d$ in an open set $O$ is defined as

$$\mathcal{P}(E,O) = \sup \Big\{ \int_E \operatorname{div} f(x)\, d\mathcal{L}^d(x) : f \in C_c^\infty(O, B(0,1)) \Big\}$$

where $C_c^\infty(O, B(0,1))$ is the set of the $C^\infty$ vector functions from $\mathbb{R}^d$ to $B(0,1)$ having a compact support included in $O$ and div is the usual divergence operator, defined for a $C^1$ vector function $f$ with scalar components $(f_1, \ldots, f_d)$ as

$$\operatorname{div} f = \frac{\partial f_1}{\partial x_1} + \cdots + \frac{\partial f_d}{\partial x_d}.$$

We define the perimeter $\mathcal{P}(E)$ of a set $E$ as $\mathcal{P}(E) = \mathcal{P}(E, \mathbb{R}^d)$. The set $E$ is said to have finite perimeter in an open set $O$ if $\mathcal{P}(E,O)$ is finite. The set $E$ is said to be of locally finite perimeter or to be a Caccioppoli set if $\mathcal{P}(E,O)$ is finite for every bounded open set $O$ of $\mathbb{R}^d$. The set $E$ is of finite perimeter if $\mathcal{P}(E) = \mathcal{P}(E, \mathbb{R}^d)$ is finite.

A set $E$ has finite perimeter in an open set $O$ if and only if its characteristic function $\chi_E$ is a function of bounded variation in $O$. The distributional derivative $\nabla\chi_E$ of $\chi_E$ is then a vector Radon measure and $\mathcal{P}(E,O) = ||\nabla\chi_E||(O)$, where $||\nabla\chi_E||$ is the total variation measure of $\nabla\chi_E$.

The perimeter $\mathcal{P}$ extends the classical isotropic notion of area. Indeed, supposing that $E$ has a smooth boundary, we have by the classical Gauss–Green theorem

$$\begin{aligned}
\mathcal{P}(E) &= \sup \Big\{ \int_E \operatorname{div} f(x)\, d\mathcal{L}^d(x) : f \in C_c^\infty(\mathbb{R}^d, B(0,1)) \Big\} \\
&= \sup_{f \in C_c^\infty(\mathbb{R}^d, B(0,1))} \int_{\partial E} f(x) \cdot \nu_E(x)\, d\mathcal{H}^{d-1}(x) \\
&= \int_{\partial E} \sup_{f \in C_c^\infty(\mathbb{R}^d, B(0,1))} f(x) \cdot \nu_E(x)\, d\mathcal{H}^{d-1}(x).
\end{aligned}$$

Here we interchange the integral and the supremum without much care; but this can be done in a neat way. The boundary $\partial E$ being smooth, the supremum over the smooth vector fields $f$ with values in $B(0,1)$ is equal to the supremum over much less regular vector fields. Thus, modulo some technicalities that we omit here, we have

$$\begin{aligned}
\mathcal{P}(E) &= \int_{\partial E} \sup_{y \in B(0,1)} y \cdot \nu_E(x)\, d\mathcal{H}^{d-1}(x) \\
&= \int_{\partial E} |\nu_E(x)|_2\, d\mathcal{H}^{d-1}(x) = \mathcal{H}^{d-1}(\partial E).
\end{aligned}$$

**The reduced boundary.** Let $E$ be a Caccioppoli set. Its reduced boundary $\partial^* E$ consists of the points $x$ such that

- $||\nabla\chi_E||(B(x,r)) > 0$ for any $r > 0$,
- if $\nu_r(x) = -\nabla\chi_E(B(x,r))/||\nabla\chi_E||(B(x,r))$ then, as $r$ goes to 0, $\nu_r(x)$ converges towards a limit $\nu_E(x)$ such that $|\nu_E(x)|_2 = 1$.

For a point $x$ belonging to $\partial^* E$, the vector $\nu_E(x)$ is called the generalized exterior normal to $E$ at $x$. A unit vector $\nu$ is called the measure theoretic exterior normal to $E$ at $x$ if

$$\lim_{r\to 0} r^{-d}\mathcal{L}^d(B_-(x,r,\nu)\setminus E) = 0\,,$$
$$\lim_{r\to 0} r^{-d}\mathcal{L}^d(B_+(x,r,\nu)\cap E) = 0\,.$$

At each point $x$ of the reduced boundary $\partial^* E$ of $E$, the generalized exterior normal $\nu_E(x)$ is also the measure theoretic exterior normal to $E$ at $x$ and moreover for any $\varepsilon > 0$,

$$\lim_{r\to 0} \frac{1}{\alpha_{d-1}r^{d-1}}\mathcal{H}^{d-1}(\partial^* E\cap\{\,y\in B(x,r) : |(y-x)\cdot\nu_E(x)|\leq\varepsilon|y-x|_2\,\}) = 1$$

and for $\mathcal{H}^{d-1}$ almost all $x$ in $\partial^* E$,

$$\lim_{r\to 0} \frac{1}{\alpha_{d-1}r^{d-1}}\mathcal{H}^{d-1}(B(x,r)\cap\partial^* E) = 1\,.$$

The map $x\in\partial^* E\mapsto\nu_E(x)\in S^{d-1}$ is $||\nabla\chi_E||$ measurable. For any Borel set $A$ of $\mathbb{R}^d$,

$$||\nabla\chi_E||(A) = \mathcal{H}^{d-1}(A\cap\partial^* E)\,,$$
$$\nabla\chi_E(A) = \int_{A\cap\partial^* E} -\nu_E(x)\,d\mathcal{H}^{d-1}(x)\,.$$

Let $f:\partial^* E\to\mathbb{R}$ be a $||\nabla\chi_E||$ measurable bounded function. By the Besicovitch differentiation theorem (see theorem 13.4) applied to the measure $||\nabla\chi_E||$, for $\mathcal{H}^{d-1}$ almost all $x$ in $\partial^* E$,

$$\lim_{r\to 0} \frac{1}{\alpha_{d-1}r^{d-1}}\int_{B(x,r)\cap\partial^* E} f(y)\,d\mathcal{H}^{d-1}(y) = f(x)\,.$$

We state next a useful result (see [77], remark 2.14). If $E$ is a Caccioppoli set, then for any $x$ in $\mathbb{R}^d$ and for $\mathcal{H}^1$ almost all positive radius $r$,

$$\mathcal{P}(E\cap B(x,r)) = \mathcal{P}(E,\overset{\circ}{B}(x,r)) + \mathcal{H}^{d-1}(E\cap\partial B(x,r))\,.$$

Next we recall the general isoperimetric inequality and the Gauss–Green theorem.

**Isoperimetric inequality.** There exist two constants $b_{\text{iso}}, c_{\text{iso}}$ depending only upon the dimension such that, for any Caccioppoli set $E$, any ball $B(x,r)$,

$$\min\big(\mathcal{L}^d(E\cap B(x,r)), \mathcal{L}^d((\mathbb{R}^d\setminus E)\cap B(x,r))\big) \leq b_{\text{iso}}\,\mathcal{P}(E,\mathring{B}(x,r))^{d/d-1},$$
$$\min\big(\mathcal{L}^d(E), \mathcal{L}^d(\mathbb{R}^d\setminus E)\big) \leq c_{\text{iso}}\,\mathcal{P}(E)^{d/d-1}.$$

**Gauss–Green theorem.** For any compactly supported $C^1$ vector field $f$ from $\mathbb{R}^d$ to $\mathbb{R}^d$, any Caccioppoli set $E$,

$$\int_E \operatorname{div} \mathrm{f}(x)\, d\mathcal{L}^d(x) = \int_{\partial^* E} f(x)\cdot \nu_E(x)\, d\mathcal{H}^{d-1}(x).$$

## 13.4 Two technical results

Brothers and Morgan's proof of the Wulff theorem [33] relies on the following extension of the Gauss–Green theorem and the technical lemma 13.7.

**Theorem 13.6.** *Let $E$ be a Borel subset of $\mathbb{R}^d$ having finite perimeter and finite volume. Let $F=(F_1,\dots,F_d)$ be a bounded Borel vector field which is defined $\mathcal{L}^d$ almost everywhere on $E$ and $\mathcal{H}^{d-1}$ almost everywhere on $\partial^* E$. We suppose also that for $(x_1,\dots,x_d)\in E$, for $k=1,\dots,d$, the map*

$$y\mapsto F_k(x_1,\dots,x_{k-1},y,x_{k+1},\dots,x_d)$$

*is continuous non-decreasing and locally Lipschitz on its domain of definition. Then the identity of the Gauss–Green theorem holds with $F$ and $E$.*

For $v$ in the unit sphere $S^{d-1}$ of $\mathbb{R}^d$, we denote by $\pi_v$ the orthogonal projection of $\mathbb{R}^d$ onto the hyperplane $H(v)$ orthogonal to $v$.

**Lemma 13.7.** *Let $E$ be a Borel subset of $\mathbb{R}^d$ having finite perimeter and finite volume. For $v\in S^{d-1}$, we define*

$$E(v) = \big\{\,\xi\in H(v) : 0 < \mathcal{H}^1\big(E\cap\pi_v^{-1}(\xi)\big) < +\infty\,\big\}.$$

*For $\mathcal{H}^{d-1}$ almost all $v\in S^{d-1}$,*

$$\mathcal{H}^{d-1}\big(\pi_v(\partial^* E)\Delta E(v)\big) = 0,$$

*and the following property holds: for any Borel subset $A$ of $\partial^* E$,*

$$\mathcal{H}^{d-1}\big(\pi_v(A)\big) > 0 \qquad\Longleftrightarrow\qquad \mathcal{H}^{d-1}(A) > 0.$$

Thierry Quentin de Gromard has rewritten in the BV framework the original proofs given in [33] of theorem 13.6 and lemma 13.7. The rest of the section is devoted to these proofs. We start with several results leading to a better understanding of the structure of the sets of finite perimeter.

*Notation.* For $x = (x_1, \dots, x_d) \in \mathbb{R}^d$, we set $y = (x_2, \dots, x_d) \in \mathbb{R}^{d-1}$, so that $x = (x_1, y)$; in the sequel, the variable $y$ is a generic point of $\mathbb{R}^{d-1}$. For a subset $E$ of $\mathbb{R}^d$ and $y \in \mathbb{R}^{d-1}$, we define

$$E(y) = \{x_1 \in \mathbb{R} : (x_1, y) \in E\},$$

and for $u$ a function defined on a subset of $\mathbb{R}^d$, we denote by $u^y$ the partial function $x_1 \in \mathbb{R} \mapsto u(x_1, y)$.
Let $\Omega$ be an open subset of $\mathbb{R}^d$. For $u \in L^1_{\text{loc}}(\Omega)$, we define

$$\int_\Omega |D_1 u| = \sup\Big\{ \int u(x) \frac{\partial \phi}{\partial x_1}(x)\, dx : \phi \in C^1_c(\Omega),\ ||\phi||_\infty \leq 1 \Big\}.$$

For $u \in BV(\Omega)$, this quantity is finite and $|D_1 u|$ is a Radon measure on $\Omega$. The next proposition is due to E. De Giorgi and M. Miranda.

**Proposition 13.8.** *Let $\Omega$ be an open subset of $\mathbb{R}^d$ and let $u \in L^1_{loc}(\Omega)$. Then the map $y \in \mathbb{R}^{d-1} \mapsto \int_{\Omega(y)} |D_1 u^y|$ is measurable and*

$$\int_\Omega |D_1 u| = \int dy \int_{\Omega(y)} |D_1 u^y|.$$

*Proof.* We give the sketch of the proof; for additional details, see [11], chapter 3. Let $y \in \mathbb{R}^{d-1}$ be such that $u^y \in L^1_{\text{loc}}(\Omega(y))$. Using the above definition, we see that

$$\int_{\Omega(y)} |D_1 u^y| = \lim_{j \to \infty} \int u(x_1, y) \frac{\partial \phi_j}{\partial x_1}(x_1, y)\, dx_1,$$

where $(\phi_j)_{j \in \mathbb{N}}$ is a sequence of functions in $C^1_c(\Omega)$. This implies the measurability of the map $y \in \mathbb{R}^{d-1} \mapsto \int_{\Omega(y)} |D_1 u^y|$. For $u \in C^1(\Omega)$, we have

$$\begin{aligned}\int_\Omega |D_1 u| &= \int_\Omega |D_1 u(x)|\, dx \\ &= \int dy \int_{\Omega(y)} |D_1 u(x_1, y)|\, dx_1 = \int dy \int_{\Omega(y)} |D_1 u^y|.\end{aligned}$$

Let now $u \in L^1_{\text{loc}}(\Omega)$. For $\phi \in C^1_c(\Omega)$ such that $||\phi||_\infty \leq 1$, we have by Fubini's theorem

$$\int u(x) \frac{\partial \phi}{\partial x_1}(x)\, dx = \int dy \int_{\mathbb{R}} u(x_1, y) \frac{\partial \phi}{\partial x_1}(x_1, y)\, dx_1 \leq \int dy \int_{\Omega(y)} |D_1 u^y|,$$

whence, taking the supremum over $\phi$,

$$\int_\Omega |D_1 u| \leq \int dy \int_{\Omega(y)} |D_1 u^y|.$$

Conversely, there exists a sequence $(u_k)_{k\in\mathbb{N}}$ of functions in $C^1(\Omega)$ converging towards $u$ in $L^1_{\mathrm{loc}}(\Omega)$ and such that

$$\lim_{k\to\infty}\int_\Omega |D_1 u_k| = \int_\Omega |D_1 u|\,.$$

Next, we have

$$\forall k\in\mathbb{N}\qquad \int_\Omega |D_1 u_k| = \int dy \int_{\Omega(y)} |D_1 u_k^y|\,.$$

Up to the extraction of a subsequence, we can assume that, for almost all $y\in\mathbb{R}^{d-1}$, the sequence $(u_k^y)_{k\in\mathbb{N}}$ converges towards $u^y$ in $L^1_{\mathrm{loc}}(\Omega(y))$. By lower semicontinuity, we obtain

$$\int_{\Omega(y)} |D_1 u^y| \leq \liminf_{k\to\infty}\int_{\Omega(y)} |D_1 u_k^y| \qquad \text{for almost all } y\in\mathbb{R}^{d-1}\,.$$

By Fatou's lemma,

$$\int dy \int_{\Omega(y)} |D_1 u^y| \leq \liminf_{k\to\infty}\int dy\int_{\Omega(y)} |D_1 u_k^y| = \int_\Omega |D_1 u|$$

as requested. □

The following two propositions are due to A.I. Vol'pert.

**Proposition 13.9.** *Let $u\in BV(\Omega)$. For almost all $y\in\mathbb{R}^{d-1}$, the map $u^y$ belongs to $BV(\Omega(y))$ and for any bounded Borel function $g$ on $\Omega$, the maps*

$$y\mapsto \int g^y|D_1 u^y|\,,\qquad y\mapsto\int g^y D_1 u^y$$

*are defined for almost all $y\in\mathbb{R}^{d-1}$, they are measurable and moreover*

$$\int_\Omega g|D_1 u| = \int dy\int_{\Omega(y)} g^y |D_1 u^y|\,,\qquad \int_\Omega g D_1 u = \int dy \int_{\Omega(y)} g^y D_1 u^y\,.$$

The maps are set equal to 0 whenever $\Omega(y)=\emptyset$. To prove the measurability, we consider first the case where $g$ is the indicator function of a Borel subset of $\Omega$; we conclude then with a standard approximation argument. The integral identities can be checked on open sets and then extended to bounded Borel functions. For a detailed proof, we refer to [11], section 2.5.

**Proposition 13.10.** *Let $E$ be a Borel subset of $\mathbb{R}^d$ having finite volume and finite perimeter. For almost all $y\in\mathbb{R}^{d-1}$, the set $E(y)$ is equal, up to a set of null $\mathcal{H}^1$ measure, to a finite union of disjoint intervals whose endpoints belong to $\partial^* E$.*

*Proof.* We apply proposition 13.9 to the indicator function of $E$: for almost all $y \in \mathbb{R}^{d-1}$, the indicator function $1_{E(y)}$ belongs to $BV(\mathbb{R})$, whence $E(y)$ is equal to a finite union of disjoint intervals, up to a $\mathcal{H}^1$ negligible set. This union is not empty whenever $0 < \mathcal{H}^1(E(y)) < +\infty$. Let $F_1$ be the set of the points $y \in \mathbb{R}^{d-1}$ such that the preceding assertions are true. Let $C$ be the set of the endpoints of all these intervals, when $y$ varies over $F_1$. From proposition 13.9, we have

$$0 = \int_{C\setminus\partial^*E} |D_1 1_E| = \int_{F_1} dy \,\big|\{\, t \in \mathbb{R} : (t,y) \in C \setminus \partial^*E \,\}\big|$$
$$\geq \mathcal{H}^{d-1}(\{\, y \in F_1 : \exists t \in \mathbb{R} \quad (t,y) \in C \setminus \partial^*E \,\}) \,.$$

Thus, for almost all $y \in F_1$, the endpoints of the intervals associated to $E(y)$ belong to $\partial^*E$. □

**Corollary 13.11.** *Let $\pi_1$ be the orthogonal projection in $\mathbb{R}^d$ along $(1,0,\dots,0)$. Let $E$ be a Borel subset of $\mathbb{R}^d$ having finite volume and finite perimeter. Then*

$$\mathcal{H}^{d-1}\Big(\{\, y \in \mathbb{R}^{d-1} : 0 < \mathcal{H}^1(E(y)) < +\infty \,\} \setminus \pi_1(\partial^*E)\Big) = 0 \,.$$

We have now all the tools necessary to complete the proofs of theorem 13.6 and lemma 13.7.

*Proof of theorem 13.6.* Let $\Gamma_1$ be the set of the points $y \in \mathbb{R}^{d-1}$ satisfying the conclusion of proposition 13.10. The partial derivative $\partial F_1/\partial x_1(x_1,y)$ exists for almost all $x_1 \in E(y)$ and $y \in \Gamma_1$, it is non–negative and

$$\int_E \frac{\partial F_1}{\partial x_1}(x)\, dx = \int_{\Gamma_1} dy \int_{E(y)} \frac{\partial F_1}{\partial x_1}(x_1,y)\, dx_1 \,.$$

For $y \in \Gamma_1$, we have

$$E(y) = \bigcup_{1 \leq i \leq m} [a_i, b_i]$$

and since the map $F_1^y$ is locally Lipschitz, it is absolutely continuous and

$$\forall i \in \{\, 1 \cdots m \,\} \qquad \int_{a_i}^{b_i} \frac{\partial F_1}{\partial x_1}(x_1,y)\, dx_1 = F_1(b_i,y) - F_1(a_i,y) \,.$$

Moreover

$$\sum_{1 \leq i \leq m} F_1(b_i,y) - F_1(a_i,y) = -\int F_1^y D_1 1_{E(y)} \,.$$

Putting together the previous identities and applying proposition 13.8, we obtain

$$\int_E \frac{\partial F_1}{\partial x_1}(x)\, dx = -\int F_1 D_1 1_E \,.$$

This identity holds also for the other axis directions. Yet $-D1_E = 1_{\partial^* E}\nu_E \mathcal{H}^{d-1}$ (see [11], section 3.5). We recover finally the identity of the Gauss–Green theorem by summing these $d$ identities. □

*Proof of lemma 13.7.* Let $v \in S^{d-1}$ and let

$$F(v) = \{\, x \in \partial^* E : v \cdot \nu_E(x) = 0 \,\}.$$

By Fubini's theorem,

$$\begin{aligned}\int_{S^{d-1}} \mathcal{H}^{d-1}(F(v))\, d\mathcal{H}^{d-1}(v) &= \int_{S^{d-1}} d\mathcal{H}^{d-1}(v) \int_{\partial^* E} 1_{v \cdot \nu_E(x)=0}\, d\mathcal{H}^{d-1}(x) \\ &= \int_{\partial^* E} d\mathcal{H}^{d-1}(x) \int_{S^{d-1}} 1_{v \cdot \nu_E(x)=0}\, d\mathcal{H}^{d-1}(v).\end{aligned}$$

For any $x \in \partial^* E$, we have $\mathcal{H}^{d-1}(\{\, v \in S^{d-1} : v \cdot \nu_E(x) = 0 \,\}) = 0$, hence the above integral is equal to 0 and $\mathcal{H}^{d-1}(F(v)) = 0$ for $\mathcal{H}^{d-1}$ almost all $v \in S^{d-1}$. Let

$$S_E = \{\, v \in S^{d-1} : \mathcal{H}^{d-1}(F(v)) = 0 \,\}$$

and let $v \in S_E$. By the definition of $S_E$, we have $|v \cdot \nu_E(x)| > 0$ for $\mathcal{H}^{d-1}$ almost all $x \in \partial^* E$. Let $A$ be a Borel subset of $\partial^* E$ such that

$$\int_A |v \cdot \nu_E(x)|\, d\mathcal{H}^{d-1}(x) = 0.$$

This implies that for any $k \in \mathbb{N}$, the set $\{\, x \in A : |v \cdot \nu_E(x)| > 1/k \,\}$ has a null $\mathcal{H}^{d-1}$ measure. Yet, up to a set of null $\mathcal{H}^{d-1}$ measure, the set $A$ is the union of these sets, therefore $\mathcal{H}^{d-1}(A) = 0$.

Let $v \in S^{d-1}$. We denote by $D_v$ the derivative in the direction of $v$. For $B$ a subset of $\mathbb{R}^d$ and $\xi \in \mathbb{R}^d$ such that $\xi \cdot v = 0$, we define

$$B_v(\xi) = \{\, t \in \mathbb{R} : \exists y \in B \cap \pi_v^{-1}(\xi) \quad y \cdot v = t \,\}.$$

Recall that $D1_E = -1_{\partial^* E}\nu_E \mathcal{H}^{d-1}$. By considering the direction of $v$ to be the one of the first coordinate and applying proposition 13.8, we obtain that for any Borel subset $A$ of $\partial^* E$,

$$\int_A |v \cdot \nu_E(x)|\, d\mathcal{H}^{d-1}(x) = \int_A |D_v 1_E| = \int_{E(v)\cap \pi_v(A)} d\xi \int_{A_v(\xi)} |D_v 1_{E_v(\xi)}|.$$

Let next $v \in S_E$ and let $A = \partial^* E \setminus \pi_v^{-1}(E(v))$. Then $\pi_v(A) = \pi_v(\partial^* E) \setminus E(v)$ and we deduce from the above formula that

$$\int_A |v \cdot \nu_E(x)|\, d\mathcal{H}^{d-1}(x) = 0.$$

Since $v \in S_E$, then $\mathcal{H}^{d-1}(A) = 0$ and $\mathcal{H}^{d-1}(\pi_v(A)) = 0$. We have thus proved that $\mathcal{H}^{d-1}\big(\pi_v(\partial^* E) \setminus E(v)\big) = 0$. By corollary 13.11 applied in the direction of $v$, we have

$$\forall v \in S^{d-1} \qquad \mathcal{H}^{d-1}\big(E(v) \setminus \pi_v(\partial^* E)\big) = 0\,.$$

We prove the final part of the lemma. Let $A$ be a Borel subset of $\partial^* E$ such that $\mathcal{H}^{d-1}(\pi_v(A)) = 0$ for some $v \in S_E$. The above integral identity yields that $\int_A |v \cdot \nu_E(x)| \, d\mathcal{H}^{d-1}(x) = 0$ and therefore $\mathcal{H}^{d-1}(A) = 0$. For any $v \in S^{d-1}$ and any Borel subset $A$ of $\partial^* E$ such that $\mathcal{H}^{d-1}(A) = 0$, we have $\mathcal{H}^{d-1}(\pi_v(A)) = 0$ as well. $\square$

# 14

# Surface energy

The results of this chapter and the following one are valid for any function $\tau$ from $S^{d-1}$ to $\mathbb{R}^+$ satisfying the following hypothesis.

**Hypothesis on $\tau$.** We suppose that:
- the map $\tau$ does not vanish on the sphere $S^{d-1}$.
- the homogeneous extension $\tau_0$ of $\tau$ to $\mathbb{R}^d$ defined by $\tau_0(0) = 0$ and

$$\forall w \in \mathbb{R}^d \setminus \{0\} \qquad \tau_0(w) = |w|_2 \tau(w/|w|_2)$$

is a convex function.

The above hypothesis implies in particular that $\tau_0$ is continuous (because it is convex and finite everywhere), hence it reaches its minimum and its maximum over the compact sphere $S^{d-1}$. Therefore the minimum

$$\tau_{\min} = \inf \left\{ \tau(\nu) : \nu \in S^{d-1} \right\}$$

of $\tau$ on $S^{d-1}$ is positive, while the maximum

$$\tau_{\max} = \sup \left\{ \tau(\nu) : \nu \in S^{d-1} \right\}$$

is finite. By corollary 11.7 and proposition 11.9, the surface tension $\tau$ extracted from our models satisfies the above hypothesis. The Wulff crystal of $\tau$ is the set

$$\mathcal{W}_\tau = \{ x \in \mathbb{R}^d : x \cdot w \leq \tau(w) \text{ for all } w \text{ in } S^{d-1} \}.$$

This definition readily implies that $\mathcal{W}_\tau$ is closed and convex. If $|x|_2 \leq \tau_{\min}$ then $x \cdot \nu \leq \tau(\nu)$ for $\nu \in S^{d-1}$, therefore $B(0, \tau_{\min}) \subset \mathcal{W}_\tau$. For $x \in \mathcal{W}_\tau$, we have

$$|x|_2 = \sup_{\nu \in S^{d-1}} \nu \cdot x \leq \sup_{\nu \in S^{d-1}} \tau(\nu) = \tau_{\max}$$

hence $\mathcal{W}_\tau \subset B(0, \tau_{\max})$. Thus the crystal $\mathcal{W}_\tau$ is bounded and contains the origin 0 in its interior.

**Proposition 14.1.** *The surface tension $\tau$ is the support function of its Wulff crystal $\mathcal{W}_\tau$, that is,*

$$\forall \nu \in S^{d-1} \qquad \tau(\nu) \,=\, \sup\{\, x \cdot \nu : x \in \mathcal{W}_\tau \,\}\,.$$

*The crystal $\mathcal{W}_\tau$ admits a unit outward normal vector $\nu_{\mathcal{W}_\tau}(x)$ at $\mathcal{H}^{d-1}$ almost all points $x \in \partial \mathcal{W}_\tau$. Let $x \in \partial \mathcal{W}_\tau$ be such a point. Then*

$$\tau(\nu_{\mathcal{W}_\tau}(x)) \,=\, x \cdot \nu_{\mathcal{W}_\tau}(x)$$

*and if $\nu \in S^{d-1}$ is such that $\tau(\nu) = x \cdot \nu$, then $\nu = \nu_{\mathcal{W}_\tau}(x)$.*

*Proof.* Let $\tau_0^*$ be the Fenchel transform of $\tau_0$, i.e.,

$$\forall x \in \mathbb{R}^d \qquad \tau_0^*(x) \,=\, \sup_{y \in \mathbb{R}^d} \big(x \cdot y - \tau_0(y)\big)\,.$$

If $x \in \mathcal{W}_\tau$ and $y \neq 0$, then

$$x \cdot y - \tau_0(y) \,=\, |y|_2 \Big(x \cdot \frac{y}{|y|_2} - \tau\Big(\frac{y}{|y|_2}\Big)\Big) \,\leq\, 0$$

so that $\tau_0^*(x) = 0$. If $x \notin \mathcal{W}_\tau$, then there exists $\nu \in S^{d-1}$ such that $x \cdot \nu > \tau(\nu)$ and

$$\forall \lambda > 0 \qquad \tau_0^*(x) \,\geq\, \lambda\big(x \cdot \nu - \tau(\nu)\big)\,,$$

so that $\tau_0^*(x) = +\infty$. Since $\tau_0$ is convex and continuous, it coincides with its bipolar, i.e.,

$$\forall x \in \mathbb{R}^d \qquad \tau_0(x) \,=\, \sup_{y \in \mathbb{R}^d} \big(x \cdot y - \tau_0^*(y)\big) \,=\, \sup_{y \in \mathcal{W}_\tau} x \cdot y\,.$$

Since $\mathcal{W}_\tau$ is convex and bounded, then at $\mathcal{H}^{d-1}$ almost all points $x \in \partial \mathcal{W}_\tau$, it admits a unit outward normal vector $\nu_{\mathcal{W}_\tau}(x)$ (this is a consequence of the differentiability properties of convex functions, see [113], theorem 25.5). Let $x$ be such a point. Since $x \in \mathcal{W}_\tau$, then $x \cdot \nu_{\mathcal{W}_\tau}(x) \leq \tau(\nu_{\mathcal{W}_\tau}(x))$. Moreover the crystal $\mathcal{W}_\tau$ is included in the half-space $\{\, y \in \mathbb{R}^d : y \cdot \nu_{\mathcal{W}_\tau}(x) \leq x \cdot \nu_{\mathcal{W}_\tau}(x) \,\}$. Therefore

$$\tau(\nu_{\mathcal{W}_\tau}(x)) \,=\, \sup\,\{\, y \cdot \nu_{\mathcal{W}_\tau}(x) : y \in \mathcal{W}_\tau \,\} \,\leq\, x \cdot \nu_{\mathcal{W}_\tau}(x)$$

and we conclude that $\tau(\nu_{\mathcal{W}_\tau}(x)) \,=\, x \cdot \nu_{\mathcal{W}_\tau}(x)$. Let $\nu \in S^{d-1}$. If the hyperplane $\{\, y \in \mathbb{R}^d : y \cdot \nu = x \cdot \nu \,\}$ meets the interior of $\mathcal{W}_\tau$, then

$$\tau(\nu) \,=\, \sup\,\{\, y \cdot \nu : y \in \mathcal{W}_\tau \,\} \,>\, x \cdot \nu\,.$$

Thus the equality $\tau(\nu) = x \cdot \nu$ implies that $\nu = \nu_{\mathcal{W}_\tau}(x)$. □

## 14.1 Definition

With the help of the surface tension $\tau$ defined in propositions 11.2, 11.4, 11.5, or equivalently its Wulff crystal $\mathcal{W}_\tau$, we build a surface energy functional defined on the collection $\mathcal{B}(\mathbb{R}^d)$ of the Borel sets of $\mathbb{R}^d$.

**Definition 14.2.** *The surface energy $\mathcal{I}(A,O)$ of a Borel set $A$ of $\mathbb{R}^d$ in an open set $O$ is defined as*

$$\mathcal{I}(A,O) = \sup \Big\{ \int_A \operatorname{div} f(x)\, d\mathcal{L}^d(x) : f \in C^1_c(O,\mathcal{W}_\tau) \Big\}$$

*where $C^1_c(O,\mathcal{W}_\tau)$ is the set of the $C^1$ vector functions defined on $\mathbb{R}^d$ with values in $\mathcal{W}_\tau$ having compact support included in $O$ and div is the usual divergence operator. We define also $\mathcal{I}(A) = \mathcal{I}(A,\mathbb{R}^d)$.*

Let $\tau_{\text{min}}$ and $\tau_{\text{max}}$ be the infimum and the supremum of $\tau$ over $S^{d-1}$. Then $B(0,\tau_{\text{min}}) \subset \mathcal{W}_\tau \subset B(0,\tau_{\text{max}})$ so that for every $A \in \mathcal{B}(\mathbb{R}^d)$ and every open set $O$, we have

$$\tau_{\text{min}}\mathcal{P}(A,O) \leq \mathcal{I}(A,O) \leq \tau_{\text{max}}\mathcal{P}(A,O)$$

where $\mathcal{P}$ is the classical perimeter (see section 13.3). By proposition 11.9, we have $0 < \tau_{\text{min}} \leq \tau_{\text{max}} < \infty$. Thus a set $A$ has finite surface energy in an open set $O$ if and only if it has finite perimeter in $O$. We show now that the surface energy is the surface integral of $\tau$ on the reduced boundary.

**Proposition 14.3.** *The surface energy $\mathcal{I}(A,O)$ of a Borel set $A$ of $\mathbb{R}^d$ of finite perimeter in an open set $O$ is equal to*

$$\mathcal{I}(A,O) = \int_{\partial^* A \cap O} \tau(\nu_A(x))\, d\mathcal{H}^{d-1}(x)\,.$$

*Proof.* Let $A$ be a Borel subset of $\mathbb{R}^d$ of finite perimeter in $O$, i.e., $\mathcal{P}(A,O)$ is finite and is equal to $\mathcal{H}^{d-1}(\partial^* A \cap O)$. By the Gauss–Green theorem (see section 13.3), for any function $f$ in $C^1_c(O,\mathbb{R}^d)$,

$$\int_A \operatorname{div}\, \mathrm{f}(x)\, d\mathcal{L}^d(x) = \int_{\partial^* A} f(x)\cdot \nu_A(x)\, d\mathcal{H}^{d-1}(x)\,.$$

Taking the supremum over all functions $f$ in $C^1_c(O,\mathcal{W}_\tau)$ yields

$$\mathcal{I}(A,O) = \sup\Big\{ \int_{\partial^* A} f(x)\cdot \nu_A(x)\, d\mathcal{H}^{d-1}(x) : f \in C^1_c(O,\mathcal{W}_\tau) \Big\}\,.$$

Thus, using proposition 14.1,

$$\mathcal{I}(A,O) \leq \int_{\partial^* A \cap O} \sup_{y \in \mathcal{W}_\tau} y\cdot \nu_A(x)\, d\mathcal{H}^{d-1}(x) = \int_{\partial^* A \cap O} \tau(\nu_A(x))\, d\mathcal{H}^{d-1}(x)\,.$$

Conversely, let $\varepsilon$ belong to $]0,1/2[$. For $\mathcal{H}^{d-1}$ almost all $x$ in $\partial^*A\cap O$,

$$\lim_{r\to 0}\ (\alpha_{d-1}r^{d-1})^{-1}\mathcal{H}^{d-1}(B(x,r)\cap\partial^*A) = 1\,.$$

Let $\partial^{**}A$ be the points of $\partial^*A\cap O$ where the above property holds. For any $x$ in $\partial^{**}A$, there exists a positive $r_1(x,\varepsilon)$ such that $B(x,r_1(x,\varepsilon))\subset O$ and

$$\forall r < r_1(x,\varepsilon)\qquad \big|\mathcal{H}^{d-1}(B(x,r)\cap\partial^*A)-\alpha_{d-1}r^{d-1}\big| \le \varepsilon\,\alpha_{d-1}r^{d-1}\,.$$

We use next an idea of De Giorgi. By the Egoroff theorem ([114], Chapter 3, Exercise 16), there exists a compact set $C$ included in $\partial^{**}A$ such that $\mathcal{H}^{d-1}(\partial^*A\cap O\setminus C)<\varepsilon$ and $\nabla\chi_A(B(x,r))/||\nabla\chi_A||(B(x,r))$ converges uniformly on $C$ towards $\nu_A(x)$ as $r$ goes to 0; then the restriction of $\nu_A$ to $C$ is continuous. Since $\tau$ is also continuous, then for any $x$ in $C$, there exists $r_2(x,\varepsilon)$ positive such that

$$\forall y\in C\cap B(x,r_2(x,\varepsilon))\qquad |\nu_A(y)-\nu_A(x)|_2\le\varepsilon\,,\quad |\tau(\nu_A(y))-\tau(\nu_A(x))|_2\le\varepsilon\,.$$

The family of balls $B(x,r)$, $x\in C$, $r<\min(r_1(x,\varepsilon),r_2(x,\varepsilon))$, is a Vitali relation for $C$. By the Vitali covering theorem for $\mathcal{H}^{d-1}$ (theorem 13.3), we may select from this collection of balls a finite or countable collection of disjoint balls $B(x_i,r_i)$, $i\in I$, such that

$$\text{either}\qquad \mathcal{H}^{d-1}\Big(C\setminus\bigcup_{i\in I}B(x_i,r_i)\Big)=0\qquad\text{or}\qquad \sum_{i\in I}r_i^{d-1}=\infty\,.$$

Because for each $i$ in $I$, $r_i$ is smaller than $r_1(x,\varepsilon)$,

$$(1-\varepsilon)\sum_{i\in I}\alpha_{d-1}r_i^{d-1}\le\mathcal{H}^{d-1}(\partial^*A\cap O)=\mathcal{P}(A,O)<\infty$$

and therefore the first case occurs, so that we may further select a finite subset $J$ of $I$ such that

$$\mathcal{H}^{d-1}\Big(C\setminus\bigcup_{i\in J}B(x_i,r_i)\Big)<\varepsilon\,.$$

Since $\tau$ is the support function of $\mathcal{W}_\tau$, then for each $i$ in $J$ there exists a vector $y_i$ in $\mathcal{W}_\tau$ such that $y_i\cdot\nu_A(x_i)=\tau(\nu_A(x_i))$. The balls $B(x_i,r_i)$, $i\in J$, being closed and disjoint, certainly there exists a function $f$ in $C^1_c(O,\mathcal{W}_\tau)$ such that

$$\forall i\in J\quad\forall x\in B(x_i,r_i)\qquad f(x)=y_i\,.$$

For such a function $f$,

$$\begin{aligned}&\Big|\int_{\partial^*A}f(x)\cdot\nu_A(x)\,d\mathcal{H}^{d-1}(x)-\int_{\partial^*A\cap O}\tau(\nu_A(x))\,d\mathcal{H}^{d-1}(x)\Big|\\&\quad\le 2||\tau||_\infty\mathcal{H}^{d-1}(\partial^*A\cap O\setminus C)+\Big|\int_C\big(f(x)\cdot\nu_A(x)-\tau(\nu_A(x))\big)\,d\mathcal{H}^{d-1}(x)\Big|\end{aligned}$$

$$\leq 2||\tau||_\infty \Big(\mathcal{H}^{d-1}(\partial^* A \cap O \setminus C) + \mathcal{H}^{d-1}\Big(C \setminus \bigcup_{i \in J} B(x_i, r_i)\Big)\Big)$$
$$+ \sum_{i \in J} \Big| \int_{B(x_i,r_i) \cap C} (y_i \cdot \nu_A(x) - \tau(\nu_A(x)))\, d\mathcal{H}^{d-1}(x) \Big| .$$

For $i$ in $J$ and $x$ in $B(x_i, r_i) \cap C$, because $r_i < \min(r_1(x_i, \varepsilon), r_2(x_i, \varepsilon))$,

$$|y_i \cdot \nu_A(x) - \tau(\nu_A(x))| \leq$$
$$|y_i \cdot (\nu_A(x) - \nu_A(x_i))| + |\tau(\nu_A(x_i)) - \tau(\nu_A(x))| \leq (||\tau||_\infty + 1)\, \varepsilon .$$

Integrating these inequalities,

$$\Big| \int_{\partial^* A} f(x) \cdot \nu_A(x)\, d\mathcal{H}^{d-1}(x) - \int_{\partial^* A \cap O} \tau(\nu_A(x))\, d\mathcal{H}^{d-1}(x) \Big|$$
$$\leq 4\varepsilon ||\tau||_\infty + \varepsilon\, (||\tau||_\infty + 1) \mathcal{H}^{d-1}(\partial^* A \cap O) ,$$

whence

$$\mathcal{I}(A, O) \geq \int_{\partial^* A \cap O} \tau(\nu_A(x))\, d\mathcal{H}^{d-1}(x) - 4\varepsilon ||\tau||_\infty - \varepsilon\, (||\tau||_\infty + 1) \mathcal{P}(A, O) .$$

Letting $\varepsilon$ go to zero, we obtain the converse inequality and the claim of the proposition. □

The next lemma bounds the surface energy of a Caccioppoli set intersected with a ball.

**Lemma 14.4.** *Let $A$ be a Caccioppoli set. For $x$ in $\mathbb{R}^d$, for $\mathcal{H}^1$ almost all positive radius $r$,*

$$\mathcal{I}(A \cap B(x,r)) \leq \mathcal{I}(A, \overset{o}{B}(x,r)) + ||\tau||_\infty \mathcal{H}^{d-1}(A \cap \partial B(x,r)) .$$

*Proof.* Let $r > 0$. Since

$$\partial^*(A \cap B(x,r)) \cap \overset{o}{B}(x,r) = \partial^* A \cap \overset{o}{B}(x,r)$$

and $\nu_{A \cap B(x,r)}(y) = \nu_A(y)$ for $y$ in $\partial^* A \cap \overset{o}{B}(x,r)$, using proposition 14.3,

$$\mathcal{I}(A \cap B(x,r)) = \int_{\partial^*(A \cap B(x,r))} \tau(\nu_{A \cap B(x,r)}(y))\, d\mathcal{H}^{d-1}(y)$$
$$\leq \mathcal{I}(A, \overset{o}{B}(x,r)) + ||\tau||_\infty \mathcal{H}^{d-1}(\partial^*(A \cap B(x,r)) \cap \partial B(x,r)) .$$

Next, by the identity recalled at the end of section 13.3, for $\mathcal{H}^1$ almost all positive $r$,

$$\mathcal{H}^{d-1}(\partial^*(A \cap B(x,r)) \cap \partial B(x,r)) = \mathcal{P}(A \cap B(x,r)) - \mathcal{P}(A, \overset{o}{B}(x,r))$$
$$= \mathcal{H}^{d-1}(A \cap \partial B(x,r)) .$$

Combining these inequalities, the result follows. □

## 14.2 Lowersemicontinuity and compactness

Let $O$ be an open subset of $\mathbb{R}^d$. For a fixed function $f$ in $C^1_c(O,\mathcal{W}_\tau)$, the map

$$A \in \mathcal{B}(\mathbb{R}^d) \mapsto \int_A \operatorname{div} f(x)\, d\mathcal{L}^d(x)$$

is continuous for the $L^1$ convergence of sets, or the convergence with respect to the volume: we say that a sequence of Borel sets $(E_n)_{n\in\mathbb{N}}$ converges towards $E \in \mathcal{B}(\mathbb{R}^d)$ if $\mathcal{L}^d(E_n \Delta E)$ converges to 0 as $n$ goes to $\infty$, where $\Delta$ is the symmetric difference operator. The surface energy $\mathcal{I}(\cdot,O)$, being the supremum of all these maps, is lower semicontinuous: if $(E_n)_{n\in\mathbb{N}}$ is a sequence of sets in $\mathcal{B}(\mathbb{R}^d)$ converging towards a set $E$ in $L^1$, then

$$\mathcal{I}(E,O) \leq \liminf_{n\to\infty} \mathcal{I}(E_n,O)\,.$$

**Theorem 14.5.** *For any bounded open subset $U$ of $\mathbb{R}^d$, the functional $\mathcal{I}(\cdot,U)$ is a good rate function on $\mathcal{B}(U)$ endowed with the topology of $L^1$ convergence, i.e., for any $\lambda$ in $\mathbb{R}^+$, the level set $\{\, E \in \mathcal{B}(U) : \mathcal{I}(E,U) \leq \lambda \,\}$ is compact.*

*Proof.* For any bounded open subset $U$ of $\mathbb{R}^d$ and any $\lambda > 0$, the collection of sets $\big\{\, E \in \mathcal{B}(U) : \mathcal{P}(E) \leq \lambda \,\big\}$ is compact for the topology $L^1$. The original formulation of this property can be found in [57], Teorema 2.4, or [55], Teorema I. This compactness property is also an immediate consequence of the compactness theorem stated in [101], chapter 2, p.70. Modern presentations are formulated through functions of bounded variation: if $O$ is an open bounded subset of $\mathbb{R}^d$ with sufficiently regular boundary (say $C^1$), then a set of functions in $L^1(O)$ uniformly bounded in BV–norm is relatively compact in $L^1(O)$ (see any of the following references: [67], Section 5.2.3, [77], theorem 1.19, [133], corollary 5.3.4). To deduce the compactness result on sets of finite perimeter, we choose an open bounded subset $O$ of $\mathbb{R}^d$ with regular boundary containing $U$ in its interior. We embed $\mathcal{B}(U)$ in $L^1(O)$ by associating to a Borel set $E$ of $\mathcal{B}(U)$ its characteristic function $\chi_E$ and we simply remark that the set $\{\, \chi_E : E \in \mathcal{B}(U) \,\}$ is a closed subset of $L^1(O)$.

This compactness property together with the inequality stated just before proposition 14.3 and the lower semicontinuity of $\mathcal{I}$ imply that it is a good rate function on $\mathcal{B}(U)$. □

## 14.3 Covering

We now state an approximation result used for proving the large deviation local upper bound.

**Lemma 14.6.** *Let $A$ be a Caccioppoli set having finite perimeter in an open subset $O$ of $\mathbb{R}^d$. For any positive $\varepsilon, \delta$, there exists a finite collection of disjoint* *balls $B(x_i, r_i)$, $i \in I$, such that: for any $i$ in $I$, $x_i$ belongs to $\partial^* A$, $r_i$ belongs to $]0,1[$, $B(x_i, r_i)$ is included in $O$,*

$$\mathcal{L}^d\big((A \cap B(x_i,r_i)) \Delta B_-(x_i, r_i, \nu_A(x_i))\big) \leq \delta\, \alpha_d\, r_i^d\,,$$
$$\Big|\mathcal{I}(A,O) - \sum_{i \in I} \alpha_{d-1} r_i^{d-1} \tau(\nu_A(x_i))\Big| \leq \varepsilon\,.$$

*Proof.* Let $\varepsilon, \delta$ be positive, with $\varepsilon < 1/2$. Because a generalized normal vector is also a measure theoretic normal, for any $x$ in $\partial^* A$, there exists a positive $r_1(x,\delta)$ such that, for any $r < r_1(x,\delta)$,

$$\mathcal{L}^d\big((A \cap B(x,r)) \Delta B_-(x, r, \nu_A(x))\big) \leq \delta\, \alpha_d\, r^d\,.$$

The map $x \in \partial^* A \mapsto \nu_A(x) \in S^{d-1}$ is measurable with respect to the measure $\mathcal{H}^{d-1}|_{\partial^* A}$. Using the results on Caccioppoli sets stated in section 13.3, for $\mathcal{H}^{d-1}$ almost all $x$ in $\partial^* A$,

$$\lim_{r \to 0} \quad \frac{1}{\alpha_{d-1} r^{d-1}} \mathcal{H}^{d-1}(B(x,r) \cap \partial^* A) = 1\,,$$
$$\lim_{r \to 0} \quad \frac{1}{\alpha_{d-1} r^{d-1}} \int_{B(x,r) \cap \partial^* A} \tau(\nu_A(y))\, d\mathcal{H}^{d-1}(y) = \tau(\nu_A(x))\,.$$

Let $\partial^{**} A$ be the set of the points of $\partial^* A$ where the two preceding identities hold simultaneously. Clearly $\mathcal{H}^{d-1}(\partial^* A \setminus \partial^{**} A) = 0$. For any $x$ in $\partial^{**} A$, there exists a positive $r_2(x,\varepsilon)$ such that, for any $r < r_2(x,\varepsilon)$,

$$|\mathcal{H}^{d-1}(B(x,r) \cap \partial^* A) - \alpha_{d-1} r^{d-1}| \leq \varepsilon\, \alpha_{d-1} r^{d-1}\,,$$
$$\Big|\frac{1}{\alpha_{d-1} r^{d-1}} \int_{B(x,r) \cap \partial^* A} \tau(\nu_A(y))\, d\mathcal{H}^{d-1}(y) - \tau(\nu_A(x))\Big| \leq \varepsilon\,.$$

The family of balls

$$B(x,r)\,, \quad x \in \partial^{**} A \cap O\,, \quad r < \min\big(r_1(x,\delta), r_2(x,\varepsilon), 1, \mathrm{dist}(x, \partial O)\big)\,,$$

is a Vitali relation for $\partial^{**} A \cap O$. By the Vitali covering theorem for $\mathcal{H}^{d-1}$ (theorem 13.3), we may select from this collection of balls a finite or countable collection of disjoint balls $B(x_i, r_i)$, $i \in I$, such that

$$\text{either} \quad \mathcal{H}^{d-1}\Big((\partial^{**} A \cap O) \setminus \bigcup_{i \in I} B(x_i, r_i)\Big) = 0 \quad \text{or} \quad \sum_{i \in I} r_i^{d-1} = \infty\,.$$

By hypothesis, $\mathcal{P}(A,O) = \mathcal{H}^{d-1}(\partial^* A \cap O)$ is finite. For $i$ in $I$, the radius $r_i$ is smaller than $r_2(x_i, \varepsilon)$, whence

$$(1-\varepsilon) \sum_{i \in I} \alpha_{d-1} r_i^{d-1} \leq \mathcal{H}^{d-1}(\partial^{**} A \cap O) < \infty$$

and therefore the first case occurs, so that we may select a finite subset $J$ of $I$ such that

$$\mathcal{H}^{d-1}\Big((\partial^{**}A\cap O)\setminus\bigcup_{i\in J}B(x_i,r_i)\Big) < \varepsilon\,\mathcal{H}^{d-1}(\partial^{**}A\cap O)\,.$$

We claim that the collection of balls $B(x_i,r_i), i\in J$, enjoys the desired properties. Indeed, there is only the last condition to be checked:

$$\Big|\mathcal{I}(A,O)-\sum_{i\in J}\alpha_{d-1}r_i^{d-1}\tau(\nu_A(x_i))\Big| \leq \int_{(\partial^{**}A\cap O)\setminus\bigcup_{i\in J}B(x_i,r_i)}\tau(\nu_A(x))\,d\mathcal{H}^{d-1}(x)$$
$$+\sum_{i\in J}\Big|\int_{\partial^{**}A\cap B(x_i,r_i)}\tau(\nu_A(x))\,d\mathcal{H}^{d-1}(x)-\alpha_{d-1}r_i^{d-1}\tau(\nu_A(x_i))\Big|\,.$$

The first integral on the right–hand side is less than $\varepsilon\,\mathcal{H}^{d-1}(\partial^{**}A\cap O)||\tau||_\infty$. For $i$ in $J$,

$$\Big|\int_{\partial^{**}A\cap B(x_i,r_i)}\tau(\nu_A(x))\,d\mathcal{H}^{d-1}(x)-\alpha_{d-1}r_i^{d-1}\tau(\nu_A(x_i))\Big|$$
$$\leq \varepsilon\,\alpha_{d-1}r_i^{d-1} \leq 2\varepsilon\mathcal{H}^{d-1}(B(x_i,r_i)\cap\partial^{**}A)\,,$$

whence by summing over $i$ in $J$,

$$\sum_{i\in J}\Big|\int_{\partial^{**}A\cap B(x_i,r_i)}\tau(\nu_A(x))\,d\mathcal{H}^{d-1}(x)-\alpha_{d-1}r_i^{d-1}\tau(\nu_A(x_i))\Big| \leq 2\varepsilon\,\mathcal{H}^{d-1}(\partial^{**}A\cap O).$$

Putting these inequalities together, we obtain

$$\Big|\mathcal{I}(A,O)-\sum_{i\in J}\alpha_{d-1}r_i^{d-1}\tau(\nu_A(x_i))\Big| \leq \varepsilon\,\mathcal{P}(A,O)\,(||\tau||_\infty+2)\,.$$

Since $(||\tau||_\infty+2)\mathcal{P}(A,O)$ does not depend on $\varepsilon$, we have the required estimate. □

In the FK percolation or Ising setting, we will work within the $d$–dimensional unit cube $Q=[-1/2,1/2]^d$ and we must take care of what happens close to the boundary $\partial Q$. Thus we will rely on the following variant of lemma 14.6.

**Lemma 14.7.** *Let $A$ be a subset of $Q$ having finite perimeter. For any positive $\varepsilon,\delta$, there exists a finite collection of disjoint balls $B(x_i,r_i)$, $i\in I\cup I^{\mathrm{bd}}$, such that:*

- $\forall i\in I$  $x_i\in\partial^*A\cap\overset{o}{Q}$ *and* $B(x_i,r_i)\subset\overset{o}{Q}$ .
- $\forall i\in I^{\mathrm{bd}}$  $x_i\in\partial^*A\cap\partial Q$ *and* $B_-(x_i,r_i,\nu_Q(x_i))\subset Q$.
- $\forall i\in I\cup I^{\mathrm{bd}}$  $\mathcal{L}^d((A\cap B(x_i,r_i))\Delta B_-(x_i,r_i,\nu_A(x_i)))\leq\delta\,\alpha_d\,r_i^d$ .

*Finally*

$$\Big|\mathcal{I}(A) - \sum_{i\in I\cup I^{\mathrm{bd}}} \alpha_{d-1} r_i^{d-1}\tau(\nu_A(x_i))\Big| \leq \varepsilon\,.$$

Whenever $A$ is a set of finite perimeter included in $Q$, the family of balls $B(x,r)$ such that

- either $\big(x\in\partial^* A\cap \overset{\circ}{Q}$ and $B(x,r)\subset\overset{\circ}{Q}\big)$
- or $\big(x\in\partial^* A\cap\partial Q$ and $B_-(x,r,\nu_Q(x))\subset Q\big)$

is a Vitali relation for $\partial^* A$, hence the proof runs exactly in the same way as for lemma 14.6.

## 14.4 Polyhedral approximation

In this section we state an important approximation result which is the key to the proof of the large deviation lower bound.

**Proposition 14.8.** *Every bounded Caccioppoli set $A$ can be approximated by a sequence $(A_n)_{n\in\mathbb{N}}$ of bounded $C^\infty$ sets such that*

$$\lim_{n\to\infty}\mathcal{L}^d(A_n\Delta A) = 0\,,\qquad \lim_{n\to\infty}\mathcal{I}(A_n) = \mathcal{I}(A)\,.$$

*Proof.* This result is a slight generalization of the corresponding result for the perimeters ([77], theorem 1.24). Because the proof is quite long, we only sketch the essential points: it consists in adapting some minor details of ([77], theorem 1.24). One needs a slight extension of the coarea formula [129]. Let $f$ be a function of bounded variation. We define the surface energy $\mathcal{I}(f)$ by replacing $\chi_A$ by $f$ in the definition of $\mathcal{I}$:

$$\mathcal{I}(f) = \sup\Big\{\int f(x)\,\mathrm{div}\,g(x)\,d\mathcal{L}^d(x) : g\in C^1_c(\mathbb{R}^d,\mathcal{W}_\tau)\Big\}.$$

We have then the coarea formula:

$$\mathcal{I}(f) = \int_{-\infty}^{+\infty}\mathcal{I}\big(\{\,x\in\mathbb{R}^d : f(x)<t\,\}\big)\,dt\,.$$

The first step consists in approximating the characteristic function $\chi_A$ by a regular function through a smoothing procedure (convolution with a mollifier) to obtain a sequence of $C^\infty$ functions $(f_n)_{n\in\mathbb{N}}$ with values in $[0,1]$ such that

$$\lim_{n\to\infty}\int|f_n(x)-\chi_A(x)|\,d\mathcal{L}^d(x) = 0\,,\qquad \lim_{n\to\infty}\mathcal{I}(f_n) = \mathcal{I}(A)\,.$$

Let us precise this procedure. We use a sequence of functions $(\varrho_n)_{n\in N}$ satisfying

$$\forall n \in \mathbb{N} \qquad \varrho_n \in C_c^\infty(\mathbb{R}^d), \quad \int \varrho_n \, d\mathcal{L}^d = 1,$$
$$\forall x \in \mathbb{R}^d \qquad \varrho_n(x) \geq 0, \quad \varrho_n(x) = 0 \quad \text{if} \quad |x| > 1/n.$$

We define

$$\forall n \in \mathbb{N} \quad \forall x \in \mathbb{R}^d \qquad f_n(x) = \varrho_n * 1_A(x) = \int_A \varrho_n(x-y) \, d\mathcal{L}^d(y).$$

For any $g \in C_c^1(\mathbb{R}^d, \mathcal{W}_\tau)$,

$$\int_A \operatorname{div} g(x) \, d\mathcal{L}^d(x) = \lim_{n\to\infty} \int f_n(x) \operatorname{div} g(x) \, d\mathcal{L}^d(x).$$

Yet

$$\forall n \in \mathbb{N} \qquad \int f_n(x) \operatorname{div} g(x) \, d\mathcal{L}^d(x) \leq \mathcal{I}(f_n),$$

whence $\mathcal{I}(A) \leq \liminf_{n\to\infty} \mathcal{I}(f_n)$. Conversely, for $n \in \mathbb{N}$,

$$\begin{aligned}
\int f_n(x) \operatorname{div} g(x) \, d\mathcal{L}^d(x) &= -\int \nabla f_n(x) \cdot g(x) \, d\mathcal{L}^d(x) \\
&= -\int \nabla_x \Big( \int_A \varrho_n(x-y) \, d\mathcal{L}^d(y) \Big) \cdot g(x) \, d\mathcal{L}^d(x) \\
&= -\int_A \int (\nabla_x \varrho_n(x-y)) \cdot g(x) \, d\mathcal{L}^d(x) \, d\mathcal{L}^d(y) \\
&= \int_A \int \nabla_y \varrho_n(x-y) \cdot g(x) \, d\mathcal{L}^d(x) \, d\mathcal{L}^d(y) \\
&= \int_A \operatorname{div}_y \Big( \int \varrho_n(x-y) g(x) \, d\mathcal{L}^d(x) \Big) \, d\mathcal{L}^d(y).
\end{aligned}$$

Since the crystal $\mathcal{W}_\tau$ is closed and convex, the vector field

$$y \in \mathbb{R}^d \mapsto \int \varrho_n(x-y) g(x) \, d\mathcal{L}^d(x)$$

belongs to $C_c^\infty(\mathbb{R}^d, \mathcal{W}_\tau)$. Therefore

$$\forall n \in \mathbb{N} \quad \forall g \in C_c^1(\mathbb{R}^d, \mathcal{W}_\tau) \qquad \int f_n(x) \operatorname{div} g(x) \, d\mathcal{L}^d(x) \leq \mathcal{I}(A).$$

It follows that $\mathcal{I}(f_n) \leq \mathcal{I}(A)$ for $n \in \mathbb{N}$ and we conclude that

$$\lim_{n\to\infty} \mathcal{I}(f_n) = \mathcal{I}(A)$$

as requested. Let $n \in \mathbb{N}$ and let

$$N_n = f_n(\{ x \in \mathbb{R}^d : \nabla f_n(x) = 0 \}).$$

By the Sard lemma, we have $\mathcal{H}^1(N_n) = 0$. Let us set $N = \bigcup_{n\in\mathbb{N}} N_n$. By the coarea formula,

$$\lim_{n\to\infty} \int_{[0,1]\setminus N} \mathcal{I}\big(\{\, x \in \mathbb{R}^d : f_n(x) < t \,\}\big)\, dt \;=\; \mathcal{I}(A)$$

and by Fatou's lemma

$$\int_{[0,1]\setminus N} \liminf_{n\to\infty}\; \mathcal{I}\big(\{\, x \in \mathbb{R}^d : f_n(x) < t \,\}\big)\, dt \;\le\; \mathcal{I}(A)\,.$$

Hence there exists $t \in \,]0,1[\setminus N$ such that

$$\liminf_{n\to\infty}\; \mathcal{I}\big(\{\, x \in \mathbb{R}^d : f_n(x) < t \,\}\big)\, dt \;\le\; \mathcal{I}(A)\,.$$

The sequence $(\{\, x \in \mathbb{R}^d : f_n(x) < t \,\})_{n\in\mathbb{N}}$ then contains a subsequence having the desired properties. □

A nice consequence of proposition 14.8 is that $\mathcal{I}$ can alternatively be defined by

$$\mathcal{I}(A) \;=\; \inf\; \Big\{\; \liminf_{n\to\infty} \int_{\partial A_n} \tau(\nu_{A_n}(x))\, d\mathcal{H}^{d-1}(x) \;\Big\}$$

where the infimum is taken over all the sequences of $C^\infty$ sets $(A_n)_{n\in\mathbb{N}}$ such that $\mathcal{L}^d(A_n \Delta A)$ converges to 0. Thus $\mathcal{I}$ is the largest lower semicontinuous functional extending the surface integral $A \mapsto \int_{\partial A} \tau(\nu_A(x))\, d\mathcal{H}^{d-1}(x)$ from the $C^\infty$ sets to the Borel sets.

To prove the large deviation lower bound, we will need another kind of approximation result, namely we will need to approximate the sets of finite perimeter with polyhedral sets. A Borel subset of $\mathbb{R}^d$ is polyhedral if its boundary is included in the union of a finite number of hyperplanes of $\mathbb{R}^d$.

**Proposition 14.9.** *Let $A$ be a set of finite perimeter in $\mathbb{R}^d$. There exists a sequence $(A_n)_{n\in\mathbb{N}}$ of polyhedral sets of $\mathbb{R}^d$ converging to $A$ for the topology $L^1$ such that $\mathcal{I}(A_n)$ converges to $\mathcal{I}(A)$ as $n$ goes to $\infty$.*

This result yields a definition of the surface energy analogous to the original definition of the perimeter proposed by Caccioppoli [35, 36], as the infimum of the limits of the perimeters of polyhedral approximations. For the particular case $\tau = 1$, this result was originally proved by De Giorgi [54]. Below we propose a different proof, with a special emphasis on the way to approximate a set limited by a hypersurface with a polyhedral set.

*Proof.* Let $A$ be a set of finite perimeter. By the isoperimetric inequality (see section 13.3), either $\mathcal{L}^d(A)$ or $\mathcal{L}^d(\mathbb{R}^d \setminus A)$ is finite. If a sequence $(A_n)_{n\in\mathbb{N}}$ converges towards $A$ in $L^1$ then the sequence $(\mathbb{R}^d \setminus A_n)_{n\in\mathbb{N}}$ converges towards $\mathbb{R}^d \setminus A$ in $L^1$. Moreover we have $\mathcal{I}(A) = \mathcal{I}(\mathbb{R}^d \setminus A)$. Hence we need only to consider the case where $\mathcal{L}^d(A)$ is finite. For $r$ positive, we set $A_r = A \cap B(0,r)$.

Because $\mathcal{L}^d(A)$ is finite, the set $A_r$ converges towards $A$ for the topology $L^1$ as $r$ goes to $\infty$. By lemma 14.4, for $\mathcal{H}^1$ almost all $r$,

$$\mathcal{I}(A_r) \leq \int_{\partial^* A \cap \overset{\circ}{B}(0,r)} \tau(\nu_{A_r}(x))\, d\mathcal{H}^{d-1}(x) \;+\; ||\tau||_\infty \mathcal{H}^{d-1}(A \cap \partial B(0,r))\,.$$

Let $\varepsilon > 0$. Since $\mathcal{L}^d(A)$ is finite, then the set $\{\, r > 0 : \mathcal{H}^{d-1}(A \cap \partial B(0,r)) \geq \varepsilon \,\}$ has finite $\mathcal{H}^1$ measure (less than $\mathcal{L}^d(A)/\varepsilon$). Thus there exist arbitrarily large values of $r$ such that the above inequality holds and $\mathcal{H}^{d-1}(A \cap \partial B(0,r)) < \varepsilon$, whence $\mathcal{I}(A_r) \leq \mathcal{I}(A) + \varepsilon ||\tau||_\infty$. For $r$ large enough, we have $\mathcal{L}^d(A \Delta A_r) < \varepsilon$ as well. Therefore we need only to prove the approximation result for bounded sets of finite perimeter. By proposition 14.8, it is then enough to consider the case of bounded $C^\infty$ sets of finite perimeter. The end of the proof uses a technical lemma, that we state and prove now.

**Lemma 14.10.** *Let $\Gamma$ be a hypersurface (that is a $C^1$ submanifold of $\mathbb{R}^d$ of codimension 1) and let $K$ be a compact subset of $\Gamma$. There exists a positive $M = M(\Gamma, K)$ such that:*

$$\forall \varepsilon > 0 \quad \exists r > 0 \quad \forall x, y \in K \qquad |x - y|_2 \leq r$$
$$\Rightarrow \quad \mathrm{dist}(y, \tan(\Gamma, x)) \leq M\, \varepsilon\, |x - y|_2\,.$$

($\tan(\Gamma, x)$ *is the tangent plane of $\Gamma$ at $x$).*

*Proof.* By a standard compactness argument, it is enough to prove the following local property:

$$\forall x \in \Gamma \quad \exists M(x) > 0 \quad \forall \varepsilon > 0 \quad \exists r(x,\varepsilon) > 0 \quad \forall y, z \in \Gamma \cap B(x, r(x,\varepsilon))$$
$$\mathrm{dist}(y, \tan(\Gamma, z)) \leq M(x)\, \varepsilon\, |y - z|_2\,.$$

Indeed, if this property holds, we cover $K$ by the open balls $\overset{\circ}{B}(x, r(x,\varepsilon)/2)$, $x \in K$, we extract a finite subcovering $\overset{\circ}{B}(x_i, r(x_i,\varepsilon)/2)$, $1 \leq i \leq k$, and we set

$$M = \max\{\, M(x_i) : 1 \leq i \leq k \,\}, \quad r = \min\{\, r(x_i,\varepsilon)/2 : 1 \leq i \leq k \,\}\,.$$

Now let $y, z$ belong to $K$ with $|y - z|_2 \leq r$. Let $i$ be such that $y$ belongs to $B(x_i, r(x_i,\varepsilon)/2)$. Since $r < r(x_i,\varepsilon)/2$, then both $y, z$ belong to the ball $B(x_i, r(x_i,\varepsilon))$ and it follows that

$$\mathrm{dist}(y, \tan(\Gamma, z)) \;\leq\; M(x_i)\, \varepsilon\, |y - z|_2 \;\leq\; M\, \varepsilon\, |y - z|_2\,.$$

The proof of the local property relies on a classical lemma of differential calculus ([94], I, 4, corollary 2), which we state next.

**Lemma 14.11.** *Let $E, F$ be two Banach spaces, let $U$ be an open subset of $E$, let $x, z$ in $U$ be such that $[x, z] \subset U$ and let $f : U \to F$ be a $C^1$ map. Then for any $y$ in $[x, z]$,*

$$|f(z) - f(x) - df(y)(z - x)| \;\leq\; |z - x| \sup\{\, ||df(\zeta) - df(y)|| : \zeta \in [x, z] \,\}\,.$$

*Proof of lemma 14.10 continued.* We turn now to the proof of the above local property. Since $\Gamma$ is a hypersurface, for any $x$ in $\Gamma$ there exists a neighbourhood $V$ of $x$ in $\mathbb{R}^d$, a diffeomorphism $f : V \to \mathbb{R}^d$ of class $C^1$ and a $d-1$ dimensional vector space $Z$ of $\mathbb{R}^d$ such that $Z \cap f(V) = f(\Gamma \cap V)$ (see for instance [69], 3.1.19). Let $A$ be a compact neighbourhood of $x$ included in $V$. Since $f$ is a diffeomorphism, the differential maps $y \in A \mapsto df(y)$, $u \in f(A) \mapsto df^{-1}(u)$ are continuous. Therefore they are bounded:

$$\exists M > 0 \quad \forall y \in A \quad ||df(y)|| \leq M\,, \qquad \forall u \in f(A) \quad ||df^{-1}(u)|| \leq M$$

(here $||df(x)|| = \sup\{\, |df(x)(y)|_2 : |y|_2 \leq 1\,\}$ is the standard operator norm). Since $f(A)$ is compact, the differential map $df^{-1}$ is uniformly continuous on $f(A)$:

$$\forall \varepsilon > 0 \quad \exists \delta > 0 \quad \forall u, v \in f(A) \quad |u-v|_2 \leq \delta \quad \Rightarrow \quad ||df^{-1}(u) - df^{-1}(v)|| \leq \varepsilon\,.$$

Let $\varepsilon$ be positive and let $\delta$ be associated to $\varepsilon$ as above. Let $\varrho$ be positive and small enough so that $\varrho < \delta/2$ and $B(f(x), \varrho) \subset f(A)$ (since $f$ is a $C^1$ diffeomorphism, then $f(A)$ is a neighbourhood of $f(x)$). Let $r$ be such that $0 < r < \varrho/M$ and $B(x, r) \subset A$. We claim that $M$ associated to $x$ and $r$ associated to $\varepsilon, x$ answer the problem. Let $y, z$ belong to $\Gamma \cap B(x, r)$. Since $[y, z] \subset B(x, r) \subset A$ and $||df(\zeta)|| \leq M$ on $A$, then

$$|f(y) - f(x)|_2 \leq M|y - x|_2 \leq Mr < \varrho\,, \quad |f(z) - f(x)|_2 < \varrho\,,$$
$$|f(y) - f(z)|_2 < \delta\,, \quad |f(y) - f(z)|_2 \leq M|y - z|_2\,.$$

We apply next the quoted lemma 14.11 to the map $f^{-1}$, the interval $[f(z), f(y)]$ (which is included in $B(f(x), \varrho) \subset f(A)$) and the point $f(z)$:

$$|y - z - df^{-1}(f(z))(f(y) - f(z))|_2 \leq$$
$$|f(y) - f(z)|_2 \; \sup\left\{\, ||df^{-1}(\zeta) - df^{-1}(f(z))|| : \zeta \in [f(z), f(y)]\,\right\}.$$

The right–hand side is less than $M|y-z|_2\,\varepsilon$. Since $z + df^{-1}(f(z))(f(y) - f(z))$ belongs to $\tan(\Gamma, z)$, we are done. □

*Proof of proposition 14.9 continued.* Now let $A$ be a bounded $C^\infty$ set of finite perimeter. In particular its boundary $\partial A$ is a hypersurface. Since $\tau$ is continuous and $\partial A$ is a hypersurface, for any $x$ in $\partial A$,

$$\lim_{r \to 0} \quad (\alpha_{d-1} r^{d-1})^{-1} \mathcal{H}^{d-1}(B(x, r) \cap \partial A) \;=\; 1\,,$$

$$\lim_{r \to 0} \quad (\alpha_{d-1} r^{d-1})^{-1} \int_{B(x,r) \cap \partial A} \tau(\nu_A(y))\, d\mathcal{H}^{d-1}(y) \;=\; \tau(\nu_A(x))\,.$$

Since $\mathcal{H}^{d-1}(\partial A)$ is finite, for any $x$ in $\partial A$, for $\mathcal{H}^1$ almost all positive $r$,

$$\mathcal{H}^{d-1}(\partial A \cap \partial B(x, r)) = 0\,.$$

Let $M$ be associated to $\partial A$ as in lemma 14.10 (notice that $\partial A$ is compact). We might assume that $M \geq 1$. Let $\varepsilon$ belong to $]0, 1/2[$. Let $r(\partial A, \varepsilon)$ be associated to $\partial A$, $\varepsilon$ as in lemma 14.10. For $x$ in $\partial A$, there exists a positive $r(x, \varepsilon)$ such that, for $r < r(x, \varepsilon)$,

$$|\mathcal{H}^{d-1}(B(x,r) \cap \partial A) - \alpha_{d-1} r^{d-1}| \leq \varepsilon \, \alpha_{d-1} r^{d-1},$$

$$\Big|(\alpha_{d-1} r^{d-1})^{-1} \int_{B(x,r)\cap\partial A} \tau(\nu_A(y)) \, d\mathcal{H}^{d-1}(y) - \tau(\nu_A(x))\Big| \leq \varepsilon.$$

The family of balls

$$B(x,r),\ x \in \partial A,\ r < \min(r(x,\varepsilon), r(\partial A, \varepsilon), \varepsilon),\ \mathcal{H}^{d-1}(\partial A \cap \partial B(x,r)) = 0,$$

is a Vitali relation for the set $\partial A$. By the standard Vitali covering theorem (theorem 13.3), we may select from this collection of balls a finite or countable collection of disjoint balls $B(x_i, r_i)$, $i \in I$, such that

$$\text{either} \qquad \mathcal{H}^{d-1}\Big(\partial A \setminus \bigcup_{i\in I} B(x_i, r_i)\Big) = 0 \qquad \text{or} \qquad \sum_{i \in I} r_i^{d-1} = \infty.$$

Because for each $i$ in $I$, $r_i$ is smaller than $r(x_i, \varepsilon)$,

$$(1-\varepsilon) \sum_{i\in I} \alpha_{d-1} r_i^{d-1} \leq \mathcal{H}^{d-1}(\partial A) < \infty$$

and therefore the first case occurs, so that we may select a finite subset $I_0$ of $I$ such that

$$\mathcal{H}^{d-1}\Big(\partial A \setminus \bigcup_{i\in I_0} B(x_i, r_i)\Big) < \varepsilon \mathcal{H}^{d-1}(\partial A).$$

By the very definition of the Hausdorff measure $\mathcal{H}^{d-1}$, there exists a collection of balls $B(y_j, s_j)$, $j \in J$, such that: for any $j$ in $J$, $y_j \in \partial A$, $0 < s_j < \varepsilon$,

$$\sum_{j \in J} \alpha_{d-1} s_j^{d-1} \leq \varepsilon(\mathcal{H}^{d-1}(\partial A) + 1),$$

$$\partial A \setminus \bigcup_{i \in I_0} \overset{\circ}{B}(x_i, r_i) \subset \bigcup_{j\in J} \overset{\circ}{B}(y_j, s_j).$$

By compactness of $\partial A$, we might assume in addition that $J$ is finite. The notation hyp, cyl and disc is defined in section 11.1. For each $i$ in $I_0$, let $P_i$ be a convex open polygon inside the plane $\tan(\partial A, x_i) = \operatorname{hyp}(x_i, \nu_A(x_i))$ such that

$$\operatorname{disc}(x_i, r_i, \nu_A(x_i)) \subset P_i \subset B(x_i, r_i(1+\varepsilon)),$$

$$\big|\mathcal{H}^{d-2}(\partial P_i) - \gamma_{d-2} r_i^{d-2}\big)\big| \leq \varepsilon \gamma_{d-2} r_i^{d-2},$$

$$|\mathcal{H}^{d-1}(P_i) - \alpha_{d-1} r_i^{d-1}| \leq \varepsilon \alpha_{d-1} r_i^{d-1},$$

where $\gamma_{d-2}$ is the $\mathcal{H}^{d-2}$ measure of the boundary of the unit $d-1$ dimensional ball. We set $\alpha = M\varepsilon\,(1+\varepsilon)$ and $F_i = \mathrm{cyl}\,(\overline{P}_i, \alpha r_i)$ for $i$ in $I_0$. For each $j$ in $J$, let $Q_j$ be a polyhedral set such that

$$B(y_j, s_j) \subset Q_j \subset B(y_j, 2s_j)\,, \qquad \mathcal{H}^{d-1}(\partial Q_j) \leq \alpha_{d-1}(2s_j)^{d-1}\,.$$

Let $T$ be the set

$$T = A \cup \bigcup_{i\in I_0} F_i \cup \bigcup_{j\in J} Q_j\,.$$

We claim that $T$ answers the problem. First $A \subset T \subset \mathcal{V}(A, 2(M+1)\,\varepsilon)$. Secondly

$$T \setminus A \subset \bigcup_{i\in I_0} F_i \cup \bigcup_{j\in J} Q_j$$

whence

$$\mathcal{L}^d(T\setminus A) \leq \sum_{i\in I_0} 2\alpha_{d-1} r_i^{d-1}(1+\varepsilon)\alpha r_i + \sum_{j\in J} \alpha_d (2s_j)^d\,.$$

Yet $\sum_{i\in I_0} \alpha_{d-1} r_i^{d-1} \leq 2\mathcal{H}^{d-1}(\partial A)$ and $\sum_{j\in J} \alpha_{d-1} s_j^{d-1} \leq \varepsilon\,(\mathcal{H}^{d-1}(\partial A)+1)$ whence

$$\mathcal{L}^d(T\setminus A) \leq 9M\mathcal{H}^{d-1}(\partial A)\,\varepsilon + 2^d \frac{\alpha_d}{\alpha_{d-1}}\,\varepsilon\,(\mathcal{H}^{d-1}(\partial A)+1)\,.$$

We show next that $T$ is polyhedral. Indeed, for $i$ in $I_0$, $r_i$ is smaller than $r(\partial A, \varepsilon)$, so that

$$\forall x \in \partial A \cap B(x_i, r_i) \qquad \mathrm{dist}(x,\, \mathrm{hyp}\,(x_i, \nu_A(x_i))) \leq M\,\varepsilon\,|x - x_i|_2$$

whence

$$\partial A \cap B(x_i, r_i) \subset \mathrm{cyl}\,(P_i, M\,\varepsilon\, r_i) \subset \overset{\circ}{F}_i\,.$$

Moreover, letting $\beta = \sqrt{1-\alpha^2}$, the disc

$$G_i = \mathrm{disc}\,(x_i - \alpha r_i \nu_A(x_i), \beta r_i, \nu_A(x_i))$$

is included in the interior of $A$. Indeed, $G_i$ is included in $B(x_i, r_i) \cap \partial F_i$ and therefore $G_i$ does not intersect $\partial A$. It is included in $\overset{\circ}{A}$ because $\nu_A(x_i)$ is the exterior normal to $A$ at $x_i$. Since in addition the sets $F_i$, $i \in I_0$, $Q_j$, $j \in J$ cover $\partial A$, then

$$\partial T \subset \bigcup_{i\in I_0} \partial F_i \setminus G_i \cup \bigcup_{j\in J} \partial Q_j\,.$$

Thus $\partial T$ is included in a finite union of hyperplanes and $T$ is polyhedral. Finally

$$
\begin{aligned}
\mathcal{I}(\partial T) &\le \sum_{i\in I_0} \mathcal{I}(\partial F_i \setminus G_i) + \sum_{j\in J} \mathcal{I}(\partial Q_j) \\
&\le \sum_{i\in I_0} \alpha_{d-1} r_i^{d-1} (2(1+\varepsilon) - \beta^{d-1}) \tau(\nu_A(x_i)) + \\
&\qquad ||\tau||_\infty \Big( \sum_{i\in I_0} (1+\varepsilon)\gamma_{d-2} 2\alpha r_i^{d-1} + \sum_{j\in J} \alpha_{d-1}(2s_j)^{d-1} \Big) \\
&\le (1+\varepsilon)\mathcal{I}(A) + 4\varepsilon \mathcal{H}^{d-1}(\partial A) + \\
&2\mathcal{H}^{d-1}(\partial A)||\tau||_\infty \Big( 1+\varepsilon-(1-\alpha^2)^{d-1} + \frac{\gamma_{d-2}}{\alpha_{d-1}} 3\alpha \Big) + ||\tau||_\infty 2^{d-1}\varepsilon(\mathcal{H}^{d-1}(\partial A)+1) \\
&\le (1+\varepsilon)\mathcal{I}(A) + 4\varepsilon \mathcal{H}^{d-1}(\partial A) + \\
&2\mathcal{H}^{d-1}(\partial A)||\tau||_\infty \Big( \varepsilon + 3(d-1)M^2\varepsilon^2 + \frac{\gamma_{d-2}}{\alpha_{d-1}} 6M\varepsilon \Big) + ||\tau||_\infty 2^{d-1}\varepsilon(\mathcal{H}^{d-1}(\partial A)+1) .
\end{aligned}
$$

Since $M$ depends on $A$ only, the set $T$ approximates $A$ as required. □

For the FK percolation model and in the Ising setting, we will work in finite volume and we will rely on the following variant of proposition 14.9.

**Corollary 14.12.** *Let $A$ be a subset of $[-1/2, 1/2]^d$ having finite perimeter. For any $\varepsilon > 0$, there exists a polyhedral set $D$ such that*

$$\overline{D} \subset ]-1/2, 1/2[^d , \qquad \mathcal{L}^d(A \Delta D) \le \varepsilon , \qquad \mathcal{I}(D) \le \mathcal{I}(A) + \varepsilon .$$

*Proof.* Let $\alpha \in ]0,1[$ be close to 1 so that

$$\mathcal{L}^d(A \Delta \alpha A) \le \varepsilon/2 , \qquad \mathcal{I}(\alpha A) \le \mathcal{I}(A) + \varepsilon/2 .$$

Notice that $\alpha A$ is now included in the box $[-\alpha/2, \alpha/2]^d$. We approximate $\alpha A$ by a $C^\infty$ set as in proposition 14.8, but with the help of a mollifier whose support is contained in a sufficiently small neighbourhood of 0, so that the approximating set is still inside the box $[-(1+\alpha)/4, (1+\alpha)/4]^d$. We finally approximate this $C^\infty$ set as in proposition 14.9 by a polyhedral set $D$ satisfying the desired inequalities. The procedure used in this approximation process can be controlled to ensure in addition that $\overline{D} \subset ]-1/2, 1/2[^d$: we need only to use sufficiently small balls in the covering argument. □

# 15

# The Wulff theorem

## 15.1 Statement of the theorem

We consider a surface tension function $\tau$ satisfying the hypothesis stated at the beginning of chapter 14, its associated Wulff crystal $\mathcal{W}$ and the corresponding surface energy $\mathcal{I}$. We study the following variational problem:

$$\text{minimize } \mathcal{I}(E) \text{ under the constraint } \mathcal{L}^d(\mathcal{W}) \leq \mathcal{L}^d(E) < +\infty\,. \qquad (W)$$

This problem is an anisotropic isoperimetric problem.

**Theorem 15.1.** *Up to translations and modulo Lebesgue negligible sets, the Wulff crystal $\mathcal{W}$ is the unique solution to the problem $(W)$. Equivalently, a Borel set $E$ is a solution to $(W)$ if and only if*

$$\exists x \in \mathbb{R}^d \qquad \mathcal{L}^d\big((x+\mathcal{W})\Delta E\big) = 0\,.$$

In case the surface tension $\tau$ is constant and equal to 1, we have the classical result: the ball is the solution to the isotropic isoperimetric problem. The proof of theorem 15.1 is postponed to sections 15.2 and 15.3.

Wulff was the first to attack the anisotropic situation, at the beginning of the twentieth century [131]. In 1944, Dinghas studied the problem within the class of convex polyhedra [60]. In 1978, Taylor [124, 125, 126] proved the general theorem using tools of geometric measure theory and the theory of currents [69]. However, the proof uses only currents of codimension one. These objects are isomorphic to the Caccioppoli sets. The Wulff isoperimetric theorem has been reworked and slightly generalized in the framework of the Caccioppoli sets by Fonseca [72] and Fonseca and Müller [73]. Dacorogna and Pfister [53] provide a specific proof in two dimensions for Jordan domains with boundaries parametrized by Sobolev functions. Brothers and Morgan [33] have found a more efficient proof of the uniqueness part in any dimension.

## 15.2 The anisotropic isoperimetric inequality

In this section, we will prove the following anisotropic isoperimetric inequality: For any Borel set $E$ in $\mathbb{R}^d$ such that $\mathcal{L}^d(E) < +\infty$, we have

$$\mathcal{L}^d(E) \le \frac{\mathcal{L}^d(\mathcal{W})}{\mathcal{I}(\mathcal{W})^{\frac{d}{d-1}}}\mathcal{I}(E)^{\frac{d}{d-1}}.$$

The proof relies on an anisotropic version of the Minkowski content.

**Lemma 15.2.** *For any bounded polyhedral set $A$ in $\mathbb{R}^d$, we have*

$$\mathcal{I}(A) = \lim_{\varepsilon \to 0} \frac{1}{\varepsilon}\big(\mathcal{L}^d(A + \varepsilon\mathcal{W}) - \mathcal{L}^d(A)\big).$$

*Proof.* By definition, the boundary of $A$ is the union of a finite number of $d-1$ dimensional bounded polyhedral sets $F_i$, $i \in I$, so that

$$\mathcal{I}(A) = \sum_{i \in I} \mathcal{H}^{d-1}(F_i)\tau(\nu_A(F_i)),$$

where $\nu_A(F_i)$ is the unit outward normal vector to $A$ along the interior points of the face $F_i$. Let $S = \partial A \setminus \partial^* A$ be the set of the singular points of $\partial A$; it is a $d-2$ dimensional set. Let $\delta > 0$. For $\varepsilon$ sufficiently small, the cylinders

$$\operatorname{cyl}\big(F_i \setminus \mathcal{V}(S,\delta), \nu_A(F_i), \varepsilon\tau(\nu_A(F_i))\big), \quad i \in I,$$

are disjoint and included in $(A + \varepsilon\mathcal{W}) \setminus A$ (for the notation cyl, see section 11.1). Indeed, let $y_i \in \mathcal{W}$ be such that $y_i \cdot \nu_A(F_i) = \tau(\nu_A(F_i))$. For $x \in F_i \setminus \mathcal{V}(S,\delta)$, we have

$$x + \varepsilon\tau(\nu_A(F_i))\nu_A(F_i) = y + \varepsilon y_i,$$

where $y$ is such that $|x - y| \le 2\varepsilon||\tau||_\infty$ and $(x - y) \cdot \nu_A(F_i) = 0$. Hence for $\varepsilon$ small enough (depending on $\delta$), the point $y$ belongs to $F_i$. It follows that

$$\mathcal{L}^d(A + \varepsilon\mathcal{W}) - \mathcal{L}^d(A) \ge \sum_{i \in I} \mathcal{H}^{d-1}\big(F_i \setminus \mathcal{V}(S,\delta)\big)\, \varepsilon\tau(\nu_A(F_i)).$$

Letting first $\varepsilon$ and then $\delta$ go to 0, we get

$$\liminf_{\varepsilon \to 0} \frac{1}{\varepsilon}\big(\mathcal{L}^d(A + \varepsilon\mathcal{W}) - \mathcal{L}^d(A)\big) \ge \mathcal{I}(A).$$

Conversely, for $\varepsilon$ small enough,

$$(A + \varepsilon\mathcal{W}) \setminus A \subset \mathcal{V}\big(S, \varepsilon(2||\tau||_\infty + 1)\big) \cup \bigcup_{i \in I} \operatorname{cyl}\big(F_i, \nu_A(F_i), \varepsilon\tau(\nu_A(F_i))\big).$$

Indeed, let $x = a + \varepsilon w$ where $a \in F_i$ for some $i \in I$ and $w \in \mathcal{W}$. Let $y$ be the orthogonal projection of $x$ on the hyperplane containing $F_i$. Then

$$|a - y| \le |\varepsilon w| < \varepsilon(||\tau||_\infty + 1),$$
$$|(x - y) \cdot \nu_A(F_i)| = |\varepsilon w \cdot \nu_A(F_i)| \le \varepsilon \tau(\nu_A(F_i)).$$

If $x$ does not belong to $\mathcal{V}\big(S, \varepsilon(2||\tau||_\infty + 1)\big)$, then $a \in F_i \setminus \mathcal{V}\big(S, \varepsilon(||\tau||_\infty + 1)\big)$ and $y \in F_i$, whence $x$ is in $\operatorname{cyl}\big(F_i, \nu_A(F_i), \varepsilon\tau(\nu_A(F_i))\big)$. Thus

$$\mathcal{L}^d(A + \varepsilon\mathcal{W}) - \mathcal{L}^d(A) \le \mathcal{L}^d\big(\mathcal{V}(S, \varepsilon(2||\tau||_\infty + 1))\big) + \sum_{i \in I} \mathcal{H}^{d-1}(F_i)\, \varepsilon\tau(\nu_A(F_i)).$$

Sending $\varepsilon$ to 0, we conclude that

$$\limsup_{\varepsilon \to 0} \frac{1}{\varepsilon}\big(\mathcal{L}^d(A + \varepsilon\mathcal{W}) - \mathcal{L}^d(A)\big) \le \mathcal{I}(A),$$

which completes the proof of the lemma. □

We first establish the isoperimetric inequality for polyhedral sets. The other key ingredient of the proof is the Brunn–Minkowski inequality (see [34], chapter 8), which says that for any Borel subsets $A, B$ of $\mathbb{R}^d$, we have

$$\mathcal{L}^d(A + B)^{\frac{1}{d}} \ge \mathcal{L}^d(A)^{\frac{1}{d}} + \mathcal{L}^d(B)^{\frac{1}{d}}.$$

Let $A$ be a polyhedral set in $\mathbb{R}^d$. We apply lemma 15.2 and the Brunn–Minkowski inequality to get:

$$\begin{aligned}\mathcal{I}(A) &= \lim_{\varepsilon \to 0} \frac{1}{\varepsilon}\big(\mathcal{L}^d(A + \varepsilon\mathcal{W}) - \mathcal{L}^d(A)\big) \\ &\ge \lim_{\varepsilon \to 0} \frac{1}{\varepsilon}\Big(\big(\mathcal{L}^d(A)^{\frac{1}{d}} + \varepsilon\mathcal{L}^d(\mathcal{W})^{\frac{1}{d}}\big)^d - \mathcal{L}^d(A)\Big) = \mathcal{L}^d(A)^{\frac{d-1}{d}}\, d\mathcal{L}^d(\mathcal{W})^{\frac{1}{d}}.\end{aligned}$$

Let now $E$ be a Borel set in $\mathbb{R}^d$ such that $\mathcal{L}^d(E) < +\infty$. By proposition 14.9, there exists a sequence $(A_n)_{n \in \mathbb{N}}$ of polyhedral sets of $\mathbb{R}^d$ converging to $E$ for the topology $L^1$ and such that $\mathcal{I}(A_n)$ converges to $\mathcal{I}(E)$ as $n$ goes to $\infty$. We apply the previous inequality to the set $A_n$ and we pass to the limit as $n$ goes to $\infty$:

$$\mathcal{I}(E) = \lim_{n \to \infty} \mathcal{I}(A_n) \ge \lim_{n \to \infty} \mathcal{L}^d(A_n)^{\frac{d-1}{d}}\, d\mathcal{L}^d(\mathcal{W})^{\frac{1}{d}} = \mathcal{L}^d(E)^{\frac{d-1}{d}}\, d\mathcal{L}^d(\mathcal{W})^{\frac{1}{d}}.$$

We finally use the identity stated in proposition 14.1 and the Gauss–Green theorem (see section 13.3) to compute $d\mathcal{L}^d(\mathcal{W})$ as follows:

$$\begin{aligned}d\mathcal{L}^d(\mathcal{W}) &= \int_{\mathcal{W}} \operatorname{div} x\, d\mathcal{L}^d(x) = \int_{\partial^*\mathcal{W}} x \cdot \nu_{\mathcal{W}}(x)\, d\mathcal{H}^{d-1}(x) \\ &= \int_{\partial^*\mathcal{W}} \tau(\nu_{\mathcal{W}}(x))\, d\mathcal{H}^{d-1}(x) = \mathcal{I}(\mathcal{W}).\end{aligned}$$

## 15.3 The proof of Brothers and Morgan

We present in this section the proof of the Wulff theorem of Brothers and Morgan [33]. Besides the main ideas, their proof requires some technicalities from geometric measure theory. In the original paper, the authors make appeal to the theory of currents to prove these technical results. Thierry Quentin de Gromard has rewritten the corresponding arguments in the BV framework (see theorem 13.6 and lemma 13.7). Following a remark of Olivier Couronné, we rely here on a slightly different version of the Gauss–Green theorem than in [33]: the reason is that the vector field $F$ built in the proof should always take its values in the Wulff crystal $\mathcal{W}$.

Let $E$ be a Borel subset of $\mathbb{R}^d$ having finite perimeter and such that $\mathcal{L}^d(E) = \mathcal{L}^d(\mathcal{W})$. We suppose that the vector $(1,0,\dots,0)$ satisfies the conclusion of lemma 13.7. For $k \in \{0\cdots d\}$, we denote by $\pi_k : \mathbb{R}^d \to \mathbb{R}^k$ the projection defined by

$$\forall (x_1,\dots,x_d) \in \mathbb{R}^d \qquad \pi_k(x_1,\dots,x_d) = (x_1,\dots,x_k)\,.$$

When $k = 0$, this means that $\pi_0$ is the map from $\mathbb{R}^d$ to $\mathbb{R}^0 = \{0\}$ defined by $\pi_0(x) = 0$ for $x \in \mathbb{R}^d$. Let us fix $k \in \{1\cdots d\}$. We define two maps: $A_k : \mathbb{R}^{k-1} \to \mathbb{R}$, $a_k : \mathbb{R}^{k-1} \times \mathbb{R} \to \mathbb{R}$ by

$$\begin{aligned} \forall \xi \in \mathbb{R}^{k-1} \qquad & A_k(\xi) = \mathcal{H}^{d-k+1}\big(E \cap \{\, \pi_{k-1}(x) = \xi \,\}\big)\,,\\ \forall \xi \in \mathbb{R}^{k-1} \quad \forall t \in \mathbb{R} \qquad & a_k(\xi,t) = \mathcal{H}^{d-k+1}\big(E \cap \{\, \pi_{k-1}(x) = \xi\,, x_k \leq t \,\}\big)\,. \end{aligned}$$

In particular, the map $A_1$ is given by $A_1(0) = \mathcal{H}^d(E)$. We set also

$$\forall \xi \in \mathbb{R}^d \qquad A_{d+1}(\xi) = \mathcal{H}^0\big(E \cap \{\, \pi_d(x) = \xi \,\}\big) = 1_E(\xi)\,.$$

By Fubini's theorem, the maps $a_k$ and $A_k$ are Borel and $A_k(\xi) < +\infty$ for $\mathcal{H}^{k-1}$ almost all $\xi$. Moreover the map $a_k$ is Lipschitz in $x_k$ and satisfies

$$\frac{\partial a_k}{\partial x_k} = A_{k+1} \quad \text{almost everywhere in } \mathbb{R}^d\,.$$

Let us set

$$E_{k-1} = \big\{\, \xi \in \mathbb{R}^{k-1} : 0 < A_k(\xi) < +\infty \,\big\}\,.$$

Since $\mathcal{L}^d(E) < +\infty$, we have by Fubini's theorem $\mathcal{L}^d\big(E \setminus \pi_{k-1}^{-1}(E_{k-1})\big) = 0$. Let $\xi \in E_{k-1}$. The map

$$t \in \mathbb{R} \mapsto \frac{a_k(\xi,t)}{A_k(\xi)}$$

is a non–decreasing Lipschitzian map onto $[0,1]$; let us define

$$\begin{aligned} \lambda_k(\xi) &= \sup\big\{\, t \in \mathbb{R} : a_k(\xi,t) = 0 \,\big\}\,,\\ \mu_k(\xi) &= \inf\big\{\, t \in \mathbb{R} : a_k(\xi,t) = A_k(\xi) \,\big\}\,. \end{aligned}$$

Let us define also $D_k$ by the identity

$$D_k \times \mathbb{R}^{d-k} = \bigcap_{1\leq i\leq k} \pi_i^{-1}(E_i)\,.$$

In particular, $D_d = E$. We make the convention that $D_0 = \{0\}$. To the Wulff crystal $\mathcal{W}$ we associate the functions $a_k^{\mathcal{W}}$, $A_k^{\mathcal{W}}$ in the same way as we built the functions $a_k$, $A_k$ associated to the set $E$ and we define the sets $\mathcal{W}_{k-1}$ as well. Since $\mathcal{W}$ is convex, then for $\xi \in \mathcal{W}_{k-1}$ the map

$$t \in \mathbb{R} \mapsto \frac{a_k^{\mathcal{W}}(\xi,t)}{A_k^{\mathcal{W}}(\xi)}$$

is strictly increasing from $[\lambda_k^{\mathcal{W}}(\xi), \mu_k^{\mathcal{W}}(\xi)]$ onto $[0,1]$ and it admits a locally Lipschitzian inverse on $]0,1[$. Let us start with $k=1$. There exists a map $F_1 : D_0 \times \mathbb{R}^d \to \mathbb{R}$ such that

$$\forall x \in D_0 \times \mathbb{R}^d \qquad \frac{a_1^{\mathcal{W}}(F_1(x))}{A_1^{\mathcal{W}}} = \frac{a_1(x_1)}{A_1}\,.$$

Moreover $F_1(x)$ depends only on $\pi_1(x) = x_1$. By induction, we see that for $k \in \{2\cdots d\}$, there exists a map $F_k : D_{k-1} \times \mathbb{R}^{d-k+1} \to \mathbb{R}$ such that

$$\forall x \in D_{k-1} \times \mathbb{R}^{d-k+1} \qquad \frac{a_k^{\mathcal{W}}\big(F_1(x),\dots,F_k(x)\big)}{A_k^{\mathcal{W}}\big(F_1(x),\dots,F_{k-1}(x)\big)} = \frac{a_k(x_1,\dots,x_k)}{A_k(x_1,\dots,x_{k-1})}\,.$$

Moreover $F_k(x)$ depends only on $\pi_k(x) = (x_1,\dots,x_k)$. For $\xi \in D_{k-1}$, the map $x_k \mapsto F_k(\xi, x_k)$ is continuous non–decreasing and locally Lipschitzian on $]\lambda_k(\xi), \mu_k(\xi)[$. We set finally

$$F = (F_1,\dots,F_d) : D_{d-1} \times \mathbb{R} \to \mathcal{W}\,.$$

The map $G = \big(a_1^{\mathcal{W}}/A_1^{\mathcal{W}},\dots,a_d^{\mathcal{W}}/A_d^{\mathcal{W}}\big) : \mathcal{W} \to [0,1]^d$ is one to one and Lipschitzian, hence it is a homeomorphism. Therefore

$$F = G^{-1} \circ \big(a_1/A_1,\dots,a_d/A_d\big) : D_{d-1} \times \mathbb{R} \to \mathcal{W}$$

is a Borel map. Differentiating the implicit equation defining $F_k$, we get

$$\frac{\partial F_k}{\partial x_k}(x_1,\dots,x_k) = \frac{A_{k+1}(x_1,\dots,x_k)\, A_k^{\mathcal{W}}(F_1,\dots,F_{k-1})}{A_k(x_1,\dots,x_{k-1})\, A_{k+1}^{\mathcal{W}}(F_1,\dots,F_k)}$$

for almost all $(x_1,\dots,x_k) \in D_{k-1} \times \mathbb{R}$. The classical inequality between the arithmetic and the geometric means implies that

$$\frac{1}{d}\sum_{1\leq k\leq d} \frac{\partial F_k}{\partial x_k} \geq \Big(\prod_{1\leq k\leq d} \frac{\partial F_k}{\partial x_k}\Big)^{\frac{1}{d}} = \Big(\frac{A_{d+1}(x_1,\dots,x_d)\, A_1^{\mathcal{W}}(0)}{A_1(0)\, A_{d+1}^{\mathcal{W}}(F_1,\dots,F_d)}\Big)^{\frac{1}{d}} = 1\,.$$

The conclusion of lemma 13.7 yields that $\mathcal{H}^{d-1}\big(\pi_{d-1}(\partial^* E)\Delta E_{d-1}\big)=0$. Since $D_{d-1}\subset E_{d-1}$ and $\mathcal{H}^{d-1}(E_{d-1}\setminus D_{d-1})=0$, then $\mathcal{H}^{d-1}\big(\pi_{d-1}(\partial^* E)\setminus D_{d-1}\big)=0$ and $\mathcal{H}^{d-1}\big(\partial^* E\setminus(D_{d-1}\times\mathbb{R})\big)=0$. In particular, the map $F$ is defined $\mathcal{H}^{d-1}$ almost everywhere on $\partial^* E$. We apply the extension of the Gauss–Green theorem to $F$ and $E$ (see theorem 13.6):

$$\begin{aligned} d\mathcal{L}^d(E) &\le \int_E \operatorname{div} F(x)\, d\mathcal{L}^d(x) \\ &= \int_{\partial^* E} F(x)\cdot \nu_E(x)\, d\mathcal{H}^{d-1}(x) \le \int_{\partial^* E} \tau(\nu_E(x))\, d\mathcal{H}^{d-1}(x) = \mathcal{I}(E)\,. \end{aligned}$$

By proposition 14.1 and the Gauss–Green theorem, we have also

$$\int_{\mathcal{W}} \operatorname{div} x\, d\mathcal{L}^d(x) = \int_{\partial^*\mathcal{W}} x\cdot\nu_{\mathcal{W}}(x)\, d\mathcal{H}^{d-1}(x) = \int_{\partial^*\mathcal{W}} \tau(\nu_{\mathcal{W}}(x))\, d\mathcal{H}^{d-1}(x)$$

whence $d\mathcal{L}^d(\mathcal{W})=\mathcal{I}(\mathcal{W})$. These identities imply already that the Wulff crystal $\mathcal{W}$ is a solution to the problem $(W)$.

Suppose now that $E$ is a solution to $(W)$. We have then $d\mathcal{L}^d(E)=d\mathcal{L}^d(\mathcal{W})$ and $\mathcal{I}(E)=\mathcal{I}(\mathcal{W})$ so that there is equality in the previous chain of inequalities. Therefore $\operatorname{div} F = d$ almost everywhere on $E$ (recall that $\partial F_k/\partial x_k\ge 0$). Moreover

$$F(x)\cdot\nu_E(x) = \tau(\nu_E(x)) \qquad \text{for } \mathcal{H}^{d-1} \text{ almost all } x\in\partial^* E\,.$$

By proposition 14.1, we have then $\nu_E(x)=\nu_{\mathcal{W}}(F(x))$ for $\mathcal{H}^{d-1}$ almost all $x$ in $\partial^* E$ such that $\mathcal{W}$ admits a normal vector at $F(x)$. Equality in the inequality between the arithmetic and the geometric means yields

$$\frac{\partial F_1}{\partial x_1} = \cdots = \frac{\partial F_d}{\partial x_d} = 1$$

for $\mathcal{L}^d$ almost all $x\in E$. Since for $k\in\{1\cdots d\}$, the map $F_k$ is locally Lipschitzian as a function of $x_k$, then there exists $c_1\in\mathbb{R}$ such that

$$\forall x\in[\lambda_1,\mu_1]\times\mathbb{R}^{d-1} \qquad F_1(x)=x_1+c_1$$

and there exist $d-1$ Borel functions $c_k:\mathbb{R}^{k-1}\to\mathbb{R}$, $2\le k\le d$, such that

$$F_k(\xi,x_k,\cdot) = x_k + c_k(\xi) \quad \text{for almost all } \xi\in D_{k-1} \text{ and } x_k\in[\lambda_k(\xi),\mu_k(\xi)]\,.$$

By repeated applications of Fubini's theorem, we see that if $B$ is a Borel subset of $E$ such that $F(B)$ is also measurable, then $\mathcal{L}^d(B)=\mathcal{L}^d(F(B))$. The structure of $F$ implies furthermore that

$$A_2(x_1) = A_2^{\mathcal{W}}(x_1+c_1) \qquad \text{for } \mathcal{H}^1 \text{ almost all } x_1\,.$$

Indeed, for any interval $I$,

$$\int_I A_2(x_1)\,dx_1 = \mathcal{L}^d\big(E\cap\pi_1^{-1}(I)\big) = \mathcal{L}^d\big(F(E\cap\pi_1^{-1}(I))\big)\,.$$

Yet

$$F(E\cap\pi_1^{-1}(I)) \subset \mathcal{W}\cap\pi_1^{-1}(I+c_1)\,,$$

thus

$$\int_I A_2(x_1)\,dx_1 \leq \int_{I+c_1} A_2^{\mathcal{W}}(x_1)\,dx_1\,.$$

This inequality is true for any interval $I$. Moreover

$$\mathcal{L}^d(E) = \int_{\mathbb{R}} A_2(x_1)\,dx_1 \geq \int_{\mathbb{R}} A_2^{\mathcal{W}}(x_1)\,dx_1 = \mathcal{L}^d(\mathcal{W})\,,$$

so that necessarily $A_2(x_1) = A_2^{\mathcal{W}}(x_1+c_1)$ for $\mathcal{H}^1$ almost all $x_1$.

Let next $A$ be a Borel subset of $\partial^*E$ such that $\mathcal{H}^{d-1}(A) > 0$. Letting $B = E\cap\big(\pi_{d-1}(A)\times\mathbb{R}\big)$, we have $\mathcal{L}^d(B)>0$ and $\mathcal{L}^d(F(B))>0$. Thus $\mathcal{H}^{d-1}\big(\pi_{d-1}(F(B))\big)>0$. Let $F^* = (F_1,\dots,F_{d-1}) : D_{d-1}\to\pi_{d-1}(\mathcal{W})$. Thanks to the structure of $F$, we have

$$\begin{aligned}\mathcal{H}^{d-1}\big(\pi_{d-1}(F(B))\big) &= \mathcal{H}^{d-1}\big(F^*(\pi_{d-1}(B))\big)\\ &= \mathcal{H}^{d-1}\big(F^*(\pi_{d-1}(A))\big) = \mathcal{H}^{d-1}\big(\pi_{d-1}(F(A))\big)\,,\end{aligned}$$

whence $\mathcal{H}^{d-1}\big(\pi_{d-1}(F(A))\big)>0$ and $\mathcal{H}^{d-1}\big(F(A)\big)>0$. Yet $F(A)\subset\partial\mathcal{W}$ and $\mathcal{H}^{d-1}(\partial\mathcal{W}\setminus\partial^*\mathcal{W})=0$, therefore $\mathcal{H}^{d-1}\big(F(A)\cap\partial^*\mathcal{W}\big)>0$. Thus

$$F(x)\in\partial^*\mathcal{W}\quad\text{and}\quad \nu_E(x)=\nu_{\mathcal{W}}(F(x))\qquad\text{for }\mathcal{H}^{d-1}\text{ almost all }x\in\partial^*E\,.$$

Let $\mathcal{B}_d$ be the set of the orthonormal bases $V=(v_1,\dots,v_d)$ of $\mathbb{R}^d$ such that $v_1$ satisfies the conclusion of lemma 13.7. Let $F_V$ be the map constructed as above with $V$ instead of the standard basis. By lemma 13.7, there exists a countable subset $\mathcal{B}_1$ of $\mathcal{B}_d$ such that the set

$$\{\,v_1 : (v_1,\dots,v_d)\in V,\ V\in\mathcal{B}_1\,\}$$

is dense in the unit sphere $S^{d-1}$. For $V\in\mathcal{B}_1$, the map $F_{1,V}$ is a translation, therefore, up to a Lebesgue negligible set, the set $E$ is bounded in the direction of the first vector of the basis $V$. By choosing $d$ independent directions in $\mathcal{B}_1$, we conclude that, modulo a Lebesgue negligible set, the set $E$ is bounded. We translate $E$ so that its mass center is the origin. Let $V\in\mathcal{B}_1$, we have then $F_{1,V}(x_1)=x_1+c_{1,V}$ for some constant $c_{1,V}$ and

$$0=\int_E x_1\,dx = \int_{\mathbb{R}} x_1 A_2(x_1)\,dx_1 = \int_{\mathbb{R}} x_1 A_2^{\mathcal{W}}(x_1+c_{1,V})\,dx_1 = -c_{1,V}\mathcal{L}^d(\mathcal{W}).$$

Therefore $F_{1,V}(x_1)=x_1$ for any $V\in\mathcal{B}_1$.

Let us define

$$N = \{ x \in \partial^* E : \forall V \in \mathcal{B}_1 \quad F_V(x) \in \partial^* \mathcal{W}, \quad \nu_E(x) = \nu_{\mathcal{W}}(F_V(x)) \} .$$

We have then $\mathcal{H}^{d-1}(\partial^* E \setminus N) = 0$. Let $\nu \in S^{d-1}$ be such that the set

$$N(\nu) = N \cap \{ x \in \partial^* E : \nu_E(x) = \nu \}$$

is not empty. Because of the definition of $N$, the set

$$N^{\mathcal{W}}(\nu) = \{ y \in \partial^* \mathcal{W} : \nu_{\mathcal{W}}(y) = \nu \}$$

is also not empty. Since $\mathcal{W}$ is convex, the closure $\overline{N^{\mathcal{W}}(\nu)}$ is convex and moreover

$$\forall y \in \overline{N^{\mathcal{W}}(\nu)} \cap \partial^* \mathcal{W} \qquad \nu_{\mathcal{W}}(y) = \nu .$$

Indeed, for $y \in \overline{N^{\mathcal{W}}(\nu)}$, the hyperplane $\{ z \in \mathbb{R}^d : z \cdot \nu = y \cdot \nu \}$ is a supporting hyperplane for $\mathcal{W}$ and if $y$ belongs to $\partial^* \mathcal{W}$, then $\mathcal{W}$ admits a unique supporting hyperplane at $y$, which is also the tangent hyperplane. Since $c_{1,V} = 0$ for any $V \in \mathcal{B}_1$, then $N(\nu)$ is included in

$$\bigcap_{V \in \mathcal{B}_1} \Big\{ x \in \mathbb{R}^d : \inf \{ y \cdot v_1 : y \in N^{\mathcal{W}}(\nu) \} \leq x \cdot v_1 \leq \sup \{ y \cdot v_1 : y \in N^{\mathcal{W}}(\nu) \} \Big\}$$

and this set is equal to $\overline{N^{\mathcal{W}}(\nu)}$. Therefore $N(\nu) \subset \overline{N^{\mathcal{W}}(\nu)}$ and

$$\forall x \in N(\nu) \cap \partial^* \mathcal{W} \qquad \nu_E(x) = \nu_{\mathcal{W}}(x) = \nu .$$

Since $\mathcal{W}$ is convex, we have $\mathcal{H}^{d-1}(\partial \mathcal{W} \setminus \partial^* \mathcal{W}) = 0$. We conclude successively that $N \subset \partial \mathcal{W}$, $\mathcal{H}^{d-1}(\partial^* E \setminus \partial^* \mathcal{W}) = 0$ and

$$\nu_E(x) = \nu_{\mathcal{W}}(x) \qquad \text{for } \mathcal{H}^{d-1} \text{ almost all } x \in \partial^* E \cap \partial^* \mathcal{W} .$$

In addition, we have $\mathcal{I}(E) \geq \mathcal{I}(\mathcal{W})$. The surface tension $\tau$ being positive, this implies that $\mathcal{H}^{d-1}(\partial^* \mathcal{W} \setminus \partial^* E) = 0$.

Let $B$ be a ball included in the interior of $\mathcal{W}$. By the local isoperimetric inequality (see section 13.3), we have either $\mathcal{L}^d(B \setminus E) = 0$ or $\mathcal{L}^d(E \cap B) = 0$. Let $B_1, \dots, B_r$ be a finite collection of open balls included in the interior of $\mathcal{W}$ such that $B_1 \cup \cdots \cup B_r$ is connected. The previous property implies that

$$\text{either} \quad \mathcal{L}^d(B_1 \cup \cdots \cup B_r \setminus E) = 0 \quad \text{or} \quad \mathcal{L}^d((B_1 \cup \cdots \cup B_r) \cap E) = 0 .$$

Therefore we have either $\mathcal{L}^d(\mathcal{W} \setminus E) = 0$ or $\mathcal{L}^d(\mathcal{W} \cap E) = 0$. The same argument applied to $\mathbb{R}^d \setminus \mathcal{W}$ yields that either $\mathcal{L}^d(\mathbb{R}^d \setminus \mathcal{W} \setminus E) = 0$ or $\mathcal{L}^d((\mathbb{R}^d \setminus \mathcal{W}) \cap E) = 0$. Yet $\mathcal{L}^d(\mathcal{W}) \leq \mathcal{L}^d(E) < \infty$, hence the only possible case is that $\mathcal{L}^d(\mathcal{W} \Delta E) = 0$.
□

## 15.4 Stability of the Wulff crystal

We quote next a stability result for the solution of $(W)$ (see [72], proposition 5.1).

**Proposition 15.3.** *If $(E_n)_{n\in\mathbb{N}}$ is a minimizing sequence of the variational problem $(W)$, then there exists a sequence $(a_n)_{n\in\mathbb{N}}$ of vectors in $\mathbb{R}^d$ such that*

$$\lim_{n\to\infty} \mathcal{L}^d\big((E_n - a_n)\Delta\mathcal{W}\big) = 0 .$$

*Proof.* Let $(E_n)_{n\in\mathbb{N}}$ be a minimizing sequence of the variational problem $(W)$. The sequence

$$F_n = \Big(\frac{\mathcal{L}^d(\mathcal{W})}{\mathcal{L}^d(E_n)}\Big)^{\frac{1}{d}} E_n , \qquad n\in\mathbb{N}$$

is also a minimizing sequence for $(W)$, and it satisfies $\mathcal{L}^d(F_n)=\mathcal{L}^d(\mathcal{W})$ for all $n\in\mathbb{N}$. We denote by $(e_1,\dots,e_d)$ the canonical basis of $\mathbb{R}^d$. For $i\in\{1\cdots d\}$ and $x_i\in\mathbb{R}$, we define the half spaces

$$H_i^-(x_i) = \big\{\, x\in\mathbb{R}^d : x\cdot e_i < x_i \,\big\} ,$$
$$H_i^+(x_i) = \big\{\, x\in\mathbb{R}^d : x\cdot e_i > x_i \,\big\} ,$$

and the hyperplane

$$H_i(x_i) = \big\{\, x\in\mathbb{R}^d : x\cdot e_i = x_i \,\big\} .$$

For $n\in\mathbb{N}$, there exists $x_n\in\mathbb{R}^d$ such that, for $i\in\{1\cdots d\}$,

$$\mathcal{L}^d\big((F_n - x_n)\cap H_i^-(0)\big) = \frac{1}{2}\mathcal{L}^d(\mathcal{W}) = \mathcal{L}^d\big((F_n - x_n)\cap H_i^+(0)\big) .$$

We set

$$\forall n\in\mathbb{N} \qquad G_n = F_n - x_n .$$

Let us fix $i\in\{1\cdots d\}$. For almost all $x_i\in\mathbb{R}$, we have for $*=-,+$,

$$\mathcal{I}(G_n\cap H_i^*(x_i)) \le \mathcal{I}(G_n, H_i^*(x_i)) + ||\tau||_\infty \mathcal{H}^{d-1}(G_n\cap H_i(x_i)) .$$

The proof of this fact is similar to the one of lemma 14.4. Let $N_{i,n}$ be the set of the values $x_i\in\mathbb{R}$ satisfying the above inequality. Let $\delta>0$. Since

$$\mathcal{L}^d(G_n) \ge \int_0^{\mathcal{L}^d(\mathcal{W})/\delta} \mathcal{H}^{d-1}(G_n\cap H_i(x_i))\, dx_i ,$$

then there exists $x_{i,n}^+\in N_{i,n}\cap[0,\mathcal{L}^d(\mathcal{W})/\delta]$ such that $\mathcal{H}^{d-1}(G_n\cap H_i(x_{i,n}^+))\le\delta$. Now

$$\begin{aligned}\mathcal{I}(G_n) &\ge \mathcal{I}(G_n, H_i^-(x_{i,n}^+)) + \mathcal{I}(G_n, H_i^+(x_{i,n}^+))\\ &\ge \mathcal{I}(G_n\cap H_i^-(x_{i,n}^+)) + \mathcal{I}(G_n\cap H_i^+(x_{i,n}^+)) - 2\delta||\tau||_\infty .\end{aligned}$$

Let us set

$$\alpha_{i,n}^+ = \frac{\mathcal{L}^d(G_n \cap H_i^+(x_{i,n}^+))}{\mathcal{L}^d(\mathcal{W})}.$$

The definition of $G_n$ implies that $\alpha_{i,n}^+ \leq 1/2$. The anisotropic isoperimetric inequality of section 15.2 applied to the sets $G_n \cap H_i^-(x_{i,n}^+)$ and $G_n \cap H_i^+(x_{i,n}^+)$ yields

$$\alpha_{i,n}^+ \mathcal{I}(\mathcal{W})^{\frac{d}{d-1}} \leq \mathcal{I}(G_n \cap H_i^+(x_{i,n}^+))^{\frac{d}{d-1}},$$
$$(1-\alpha_{i,n}^+)\mathcal{I}(\mathcal{W})^{\frac{d}{d-1}} \leq \mathcal{I}(G_n \cap H_i^-(x_{i,n}^+))^{\frac{d}{d-1}}.$$

Reporting in the previous inequality, we obtain

$$(\alpha_{i,n}^+)^{\frac{d-1}{d}} + (1-\alpha_{i,n}^+)^{\frac{d-1}{d}} \leq \frac{\mathcal{I}(G_n) + 2\delta||\tau||_\infty}{\mathcal{I}(\mathcal{W})}.$$

We define analogously $x_{i,n}^-$ and $\alpha_{i,n}^-$ by working on the interval $[-\mathcal{L}^d(\mathcal{W})/\delta, 0]$. We have then, for $\delta > 0$ and $n \in \mathbb{N}$,

$$\mathcal{L}^d\Big(G_n \setminus \Big[-\frac{\mathcal{L}^d(\mathcal{W})}{\delta}, \frac{\mathcal{L}^d(\mathcal{W})}{\delta}\Big]^d\Big) \leq \sum_{1\leq i\leq d} (\alpha_{i,n}^- + \alpha_{i,n}^+)\mathcal{L}^d(\mathcal{W}).$$

From the previous inequalities, we deduce that

$$\limsup_{\delta\to 0}\ \limsup_{n\to\infty}\ \mathcal{L}^d\Big(G_n \setminus \Big[-\frac{\mathcal{L}^d(\mathcal{W})}{\delta}, \frac{\mathcal{L}^d(\mathcal{W})}{\delta}\Big]^d\Big) = 0$$

which yields the following tightness property:

$$\forall \varepsilon > 0 \quad \exists m \in \mathbb{N} \quad \forall n \in \mathbb{N} \qquad \mathcal{L}^d(G_n \setminus [-m,m]^d) \leq \varepsilon.$$

Moreover, we have

$$\forall n, m \in \mathbb{N} \qquad \mathcal{I}(G_n, ]-m,m[^d) \leq \mathcal{I}(G_n) \leq \mathcal{I}(E_n).$$

By theorem 14.5, the sequence $(G_n \cap ]-m,m[^d)_{n\in\mathbb{N}}$ admits a subsequence converging towards a Borel subset of $]-m,m[^d$. With the help of a standard diagonal argument, we see that there exists a Borel set $G$ in $\mathbb{R}^d$ and a subsequence $(G_{\phi(n)})_{n\in\mathbb{N}}$ such that, for any $m \in \mathbb{N}$,

$$\lim_{n\to\infty} \mathcal{L}^d\Big((G_{\phi(n)} \Delta G) \cap ]-m,m[^d\Big) = 0.$$

The tightness property then implies that

$$\lim_{n\to\infty} \mathcal{L}^d(G_{\phi(n)} \Delta G) = 0.$$

We have already proved that the sequence $(G_n)_{n\in\mathbb{N}}$ is relatively compact in $\mathcal{B}(\mathbb{R}^d)$ endowed with the topology of $L^1$ convergence. We examine next the possible limits of a subsequence. The set $G$ satisfies $\mathcal{L}^d(G) = \mathcal{L}^d(\mathcal{W})$, and since $\mathcal{I}$ is lower semicontinuous (see section 14.2), we have $\mathcal{I}(G) \leq \mathcal{I}(\mathcal{W})$. Thus $G$ is a solution to the problem (W). Since we have also

$$\forall i \in \{1 \cdots d\} \qquad \mathcal{L}^d\big(G \cap H_i^-(0)\big) = \frac{1}{2}\mathcal{L}^d(\mathcal{W}) = \mathcal{L}^d\big(G \cap H_i^+(0)\big),$$

we conclude finally with the help of theorem 15.1 that $\mathcal{L}^d(G \Delta \mathcal{W}) = 0$. The only possible limit of a subsequence of $(G_n)_{n \in \mathbb{N}}$ is $\mathcal{W}$, therefore the whole sequence is converging towards $\mathcal{W}$. □

**Corollary 15.4.** *(Stability of the Wulff crystal) For any positive $\delta$, there exists a positive $\eta$ such that*

$$\inf\Big\{\mathcal{I}(E) : E \in \mathcal{B}(\mathbb{R}^d),\, \mathcal{L}^d(\mathcal{W}) - \eta \leq \mathcal{L}^d(E) < \infty,\, \inf_{x \in \mathbb{R}^d} \mathcal{L}^d\big(E \Delta (x + \mathcal{W})\big) \geq \delta\Big\} \geq \mathcal{I}(\mathcal{W}) + \eta\,.$$

*Proof.* If the result was false, there would exist a positive $\delta$ and a sequence $(E_n)_{n \in \mathbb{N}}$ such that

$$\liminf_{n \to \infty} \mathcal{L}^d(E_n) \geq \mathcal{L}^d(\mathcal{W}), \qquad \lim_{n \to \infty} \mathcal{I}(E_n) = \mathcal{I}(\mathcal{W})$$

and for all $n$ in $\mathbb{N}$

$$\forall x \in \mathbb{R}^d \qquad \mathcal{L}^d\big(E_n \Delta (x + \mathcal{W})\big) \geq \delta\,.$$

For $n$ in $\mathbb{N}$, let $\lambda_n = (\mathcal{L}^d(\mathcal{W})/\mathcal{L}^d(E_n))^{1/d}$. On one hand, $\limsup_{n \to \infty} \lambda_n \leq 1$; on the other hand, by theorem 15.1 or simply the anisotropic inequality proved in section 15.2, $(\lambda_n)^{d-1}\mathcal{I}(E_n) = \mathcal{I}(\lambda_n E_n) \geq \mathcal{I}(\mathcal{W})$, so that $\liminf_{n \to \infty} \lambda_n \geq 1$. Therefore $\lim_{n \to \infty} \lambda_n = 1$. It follows that $(\lambda_n E_n)_{n \in \mathbb{N}}$ is a minimizing sequence of the problem $(W)$. Yet, for $x$ in $\mathbb{R}^d$, using the hypothesis on $E_n$ and the convexity of $\mathcal{W}$,

$$\begin{aligned}\mathcal{L}^d\big((\lambda_n E_n) \Delta (\lambda_n x + \mathcal{W})\big) &\geq (\lambda_n)^d \mathcal{L}^d\big(E_n \Delta (x + \mathcal{W})\big) - \mathcal{L}^d(\mathcal{W} \Delta \lambda_n \mathcal{W}) \\ &\geq (\lambda_n)^d \delta - |1 - (\lambda_n)^d| \mathcal{L}^d(\mathcal{W})\,.\end{aligned}$$

Taking the infimum over $x$ in $\mathbb{R}^d$ and passing to the limit, we get

$$\liminf_{n \to \infty} \inf_{x \in \mathbb{R}^d} \mathcal{L}^d\big((\lambda_n E_n) \Delta (x + \mathcal{W})\big) \geq \delta\,.$$

This stands in contradiction with proposition 15.3. □

# Part VII

# Final steps of the proofs

## 16

# LDP for the cluster shapes

This chapter is devoted to the proof of the large deviation principle stated in theorem 7.3. We work here within the topological vector space $\mathcal{M}(\mathbb{R}^d)$ of the signed Borel measures on $\mathbb{R}^d$ endowed with the weak topology. The collection consisting of the finite intersections of the sets

$$\big\{\,\nu\in\mathcal{M}(\mathbb{R}^d):|\nu(f)|\leq\eta\,\big\}\,,\quad \eta>0\,,\quad f\in C_c(\mathbb{R}^d,\mathbb{R})$$

is a basis of neighbourhoods of the origin.

We prove first that $\mathcal{J}$ is lower semicontinuous. Let $(\nu_n)_{n\in\mathbb{N}}$ be a sequence of Borel measures on $\mathbb{R}^d$ converging weakly towards $\nu$. Let

$$\alpha\;=\;\liminf_{n\to\infty}\mathcal{J}(\nu_n)$$

and suppose that $\alpha<+\infty$. We have to prove that $\mathcal{J}(\nu)\leq\alpha$. Let $\beta>\alpha$. Up to the extraction of a subsequence, we can assume that $\mathcal{J}(\nu_n)\leq\beta$ for $n\in\mathbb{N}$. This implies that for each $n\in\mathbb{N}$, there exists a Borel subset $A_n$ of $\mathbb{R}^d$ such that $\nu_n$ has density $\theta 1_{A_n}$ with respect to the Lebesgue measure and $\mathcal{I}(A_n)\leq\beta$. In particular, we have

$$\forall n,m\in\mathbb{N}\qquad \mathcal{I}(A_n,\overset{\circ}{B}(0,m))\;\leq\;\beta\,.$$

By theorem 14.5, the sequence $(A_n\cap B(0,m))_{n\in\mathbb{N}}$ admits a subsequence converging towards a subset of $B(0,m)$. With the help of a standard diagonal argument, we see that there exists a Borel set $A$ and a subsequence $(A_{\phi(n)})_{n\in\mathbb{N}}$ such that, for any $m\in\mathbb{N}$,

$$\lim_{n\to\infty}\mathcal{L}^d\Big((A_{\phi(n)}\Delta A)\cap B(0,m)\Big)\;=\;0\,,\qquad \mathcal{I}(A,\overset{\circ}{B}(0,m))\leq\beta\,.$$

This implies that $\mathcal{I}(A)\leq\beta$ and that $\nu$ is the measure with density $\theta 1_A$ with respect to the Lebesgue measure. Therefore $\mathcal{J}(\nu)\leq\beta$. This being true for any $\beta>\alpha$, we conclude indeed that $\mathcal{J}(\nu)\leq\alpha$ and that $\mathcal{J}$ is lower semicontinuous.

Our next goal is to prove that the random measure $\mathcal{C}_n$ is $\mathcal{J}$–tight in the sense of definition 6.3: there exists a positive constant $c$ such that

$$\forall \eta, \lambda > 0\,, \quad \forall f \in C_c(\mathbb{R}^d, \mathbb{R})\,,$$
$$\limsup_{n\to\infty} \frac{1}{n^{d-1}} \ln P\big(\forall \nu \in \mathcal{J}^{-1}([0,\lambda]) \quad |\mathcal{C}_n(f) - \nu(f)| > \eta\big) \leq -c\lambda\,.$$

We then prove the local large deviation upper bound and finally the large deviation lower bound.

## 16.1 Coarse grained image

In order to prove the $\mathcal{J}$–tightness, we build an auxiliary random measure $\widetilde{\mathcal{C}}_n$ which is exponentially contiguous to $\mathcal{C}_n$ and we provide suitable probabilistic estimates on the surface energy of $\widetilde{\mathcal{C}}_n$. The construction will depend on a parameter $K \in \mathbb{N}$. We work with the lattice rescaled by a factor $K$ and we build a coarse grained image of $\mathcal{C}_n$. We first define

$$\underline{\mathcal{C}}_n = \big\{\, \underline{x} \in \mathbb{Z}^d : \mathcal{C}_n(B_n(\underline{x})) > 0 \,\big\}\,.$$

Notice that $\underline{\mathcal{C}}_n$ is $\mathbb{L}^d$–connected. The condition $\mathcal{C}_n(B_n(\underline{x})) > 0$ is equivalent to saying that the cluster $C(0)$ intersects the block $B(\underline{x})$. The set $\underline{\mathcal{C}}_n$ contains a large number of small holes which do not contribute to create any surface energy. To get rid of these holes, we fill them out by the operation "fill" which we now describe. We look at the residual $\mathbb{L}^{d,\infty}$–components of $\underline{\mathcal{C}}_n$, that is the $\mathbb{L}^{d,\infty}$-connected components of $\mathbb{Z}^d \setminus \underline{\mathcal{C}}_n$. If $\operatorname{diam}_\infty C(0) \leq K \ln n$, we set $\operatorname{fill} \underline{\mathcal{C}}_n = \emptyset$; if $\operatorname{diam}_\infty C(0) > K \ln n$, we define

$$\operatorname{fill} \underline{\mathcal{C}}_n = \underline{\mathcal{C}}_n \cup \bigcup \underline{R}\,,$$

where the union runs over the finite residual $\mathbb{L}^{d,\infty}$–components $\underline{R}$ of $\underline{\mathcal{C}}_n$ such that $\operatorname{diam}_\infty \underline{R} < \ln n$. Finally, the auxiliary measure $\widetilde{\mathcal{C}}_n$ is the measure with density $\theta\, 1_{C_n}$ with respect to the Lebesgue measure $\mathcal{L}^d$, where $C_n$ is the region

$$C_n = \bigcup_{\underline{x} \in \operatorname{fill} \underline{\mathcal{C}}_n} B_n(\underline{x})\,.$$

We drop the parameter $K$ from the notation, yet the set $C_n$ and the measure $\widetilde{\mathcal{C}}_n$ depend on $K$.

**Lemma 16.1.** *For $K$ large enough, there exist two positive constants $b, c$ such that, for any bounded open subset $O$ of $\mathbb{R}^d$, there exists $n_0 \in \mathbb{N}$ depending only on the volume $\mathcal{L}^d(\mathcal{V}(O,1))$ of $\mathcal{V}(O,1)$ such that*

$$\forall n \geq \max(n_0, K) \quad \forall u > 0 \qquad P\big(\mathcal{P}(C_n, O) \geq u\big) \leq b \exp(-cun^{d-1})\,.$$

AQ: Please scale this figure

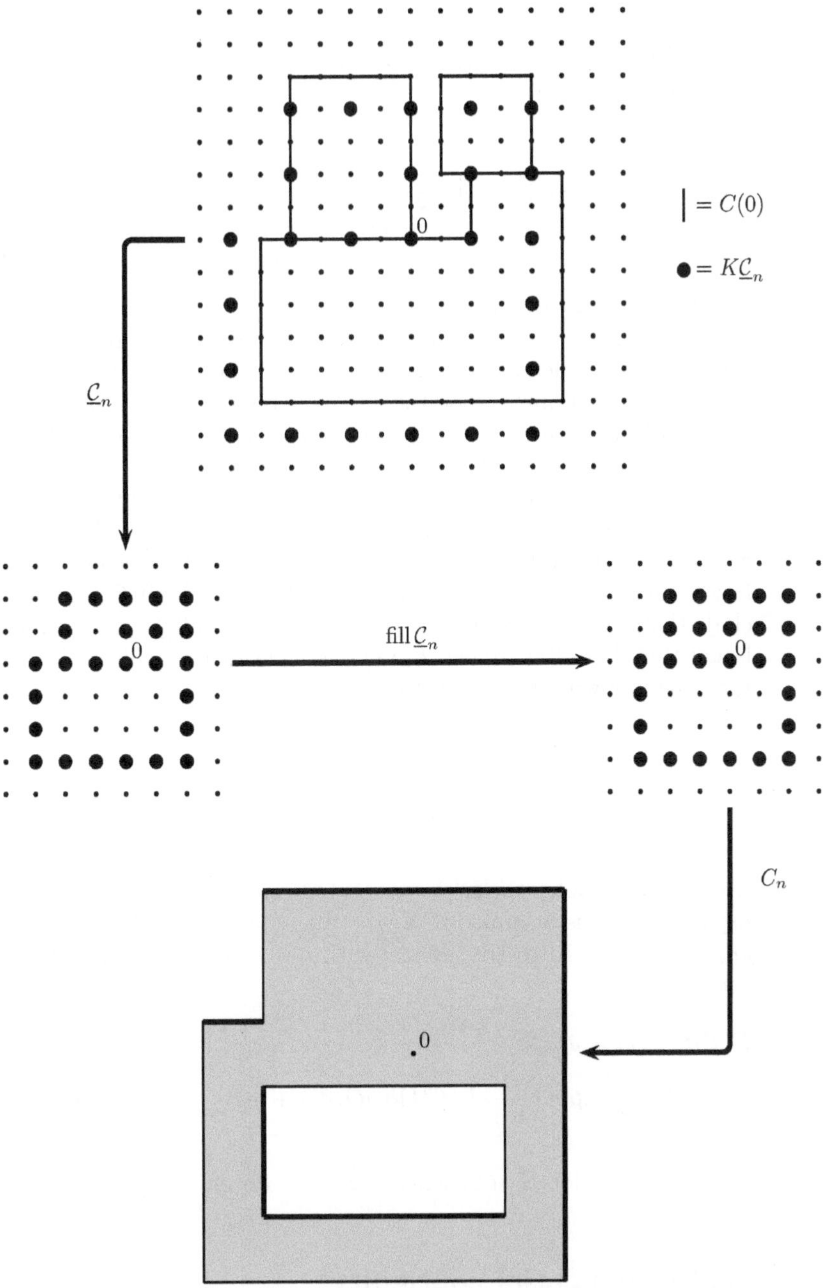


Construction of $C_n$ from $C(0)$

*Proof.* Let us set

$$\underline{O} = \{\underline{x} \in \mathbb{Z}^d : B_n(\underline{x}) \cap O \neq \emptyset\} = \{\underline{x} \in \mathbb{Z}^d : B(\underline{x}) \cap nO \neq \emptyset\}.$$

For $\underline{x} \in \mathbb{Z}^d$, the block variable $X(\underline{x})$ is the indicator function of the event $R(B'(\underline{x}), K)$, i.e.,

- Inside the event block $B'(\underline{x})$ there is exactly one crossing cluster $C(\underline{x})$, it is the only cluster in $B'(\underline{x})$ having diameter larger than or equal to $K$ and it crosses every sub–box of $B'(\underline{x})$ with diameter larger than or equal to $K$.

By construction, we have $\partial^{in}\mathrm{fill}\,\underline{\mathcal{C}_n} \subset \partial^{in}\underline{\mathcal{C}_n}$ and if $\underline{x}$ belongs to $\partial^{in}\underline{\mathcal{C}_n}$ then certainly $X(\underline{x}) = 0$: if $X(\underline{x})$ was equal to 1, the cluster $C(0)$ would intersect all the $2d$ blocks which are neighbours of $B(\underline{x})$. Any $\mathbb{L}^{d,\infty}$-connected component of $\partial^{in}\mathrm{fill}\,\underline{\mathcal{C}_n}$ contains either the external boundary of a finite residual component of $\mathrm{fill}\,\underline{\mathcal{C}_n}$ or the external boundary of the infinite residual component of $\mathrm{fill}\,\underline{\mathcal{C}_n}$. The external boundary of a finite residual component of $\mathrm{fill}\,\underline{\mathcal{C}_n}$ has diameter larger than $\ln n + 1$ and therefore has cardinality strictly larger than $\ln n$. In the case where $K \ln n < \mathrm{diam}_\infty C(0) < +\infty$, the external boundary of the unbounded residual component of $\underline{\mathcal{C}_n}$ has also cardinality strictly larger than $\ln n$. If $\mathrm{diam}_\infty C(0) \leq K \ln n$, then $\partial^{in}\mathrm{fill}\,\underline{\mathcal{C}_n}$ is empty. We conclude that any $\mathbb{L}^{d,\infty}$-connected component of $\partial^{in}\mathrm{fill}\,\underline{\mathcal{C}_n}$ has cardinality strictly larger than $\ln n$. Each block $B_n(\underline{x})$ has a perimeter $2d(K/n)^{d-1}$, therefore we have the following bound on the perimeter of $C_n$:

$$\begin{aligned}\mathcal{P}(C_n, O) &\leq 2d\Big(\frac{K}{n}\Big)^{d-1} \big|\underline{O} \cap \partial^{in}\mathrm{fill}\,\underline{\mathcal{C}_n}\big| \\ &\leq 2d\Big(\frac{K}{n}\Big)^{d-1} \Big|\big\{\,\underline{x} \in \underline{O} : |\underline{C}(\underline{x})| \geq \ln n\,\big\}\Big|,\end{aligned}$$

where $\underline{C}(\underline{x})$ denotes the $\mathbb{L}^{d,\infty}$–connected component of occupied sites containing $\underline{x}$ (a site $\underline{y}$ is occupied if $X(\underline{y}) = 0$). To estimate the right–hand side, we apply lemma 9.10 to the set $n\underline{O}$ with $t = \ln n$, $s = (n/K)^{d-1}(u/2d)$ and we get

$$P\big(\mathcal{P}(C_n, O) \geq u\big) \leq 2 \sum_{j \geq s} \exp j \left( \frac{1}{\ln n} \ln \mathcal{L}^d\big(\mathcal{V}(nO, d)\big) + \ln b + \frac{1}{3^d} \ln P(X(\underline{0}) = 0) \right).$$

By lemma 9.4, for $K$ sufficiently large, we obtain the required bound. □

## 16.2 Exponential contiguity

Let us fix a bounded uniformly continuous function $f : \mathbb{R}^d \to \mathbb{R}$. Let $x \in \mathbb{R}^d$ and let $r > 0$. Our goal is to estimate $|\mathcal{C}_n(f 1_{B(x,r)}) - \widetilde{\mathcal{C}}_n(f 1_{B(x,r)})|$. To this end, we use another block argument with the scale $L = K \ln n$. We work with the lattice rescaled by a factor $L$. Let $\varepsilon > 0$. For $\underline{y} \in \mathbb{Z}^d$, the block variable $Y(\underline{y})$ is the indicator function of the event $RVW(B(\underline{y}), B'(\underline{y}), L, \varepsilon)$ (see the definition before corollary 9.6). Let

$$\underline{A} = \{ \underline{y} \in \mathbb{Z}^d : B_n(\underline{y}) \subset B(x,r) \} .$$

We have $|\underline{A}| L^d \leq n^d \mathcal{L}^d(B(x,r))$. We suppose that $L/n$ is less than 1 and small enough so that

$$\mathcal{L}^d\Big( B(x,r) \setminus \bigcup_{\underline{y}\in\underline{A}} B_n(\underline{y}) \Big) \leq \varepsilon \mathcal{L}^d(B(x,r)) ,$$

$$\forall y, z \in \mathbb{R}^d \qquad |y - z| \leq \frac{L}{n} \quad \Rightarrow \quad |f(y) - f(z)| \leq \varepsilon ||f||_\infty .$$

We evaluate $|\mathcal{C}_n(f 1_{B(x,r)}) - \widetilde{\mathcal{C}}_n(f 1_{B(x,r)})|$ by decomposing it on blocks of size $L/n$ and we use the uniform continuity of $f$:

$$\begin{aligned} &|\mathcal{C}_n(f 1_{B(x,r)}) - \widetilde{\mathcal{C}}_n(f 1_{B(x,r)})| \\ &\qquad \leq 2\varepsilon ||f||_\infty \mathcal{L}^d(B(x,r)) + \sum_{\underline{y}\in\underline{A}} \Big| \int_{B_n(\underline{y})} f\, d\mathcal{C}_n - \int_{B_n(\underline{y})} f\, d\widetilde{\mathcal{C}}_n \Big| \\ &\qquad \leq 4\varepsilon ||f||_\infty \mathcal{L}^d(B(x,r)) + ||f||_\infty \sum_{\underline{y}\in\underline{A}} \big| \mathcal{C}_n(B_n(\underline{y})) - \widetilde{\mathcal{C}}_n(B_n(\underline{y})) \big| . \end{aligned}$$

Next we study the last sum in the above inequality. If the diameter of $C(0)$ is less than or equal to $K \ln n$, then the number of blocks contributing to the sum is less than $3^d$ and the sum is bounded by $3^d (L/n)^d$. From now on, we suppose that the diameter of $C(0)$ is strictly larger than $K \ln n$. If $\underline{y} \in \underline{A}$ is such that $B_n(\underline{y})$ does not intersect $C_n$, then $\mathcal{C}_n(B_n(\underline{y})) = \widetilde{\mathcal{C}}_n(B_n(\underline{y})) = 0$ and the corresponding term in the sum vanishes. So we need only to consider the blocks such that $B_n(\underline{y}) \cap C_n \neq \emptyset$. Let $\underline{y} \in \underline{A}$ be such that $B_n(\underline{y}) \cap C_n \neq \emptyset$. We distinguish several cases.

- If $Y(\underline{y}) = 0$, then

$$\big| \mathcal{C}_n(B_n(\underline{y})) - \widetilde{\mathcal{C}}_n(B_n(\underline{y})) \big| \leq \frac{1}{n^d} |B_n(\underline{y})| 1_{Y(\underline{y})=0} .$$

Suppose next that $Y(\underline{y}) = 1$. Several subcases arise:

- $B_n(\underline{y}) \not\subset C_n$. Then we bound

$$\big| \mathcal{C}_n(B_n(\underline{y})) - \widetilde{\mathcal{C}}_n(B_n(\underline{y})) \big| \leq \frac{1}{n^d} |B_n(\underline{y})|$$

and we notice that since $B_n(\underline{y})$ meets $C_n$, it meets also the boundary of $C_n$. Thus

$$B_n(\underline{y}) \subset \Big\{ x \in \mathbb{R}^d : \mathrm{d}_\infty(x, \partial C_n \cap B(x,r)) \leq \frac{L}{n} \Big\} .$$

Moreover

$$\begin{aligned}\mathcal{L}^d\Big(\{\,x\in\mathbb{R}^d:\mathrm{d}_\infty(x,\partial C_n\cap B(x,r))\leq\frac{L}{n}\,\}\Big)\\ &\leq\big|\partial^{in}nC_n\cap B(nx,nr+d)\big|\Big(\frac{2L+2}{n}\Big)^d\\ &\leq\mathcal{P}\big(nC_n,B(nx,nr+d)\big)\Big(\frac{2L+2}{n}\Big)^d\\ &\leq\mathcal{P}\big(C_n,B(x,r+d)\big)\frac{(3L)^d}{n}\,.\end{aligned}$$

• $B_n(\underline{y})\subset C_n$ and $\mathcal{C}_n(B_n(\underline{y}))=0$. These conditions imply that the whole block $B_n(\underline{y})$ has been added to $C_n$ by the filling operation. Yet the regions which are added by the filling operation have a diameter at most $K(\ln n-1)$, so this case cannot occur.

• $B_n(\underline{y})\subset C_n$ and $\mathcal{C}_n(B_n(\underline{y}))>0$. The definition of the block event associated to the variable $Y$ implies that

$$\big|\mathcal{C}_n(B_n(\underline{y}))-\widetilde{\mathcal{C}}_n(B_n(\underline{y}))\big|\;=\;\Big|\mathcal{C}_n(B_n(\underline{y}))-\frac{\theta}{n^d}\big|B_n(\underline{y})\big|\Big|\;\leq\;\frac{\varepsilon}{n^d}\big|B_n(\underline{y})\big|\,.$$

Summing the previous inequalities over $\underline{y}\in\underline{A}$, we get

$$|\mathcal{C}_n(f)-\widetilde{\mathcal{C}}_n(f)|\leq$$

$$||f||_\infty\mathcal{L}^d(B(x,r))\Big(5\varepsilon+\frac{1}{|\underline{A}|}\sum_{\underline{y}\in\underline{A}}1_{Y(\underline{y})=0}\Big)+||f||_\infty\frac{(3L)^d}{n}\mathcal{P}\big(C_n,B(x,r+d)\big)\,.$$

Letting $c(f,r)=||f||_\infty(\mathcal{L}^d(B(x,r))+3^d)$, we obtain

$$\begin{aligned}&P\big(|\mathcal{C}_n(f)-\widetilde{\mathcal{C}}_n(f)|\geq 7c(f,r)\varepsilon\big)\\ &\qquad\leq P\Big(\frac{1}{|\underline{A}|}\sum_{\underline{y}\in\underline{A}}1_{Y(\underline{y})=0}\geq\varepsilon\Big)+P\Big(\frac{L^d}{n}\mathcal{P}\big(C_n,B(x,r+d)\big)\geq\varepsilon\Big)\,.\end{aligned}$$

Let $\Lambda$ be a box such that $B(x,r)\subset\Lambda\subset B(x,dr)$ and let

$$\underline{\Lambda}=\{\,\underline{y}\in\mathbb{Z}^d:B_n(\underline{y})\subset\Lambda\,\}\,.$$

We have then

$$\frac{1}{|\underline{A}|}\sum_{\underline{y}\in\underline{A}}1_{Y(\underline{y})=0}\;\leq\;\frac{(2d)^d}{|\underline{\Lambda}|}\sum_{\underline{y}\in\underline{\Lambda}}1_{Y(\underline{y})=0}\,.$$

Corollary 9.6 implies that for $L$ large enough the block process satisfies the condition stated at the beginning of section 9.4 with $\delta=\varepsilon/(2d)^{d+1}$. Using the estimates of lemmas 9.11 and 16.1, we conclude that there exist $b,c>0$ and $n_0$ depending only on $f$ and $r$ such that for $n\geq\max(n_0,K)$,

$$P\big(|\mathcal{C}_n(f1_{B(x,r)}) - \widetilde{\mathcal{C}}_n(f1_{B(x,r)})| \geq 7c(f,r)\varepsilon\big)$$
$$\leq 3^d \exp\Big( -\Lambda^*\Big(\frac{\varepsilon}{(2d)^d}, \frac{\varepsilon}{(2d)^{d+1}}\Big) \Big\lfloor \frac{n^d \mathcal{L}^d(\Lambda)}{(6L)^d} \Big\rfloor \Big) + b\exp\Big( -c\varepsilon\Big(\frac{n}{L}\Big)^d \Big).$$

Moreover the constants appearing in the above inequality do not depend on $x$. We have thus proved the following result.

**Lemma 16.2.** *Let $K$ be large enough, as in lemma 16.1. For any bounded uniformly continuous function $f : \mathbb{R}^d \to \mathbb{R}$, for any $r, \varepsilon > 0$, there exist a positive constant $c(f,r,\varepsilon)$ and an integer $n(f,r,\varepsilon)$ such that*

$$\forall x \in \mathbb{R}^d \quad \forall n \geq n(f,r,\varepsilon)$$
$$P\big(|\mathcal{C}_n(f1_{B(x,r)}) - \widetilde{\mathcal{C}}_n(f1_{B(x,r)})| \geq \varepsilon\big) \leq c(f,r,\varepsilon)\exp\Big( -c(f,r,\varepsilon)\frac{n^d}{(K\ln n)^d}\Big).$$

Lemma 16.2 implies in particular that for any continuous function $f$ having a compact support,

$$\forall \varepsilon > 0 \qquad \limsup_{n\to\infty} \frac{1}{n^{d-1}} \ln P\big(|\mathcal{C}_n(f) - \widetilde{\mathcal{C}}_n(f)| > \varepsilon\big) = -\infty.$$

This estimate and lemma 16.1 imply the $\mathcal{J}$–tightness announced at the beginning of the chapter.

## 16.3 Local upper bound

The final ingredient needed to get the large deviation upper bound is the local large deviation upper bound, which we prove in this section.

**Lemma 16.3.** *Let $\nu \in \mathcal{M}(\mathbb{R}^d)$ be such that $\mathcal{J}(\nu) < \infty$. For every $\varepsilon > 0$, there exists a weak neighbourhood $\mathcal{U}$ of $\nu$ in $\mathcal{M}(\mathbb{R}^d)$ such that*

$$\limsup_{n\to\infty} \frac{1}{n^{d-1}} \ln P\big(\mathcal{C}_n \in \mathcal{U}\big) \leq -(1-\varepsilon)\,\mathcal{J}(\nu).$$

*Proof.* By definition of $\mathcal{J}$, since $\mathcal{J}(\nu) < \infty$, there exists a Borel subset $A$ of $\mathbb{R}^d$ such that $\nu$ is the measure with density $\theta\, 1_A$ with respect to the Lebesgue measure and $\mathcal{J}(\nu) = \mathcal{I}(A)$. If $\mathcal{I}(A) = 0$, there is nothing to prove. Suppose that $\mathcal{I}(A) > 0$. Let $\varepsilon > 0$ and let us set $\varepsilon' = \varepsilon(1 + 1/\mathcal{I}(A))^{-1}$. Pick $\delta_0 \in \,]0,1[$ such that $c\sqrt{2\delta_0} < \varepsilon'$ where $c = c(p,d,\zeta)$ is the constant appearing in the interface lemma 12.1. Let $B(x_i, r_i)$, $i \in I$, be a finite collection of disjoint balls associated with $A, \varepsilon'$ and $\delta_0/8$, as given in the covering lemma 14.6. Since $\mathcal{I}(A) > 0$, we can assume that there are at least two balls in the collection. For $i$ in $I$, let $f_i, g_i$ be two continuous functions having compact support and taking values in $[0,1]$ such that

$$\forall x \in \mathbb{R}^d \setminus \overset{\circ}{B}_-(x_i, r_i, \nu_A(x_i)) \qquad f_i(x) = 0\,,$$
$$\forall x \in \overline{B}_+(x_i, r_i, \nu_A(x_i)) \qquad g_i(x) = 1\,,$$
$$\Big(\frac{1}{2} - \frac{\delta_0}{8\theta}\Big)\alpha_d r_i^d \le \int f_i\, d\mathcal{L}^d \le \int g_i\, d\mathcal{L}^d \le \Big(\frac{1}{2} + \frac{\delta_0}{8\theta}\Big)\alpha_d r_i^d\,.$$

We have then

$$\nu(f_i) = \theta \int_A f_i\, d\mathcal{L}^d \ge \theta \int f_i\, d\mathcal{L}^d - \mathcal{L}^d(B_-(x_i, r_i, \nu_A(x_i)) \setminus A)$$
$$\ge \Big(\frac{\theta}{2} - \frac{\delta_0}{4}\Big)\alpha_d r_i^d\,,$$
$$\nu(g_i) = \theta \int_A g_i\, d\mathcal{L}^d \le \theta \int g_i\, d\mathcal{L}^d - \theta\mathcal{L}^d(B_+(x_i, r_i, \nu_A(x_i)) \setminus A) \le \frac{\delta_0}{4}\alpha_d r_i^d\,.$$

Let $\mathcal{U}$ be the weak neighbourhood of $\nu$ in $\mathcal{M}(\mathbb{R}^d)$ defined by

$$\mathcal{U} = \Big\{\, \varrho \in \mathcal{M}(\mathbb{R}^d) : \forall i \in I \quad \varrho(f_i) > \nu(f_i) - \frac{\delta_0}{4}\alpha_d r_i^d,\ \ \varrho(g_i) < \nu(g_i) + \frac{\delta_0}{4}\alpha_d r_i^d \Big\}.$$

Suppose that $\mathcal{C}_n \in \mathcal{U}$. Let us fix $i \in I$. The intersection of $C(0)$ with the ball $B(nx_i, nr_i)$ splits into a collection $\mathcal{C}(i)$ of $B(nx_i, nr_i)$–clusters which all intersect the boundary $\partial^{in} B(nx_i, nr_i)$. Setting

$$B_-(n,i) = B_-(nx_i, nr_i, \nu_A(x_i))\,, \quad B_+(n,i) = B_+(nx_i, nr_i, \nu_A(x_i))\,,$$

we have also

$$\mathcal{C}_n(f_i) \le \sum_{C \in \mathcal{C}(i)} \big|C \cap \overset{\circ}{B}_-(n,i)\big| \le \sum_{C \in \mathcal{C}(i)} \big|C \cap B_-(n,i)\big|\,,$$
$$\mathcal{C}_n(g_i) \ge \sum_{C \in \mathcal{C}(i)} \big|C \cap \overline{B}_+(n,i)\big| \ge \sum_{C \in \mathcal{C}(i)} \big|C \cap B_+(n,i)\big|\,.$$

The very definition of the neighbourhood $\mathcal{U}$ and these inequalities yield that the event $\{\, \mathcal{C}_n \in \mathcal{U}\,\}$ is included in the event $\mathrm{Sep}_\theta(n, x_i, r_i, \nu_A(x_i), \delta_0)$ defined before lemma 12.2. We conclude that

$$P(\mathcal{C}_n \in \mathcal{U}) \le P\Big(\bigcap_{i \in I} \mathrm{Sep}_\theta(n, x_i, r_i, \nu_A(x_i), \delta_0)\Big)\,.$$

Yet the balls $B(x_i, r_i)$, $i \in I$, are compact and disjoint, hence the events $\mathrm{Sep}_\theta(n, x_i, r_i, \nu_A(x_i), \delta_0)$, $i \in I$, are independent. Applying the interface lemma 12.1 together with lemma 12.2, we get

$$\limsup_{n\to\infty} \frac{1}{n^{d-1}} \ln P(\mathcal{C}_n \in \mathcal{U}) \le -\sum_{i \in I} \alpha_{d-1} r_i^{d-1} \tau(\nu_A(x_i))\,(1 - c\sqrt{2\delta_0})$$
$$\le -\mathcal{I}(A)\,(1-\varepsilon') + \varepsilon' = -\mathcal{J}(\nu)(1-\varepsilon)$$

and we are done. □

## 16.4 Lower bound

We finish the proof of the large deviation principle with the large deviation lower bound.

**Lemma 16.4.** *Let $d \geq 3$ and let $\nu \in \mathcal{M}(\mathbb{R}^d)$. For any weak neighbourhood $\mathcal{U}$ of $\nu$ in $\mathcal{M}(\mathbb{R}^d)$, we have*

$$\liminf_{n\to\infty} \frac{1}{n^{d-1}} \ln P\big(\mathcal{C}_n \in \mathcal{U}\big) \geq -\mathcal{J}(\nu) .$$

*In two dimensions, the same bound holds for $\nu \in \mathcal{M}(\mathbb{R}^2)$ of the form $\nu = \theta 1_A$, where $A$ is an open connected set containing* 0.

*Proof.* If $\mathcal{J}(\nu) = +\infty$, there is nothing to prove. Let $\nu \in \mathcal{M}(\mathbb{R}^d)$ be such that $\mathcal{J}(\nu) < +\infty$. By definition of $\mathcal{J}$, there exists a Borel subset $A$ of $\mathbb{R}^d$ such that $\nu$ is the measure with density $\theta\, 1_A$ with respect to the Lebesgue measure and $\mathcal{J}(\nu) = \mathcal{I}(A)$. We consider first the case where $\mathcal{L}^d(A) < +\infty$. Let $\mathcal{U}$ be a weak neighbourhood of $\nu$ and let $\varepsilon > 0$. By proposition 14.9, there exists a polyhedral set $D$ such that the measure $\psi$ with density $\theta\, 1_D$ with respect to the Lebesgue measure belongs to $\mathcal{U}$ and moreover $\mathcal{I}(D) \leq \mathcal{I}(A) + \varepsilon$. We can even assume that the set $D$ is connected and contains 0: in dimensions $d \geq 3$, we can connect together the components of $D$ and a neighbourhood of 0 with thin cylinders having small surface energy, while in dimensions $d = 2$, we suppose from the beginning that $A$ is an open connected neighbourhood of 0, and these properties can be conserved when building the approximating polyhedral set $D$. By definition of a polyhedral element, $\partial D$ is the union of a finite number of $d-1$ dimensional sets $F_1, \dots, F_s$. For $j \in \{1 \cdots s\}$, let $\nu_j$ be a unit vector normal to $F_j$. We have

$$\sum_{1\leq j\leq s} \mathcal{H}^{d-1}(F_j)\tau(\nu_j) \leq \mathcal{I}(A) + \varepsilon .$$

Moreover, for each $j$ in $\{1 \cdots s\}$, the relative boundary $\partial F_j$ has finite $d-2$ dimensional Hausdorff measure (we can achieve this by a slight perturbation of the polyhedral set if necessary).

Let $f \in C_c(\mathbb{R}^d, \mathbb{R})$. We are going to estimate the probability that $\mathcal{C}_n(f)$ and $\psi(f)$ are close. Let $\varepsilon > 0$. Since $f$ is continuous and has a compact support, it is uniformly continuous. Let $\delta \in ]0,1[$ be small enough so that

$$\mathcal{L}^d\big(\mathcal{V}(\partial D, 4d\delta)\big) \leq \varepsilon ,$$

$$\forall x, y \in \mathbb{R}^d \qquad |x-y| \leq \delta \quad \Rightarrow \quad |f(x) - f(y)| \leq \varepsilon ||f||_\infty .$$

We work with the lattice rescaled by a factor $L = \lfloor \delta n \rfloor$. For $\underline{y} \in \mathbb{Z}^d$, the block variable $Y(\underline{y})$ is the indicator function of the event $RVW(\underline{B}(\underline{y}), B'(\underline{y}), L, \varepsilon)$ (see the definition before corollary 9.6). For $\delta$ small enough and $n$ large enough, there exists a connected set $\underline{E}$ containing $\underline{0}$ and such that

$$\mathrm{dist}\Big(\bigcup_{\underline{x}\in\underline{E}} B_n(\underline{x}),\mathbb{R}^d\setminus D\Big)\geq 2d\delta\,,\quad \mathcal{L}^d\Big(D\setminus\bigcup_{\underline{x}\in\underline{E}} B_n(\underline{x})\Big)\leq\varepsilon\,,$$

and moreover $|\underline{E}|\leq c$ where $c=c(\delta)$ is a constant independent of $n$. Let $\mathcal{E}$ be the intersection of the events

$$\{\,0\longleftrightarrow\partial^{in}B'(\underline{0})\,\}\,,\qquad\{\,Y(\underline{x})=1\,\}\,,\ \underline{x}\in\underline{E}\,,\qquad \mathrm{wall}\,(F_j,n)\,,\ 1\leq j\leq s\,.$$

See lemma 12.3 for the definition of the event "wall". Let us estimate the probability of $\mathcal{E}$. Since $\mathbb{R}^d\setminus D$ is at a distance larger than $d\delta$ from the blocks $B_n(\underline{y})$, $\underline{y}\in\underline{E}$, the events $\mathrm{wall}\,(F_j,n)$, $1\leq j\leq s$, are independent from the configuration inside the blocks whose index is in $\underline{E}$. Therefore

$$P(\mathcal{E})\geq P\Big(\{\,0\longleftrightarrow\partial^{in}B'(\underline{0})\,\}\cap\bigcap_{\underline{x}\in\underline{E}}\{\,Y(\underline{x})=1\,\}\Big)P\Big(\bigcap_{1\leq j\leq s}\mathrm{wall}\,(F_j,n)\Big)\,.$$

From corollary 9.6, the probability $P(Y(\underline{x})=1)$ goes to 1 as $n$ goes to $\infty$. Moreover $P(0\longleftrightarrow\partial^{in}B'(\underline{0}))\geq\theta$, therefore

$$\liminf_{n\to\infty}P\Big(\{\,0\longleftrightarrow\partial^{in}B'(\underline{0})\,\}\cap\bigcap_{\underline{x}\in\underline{E}}\{\,Y(\underline{x})=1\,\}\Big)\geq\theta\,.$$

By the FKG inequality and lemma 12.3,

$$\liminf_{n\to\infty}\frac{1}{n^{d-1}}\ln P\Big(\bigcap_{1\leq j\leq s}\mathrm{wall}\,(F_j,n)\Big)\geq-\sum_{1\leq j\leq s}\mathcal{H}^{d-1}(F_j)\tau(\nu_j)\,.$$

Combining the previous inequalities, we get

$$\liminf_{n\to\infty}\frac{1}{n^{d-1}}\ln P(\mathcal{E})\geq-\mathcal{I}(A)-\varepsilon\,.$$

Suppose that $\mathcal{E}$ occurs. Let

$$\underline{A}=\{\,\underline{y}\in\mathbb{Z}^d:B_n(\underline{y})\cap\mathrm{supp}\,f\neq\emptyset\,\}\,.$$

Since $\delta<1$, each block of size $L$ which intersects $n\mathrm{supp}\,f$ is included in $n\mathcal{V}(\mathrm{supp}\,f,d)$, thus $|\underline{A}|L^d\leq n^d\mathcal{L}^d(\mathcal{V}(\mathrm{supp}\,f,d))$. To evaluate $|\mathcal{C}_n(f)-\psi(f)|$, we decompose it on blocks of size $L/n$ and we use the uniform continuity of $f$:

$$\begin{aligned}|\mathcal{C}_n(f)-\psi(f)|&\leq\sum_{\underline{y}\in\underline{A}}\Big|\int_{B_n(\underline{y})}f\,d\mathcal{C}_n-\int_{B_n(\underline{y})}f\,d\psi\Big|\\&\leq 2\varepsilon||f||_\infty\mathcal{L}^d(\mathcal{V}(\mathrm{supp}\,f,d))+||f||_\infty\sum_{\underline{y}\in\underline{A}}|\mathcal{C}_n(B_n(\underline{y}))-\psi(B_n(\underline{y}))|\,.\end{aligned}$$

Next we study the sum in the above inequality. Let $\underline{y}\in\underline{A}$. We distinguish several cases:

• If $\underline{y} \in \underline{E}$ then $B_n(\underline{y}) \subset D$ and

$$\big|\mathcal{C}_n(B_n(\underline{y})) - \psi(B_n(\underline{y}))\big| = \Big|\mathcal{C}_n(B_n(\underline{y})) - \frac{\theta}{n^d}|B_n(\underline{y})|\Big| \leq \frac{\varepsilon}{n^d}\big|B_n(\underline{y})\big|.$$

• If $\mathrm{dist}(B_n(\underline{y}), D) > 2d\delta$, then both terms $\mathcal{C}_n(B_n(\underline{y}))$ and $\psi(B_n(\underline{y}))$ vanish. Indeed, the occurrence of $\mathcal{E}$ precludes the existence of an open path from 0 to $\mathbb{Z}^d \setminus \mathcal{V}(D, 2d\delta)$, hence $\mathcal{C}_n(B_n(\underline{y})) = 0$.

• If $\mathrm{dist}(B_n(\underline{y}), D) \leq 2d\delta$ and $\underline{y} \notin \underline{E}$, then $B_n(\underline{y})$ is included in the set

$$\mathcal{V}(\partial D, 4d\delta) \cup D \setminus \bigcup_{\underline{x} \in \underline{E}} B_n(\underline{x})$$

whose Lebesgue measure is less than $2\varepsilon$.

On the event $\mathcal{E}$, we thus have

$$|\mathcal{C}_n(f) - \psi(f)| \leq 5\varepsilon ||f||_\infty \mathcal{L}^d(\mathcal{V}(\mathrm{supp}\, f, d)).$$

These estimates can be carried out simultaneously for a finite number of functions $f_1, \dots, f_r \in C_c(\mathbb{R}^d, \mathbb{R})$: there exists $c > 0$ (depending on $f_1, \dots, f_r$) such that, for any $\varepsilon > 0$, there exists an event $\mathcal{E}$ satisfying

$$\mathcal{E} \subset \Big\{\forall i \in \{1 \cdots r\} \quad |\mathcal{C}_n(f_i) - \psi(f_i)| \leq c\varepsilon\Big\},$$

$$\liminf_{n \to \infty} \frac{1}{n^{d-1}} \ln P(\mathcal{E}) \geq -\mathcal{J}(\nu) - \varepsilon.$$

Yet $\mathcal{U}$ is a weak neighbourhood of $\psi$, hence there exist $f_1, \dots, f_r \in C_c(\mathbb{R}^d, \mathbb{R})$ and $\varepsilon > 0$ such that

$$\Big\{\varrho \in \mathcal{M}(\mathbb{R}^d) : \forall i \in \{1 \cdots r\} \quad |\varrho(f_i) - \psi(f_i)| \leq c\varepsilon\Big\} \subset \mathcal{U}.$$

Therefore

$$\liminf_{n \to \infty} \frac{1}{n^{d-1}} \ln P\big(\mathcal{C}_n \in \mathcal{U}\big) \geq -\mathcal{J}(\nu) - \varepsilon$$

and we conclude by sending $\varepsilon$ to 0.

To finish, let us explain briefly how to handle the case where $\mathcal{L}^d(A) = +\infty$. In this situation, the polyhedral set $D$ approximating $A$ is such that $\mathbb{R}^d \setminus D$ is bounded. The main difference with the previous case is that now we would like that the cluster $C(0)$ fills the complement of $D$ instead of $D$. We can restrict ourselves to a compact set because the test functions have compact support. Let $D$ be a polyhedral set approximating $A$ as above. Let $f \in C_c(\mathbb{R}^d, \mathbb{R})$. Let $\Lambda$ be a box containing simultaneously the set $\mathbb{R}^d \setminus D$ and the support of the function $f$. Let $\Lambda' = \mathcal{V}(\Lambda, 1)$. For $\delta$ small enough and $n$ large enough, there exists a connected set $\underline{E}$ containing $\underline{0}$ and such that

$$\mathrm{dist}\Big(\bigcup_{\underline{x}\in\underline{E}} B_n(\underline{x}), D\Big) \geq 2d\delta\,, \quad \mathcal{L}^d\Big((\Lambda' \setminus D) \setminus \bigcup_{\underline{x}\in\underline{E}} B_n(\underline{x})\Big) \leq \varepsilon\,,$$

and moreover $|\underline{E}| \leq c$ where $c = c(\delta)$ is a constant independent of $n$. The rest of the argument is similar to the first case, except that one should replace $D$ by its complement $\mathbb{R}^d \setminus D$. $\square$

# 17

# Enhanced upper bound

This chapter is devoted to the proof of the enhanced large deviation upper bound stated at the end of theorem 7.3. The main difficulty to prove the enhanced upper bound is to recover some compactness, because we are working in infinite volume. On the event $\{ n^d < |C(0)| < \infty \}$, we cannot restrict the analysis to a box of linear diameter cte$\times n$, up to surface large deviation order. However, we control in section 17.3 the perimeter of the set $C_n$ and we prove that, up to surface large deviation order, the cluster $C(0)$ is contained in the ball $B(0, an^{d-1})$ for some $a > 0$. With the help of the lemma stated in the next section 17.1, we can cover $C(0)$ up to a small fractional volume by a finite number of random disjoint balls having linear diameter cte$\times n$. The volume of $C(0)$ outside these balls is controlled thanks to a local isoperimetric inequality and the bound on the perimeter of $C_n$. We condition on the collection of the random balls, and we prove in section 17.2 a large deviation principle for the random measure $\mathcal{C}_n$ restricted to these balls, which is in addition uniform with respect to the choice of the balls. Putting all these elements together, we conclude the proof in section 17.3.

## 17.1 A lemma from discrete geometry

**Lemma 17.1.** *Let $X$ be a finite subset of $\mathbb{Z}^d$. There exists a subset $E$ of $X \times \{1, \dots, 3^{|X|}\}$ such that $|E| \leq |X|$ and*

- $\forall (a,r) \in E \quad B(a,r) \cap X \neq \emptyset\,,$
- $\bigcup_{x \in X} B(x,1) \subset \bigcup_{(a,r) \in E} B(a,r)\,,$
- $\forall (a,r),\, (b,s) \in E \quad (a,r) \neq (b,s) \Rightarrow B(a,r+1) \cap B(b,s+1) = \emptyset\,.$

*Proof.* We use an algorithm to build the set $E$ starting from $X$. The algorithm works with a sequence of subsets of $X \times \{ 1, \dots, 3^{|X|} \}$ which might violate only the last condition and it stops when this condition is fulfilled. We initialize the algorithm with the set $E(1) = \{ (a,1) : a \in X \}$. The only condition which might be violated by $E(1)$ is the last one. Suppose that $E(h)$ has been built for some $h \geq 1$. If $E(h)$ answers the problem, the algorithm stops. Otherwise the last condition is violated: there exist $(a,r)$ and $(b,s)$ in $E(h)$ such that $B(a,r+1) \cap B(b,s+1) \neq \emptyset$. We define then

$$E(h+1) = \{ (a, r+2s+2) \} \cup E(h) \setminus \{ (a,r), (b,s) \} .$$

Obviously $|E(h+1)| \leq |E(h)|-1$ and $E(h+1)$ satisfies the first two conditions when $E(h)$ does. Moreover

$$\max\{ r : (a,r) \in E(h+1) \} \leq 3\max\{ r : (a,r) \in E(h) \} + 2 .$$

Necessarily the algorithm stops at some step $h$ less than $|X|$. The final set $E(h)$ answers the problem. □

## 17.2 Uniform large deviation upper bounds

We introduce some notation necessary to state the next lemma. For $U$ an open subset of $\mathbb{R}^d$, we denote by $\mathcal{B}(U)$ the collection of the Borel subsets of $U$ and by $\mathcal{M}(U)$ the vector space of the signed Borel measures on $\mathcal{B}(U)$. Let $m$ be an integer and let $U_1, \dots, U_m$ be $m$ open balls in $\mathbb{R}^d$. We consider the product space $\mathcal{M}(U_1) \times \cdots \times \mathcal{M}(U_m)$ endowed with the product of the weak topologies of each space $\mathcal{M}(U_1), \dots, \mathcal{M}(U_m)$. We define a map $\mathcal{J}^m : \mathcal{M}(U_1) \times \cdots \times \mathcal{M}(U_m) \to [0, +\infty]$ by setting

$$\mathcal{J}^m(\nu_1, \dots, \nu_m) = \mathcal{I}(E_1, U_1) + \cdots + \mathcal{I}(E_m, U_m)$$

if $\nu_1, \dots, \nu_m$ are the measures with densities $\theta 1_{E_1}, \dots, \theta 1_{E_m}$ with respect to the Lebesgue measure, where $E_1 \in \mathcal{B}(U_1), \dots, E_m \in \mathcal{B}(U_m)$; otherwise, we set $\mathcal{J}^m(\nu_1, \dots, \nu_m) = +\infty$. For any open set $O$, the map $E \in \mathcal{B}(O) \mapsto \mathcal{I}(E, O)$ is lower semicontinuous (see section 14.2), therefore $\mathcal{J}^m$ is also lower semicontinuous. Since in addition the sets $U_1, \dots, U_m$ are bounded, the compactness result theorem 14.5 implies that the level sets of $\mathcal{J}^m$ are compact, hence $\mathcal{J}^m$ is a good rate function on the space $\mathcal{M}(U_1) \times \cdots \times \mathcal{M}(U_m)$ endowed with the product of the weak topologies.
Let $T(U_1, \dots, U_m)$ be the set of the $m$–uples of integer translations sending the sets $U_1, \dots, U_m$ onto pairwise disjoint sets, that is,

$$T(U_1, \dots, U_m) = \\ \{ (x_1, \dots, x_m) \in (\mathbb{Z}^d)^m : x_1 + U_1, \dots, x_m + U_m \text{ are pairwise disjoint} \} .$$

For $\nu \in \mathcal{M}(\mathbb{R}^d)$, $U \in \mathcal{B}(\mathbb{R}^d)$ and $x \in \mathbb{R}^d$, we denote by $\nu|_{U+x}(\cdot + x)$ the measure obtained as the restriction of $\nu$ to $U + x$ and translated by $-x$, defined by

$$\forall E \in \mathcal{B}(\mathbb{R}^d) \qquad \nu|_{U+x}(E + x) = \nu\big((E + x) \cap (U + x)\big) .$$

To $U_1,\dots,U_m$ and $(x_1,\dots,x_m)$ in $T(U_1,\dots,U_m)$ we associate the map $\phi^{U_1,\dots,U_m}_{x_1,\dots,x_m}$ from $\mathcal{M}(\mathbb{R}^d)$ to $\mathcal{M}(U_1)\times\cdots\times\mathcal{M}(U_m)$ defined by

$$\forall \nu\in\mathcal{M}(\mathbb{R}^d)\qquad \phi^{U_1,\dots,U_m}_{x_1,\dots,x_m}(\nu)\,=\,\big(\nu|_{U_1+x_1}(\cdot+x_1),\dots,\nu|_{U_m+x_m}(\cdot+x_m)\big)\,.$$

We denote by $|\cdot|_1$ the norm $\big|(x_1,\dots,x_m)\big|_1\,=\,|x_1|+\cdots|x_m|\,.$

**Lemma 17.2.** **(Contiguity)** *For any uniformly continuous bounded function $f$ defined on $\mathbb{R}^d$ and $\delta>0$,*

$$\limsup_{n\to\infty}\frac{1}{n^{d-1}}\ln\sup_{x_1,\dots,x_m}P\Big(\big|\phi^{U_1,\dots,U_m}_{x_1,\dots,x_m}(\mathcal{C}_n)(f)-\phi^{U_1,\dots,U_m}_{x_1,\dots,x_m}(\widetilde{\mathcal{C}}_n)(f)\big|_1\geq\delta\Big)=-\infty,$$

*where the supremum is taken over $(x_1,\dots,x_m)$ in $T(U_1,\dots,U_m)$.*

*Proof.* We will apply lemma 16.2. For any $(x_1,\dots,x_m)$ in $T(U_1,\dots,U_m)$, for any $\delta>0$ and for $n$ large enough (depending only on $f$ and $\delta$),

$$\begin{aligned}
P\Big(\big|\phi^{U_1,\dots,U_m}_{x_1,\dots,x_m}(\mathcal{C}_n)(f)-\phi^{U_1,\dots,U_m}_{x_1,\dots,x_m}(\widetilde{\mathcal{C}}_n)(f)\big|_1\geq\delta\Big)&\\
=P\Big(\sum_{1\leq i\leq m}\Big|\int_{x_i+U_i}f(x)\,d\mathcal{C}_n(x)-\int_{x_i+U_i}f(x)\,d\widetilde{\mathcal{C}}_n(x)\Big|\geq\delta\Big)&\\
\leq\sum_{1\leq i\leq m}P(|\mathcal{C}_n(f1_{x_i+U_i})-\widetilde{\mathcal{C}}_n(f1_{x_i+U_i})|\geq\frac{\delta}{m})&\\
\leq\sum_{1\leq i\leq m}c(f,r_i,\frac{\delta}{m})\exp\Big(-c(f,r_i,\frac{\delta}{m})\exp\frac{n^d}{(K\ln n)^d}\Big)&,
\end{aligned}$$

where $r_i$ is the radius of the ball $U_i$ for $1\leq i\leq m$. The last inequalities hold as soon as $n\geq\max_{1\leq i\leq m}n(f,r_i,\delta/m)$. The constants $n(f,r_i,\delta/m)$, $c(f,r_i,\delta/m)$ are those appearing in lemma 16.2, they do not depend on $(x_1,\dots,x_m)$ in $T(U_1,\dots,U_m)$. □

**Lemma 17.3.** **($\mathcal{J}^m$–tightness)** *There exists a positive constant c such that*

$$\forall\lambda>0\qquad\limsup_{n\to\infty}\frac{1}{n^{d-1}}\ln\sup_{x_1,\dots,x_m}P\Big(\mathcal{J}^m\Big(\phi^{U_1,\dots,U_m}_{x_1,\dots,x_m}(\widetilde{\mathcal{C}}_n)\Big)\geq\lambda\Big)\,\leq\,-c\lambda\,,$$

*where the supremum is taken over $(x_1,\dots,x_m)$ in $T(U_1,\dots,U_m)$.*

*Proof.* Using the inequalities stated before proposition 14.3 and lemma 16.1,

$$\begin{aligned}
P\Big(\mathcal{J}^m\Big(\phi^{U_1,\dots,U_m}_{x_1,\dots,x_m}(\widetilde{\mathcal{C}}_n)\Big)\geq\lambda\Big)\,&\leq\,P\Big(\sum_{1\leq i\leq m}\mathcal{I}(C_n,x_i+U_i)\geq\lambda\Big)\\
&\leq\sum_{1\leq i\leq m}P\Big(\mathcal{I}(C_n,x_i+U_i)\geq\frac{\lambda}{m}\Big)\\
&\leq\sum_{1\leq i\leq m}P\Big(\mathcal{P}(C_n,x_i+U_i)\geq\frac{\lambda}{m||\tau||_\infty}\Big)\\
&\leq\,mb\exp\Big(-cn^{d-1}\frac{\lambda}{m||\tau||_\infty}\Big)\,,
\end{aligned}$$

where the last inequality holds for $K$ large enough and for $n$ larger than a value depending only on $K$ and the geometries of the sets $U_1, \cdots U_m$ (but not on $x_1, \ldots, x_m$). Thus these bounds are independent of $(x_1, \ldots, x_m)$ in $T(U_1, \ldots, U_m)$. □

**Lemma 17.4. (Local estimate)** *Let $(\nu_1, \ldots, \nu_m) \in \mathcal{M}(U_1) \times \cdots \times \mathcal{M}(U_m)$ be such that $\mathcal{J}^m(\nu_1, \ldots, \nu_m) < \infty$. For any positive $\varepsilon$, there exists a neighbourhood $\mathcal{U}$ of $(\nu_1, \ldots, \nu_m)$ such that*

$$\limsup_{n\to\infty} \frac{1}{n^{d-1}} \ln \sup_{x_1,\ldots,x_m} P\Big(\phi^{U_1,\ldots,U_m}_{x_1,\ldots,x_m}(\mathcal{C}_n) \in \mathcal{U}\Big) \le -\mathcal{J}^m(\nu_1, \ldots, \nu_m) + \varepsilon ,$$

*where the supremum is taken over $(x_1, \ldots, x_m)$ in $T(U_1, \ldots, U_m)$.*

*Proof.* We need only to consider the case where $\mathcal{J}^m(\nu_1, \ldots, \nu_m) > 0$. By the definition of $\mathcal{J}^m$, there exist $E_1 \in \mathcal{B}(U_1), \ldots, E_m \in \mathcal{B}(U_m)$ such that $\nu_1, \ldots, \nu_m$ are the measures with densities $\theta 1_{E_1}, \ldots, \theta 1_{E_m}$ with respect to the Lebesgue measure. Let $\varepsilon$ be positive. Let $\delta_0 \in ]0,1[$ be such that $c\sqrt{2\delta_0} < \varepsilon$ where $c = c(p, d, \zeta)$ is the constant appearing in the interface lemma 12.1. Let $l$ belong to $\{1, \ldots, m\}$. By lemma 14.6 applied to the set $E_l$ and the open set $U_l$, there exists a finite collection of disjoint balls $B(x_i^l, r_i^l)$, $i \in I(l)$, such that: for any $i$ in $I(l)$, $x_i^l$ belongs to $\partial^* E_l \cap U_l$, $r_i^l$ belongs to $]0,1[$, $B(x_i^l, r_i^l)$ is included in $U_l$,

$$\mathcal{L}^d\big((E_l \cap B(x_i^l, r_i^l)) \Delta B_-(x_i^l, r_i^l, \nu_{E_l}(x_i^l))\big) \le \delta_0\, \alpha_d (r_i^l)^d/8 ,$$

and moreover,

$$\Big|\mathcal{I}(E_l, U_l) - \sum_{i\in I(l)} \alpha_{d-1}(r_i^l)^{d-1}\tau(\nu_{E_l}(x_i^l))\Big| \le \varepsilon .$$

If $\mathcal{I}(E_l, U_l) = 0$, we simply take the empty collection, so that $I(l) = \emptyset$; if $\mathcal{I}(E_l, U_l) > 0$, we can assume that there are at least two balls in the collection. For $i$ in $I(l)$, let $f_i^l, g_i^l$ be two continuous functions having compact support included in $U_l$ and taking values in $[0,1]$ such that

$$\begin{gathered}
\forall x \in \mathbb{R}^d \setminus \overset{\circ}{B}_-(x_i^l, r_i^l, \nu_{E_l}(x_i^l)) \qquad f_i^l(x) = 0 ,\\
\forall x \in \overline{B}_+(x_i^l, r_i^l, \nu_{E_l}(x_i^l)) \qquad g_i^l(x) = 1 ,\\
\Big(\frac{1}{2} - \frac{\delta_0}{8\theta}\Big)\alpha_d (r_i^l)^d \le \int f_i^l \, d\mathcal{L}^d \le \int g_i^l \, d\mathcal{L}^d \le \Big(\frac{1}{2} + \frac{\delta_0}{8\theta}\Big)\alpha_d (r_i^l)^d .
\end{gathered}$$

We have then

$$\begin{aligned}
\nu(f_i^l) &= \theta \int_{E_l} f_i^l \, d\mathcal{L}^d \ge \Big(\frac{\theta}{2} - \frac{\delta_0}{4}\Big)\alpha_d (r_i^l)^d ,\\
\nu(g_i^l) &= \theta \int_{E_l} g_i^l \, d\mathcal{L}^d \le \frac{\delta_0}{4}\alpha_d (r_i^l)^d .
\end{aligned}$$

Let $\mathcal{U}(l)$ be the weak neighbourhood of $\nu_l$ in $\mathcal{M}(U_l)$ defined by

$$\mathcal{U}(l) = \Big\{ \varrho \in \mathcal{M}(\mathbb{R}^d) : \forall i \in I(l) \quad \varrho(f_i^l) > \nu(f_i^l) - \frac{\delta_0}{4}\alpha_d(r_i^l)^d\,, \\ \varrho(g_i^l) < \nu(g_i^l) + \frac{\delta_0}{4}\alpha_d(r_i^l)^d \Big\}.$$

Let us set $\mathcal{U} = \mathcal{U}_1 \times \cdots \times \mathcal{U}_m$. Suppose that $\phi^{U_1,\dots,U_m}_{x_1,\dots,x_m}(\mathcal{C}_n) \in \mathcal{U}$. Let $l$ belong to $\{1,\dots,m\}$ be such that $\mathcal{I}(E_l, U_l) > 0$ and let $i$ belong to $I(l)$. The intersection of $C(0)$ with the ball $B(nx_l + nx_i^l, nr_i^l)$ splits into a collection $\mathcal{C}(i,l)$ of $B(nx_l + nx_i^l, nr_i^l)$–clusters which all intersect the boundary $\partial^{in} B(nx_l + nx_i^l, nr_i^l)$. Setting

$$B_-(n,i,l) = B_-(nx_l + nx_i^l, nr_i^l, \nu_{E_l}(x_i^l))\,,\\ B_+(n,i,l) = B_+(nx_l + nx_i^l, nr_i^l, \nu_{E_l}(x_i^l))\,,$$

we have also

$$\int f_i^l(x - x_l)\, d\mathcal{C}_n(x) \le \sum_{C \in \mathcal{C}(i,l)} \big|C \cap \mathring{B}_-(n,i,l)\big| \le \sum_{C \in \mathcal{C}(i,l)} \big|C \cap B_-(n,i,l)\big|\,,$$
$$\int g_i^l(x - x_l)\, d\mathcal{C}_n(x) \ge \sum_{C \in \mathcal{C}(i,l)} \big|C \cap \overline{B}_+(n,i,l)\big| \ge \sum_{C \in \mathcal{C}(i,l)} \big|C \cap B_+(n,i,l)\big|\,.$$

The very definition of the neighbourhood $\mathcal{U}_l$ and these inequalities yield that $\{\phi^{U_1,\dots,U_m}_{x_1,\dots,x_m}(\mathcal{C}_n) \in \mathcal{U}\}$ is included in the event $\mathrm{Sep}_\theta(n, x_l + x_i^l, r_i^l, \nu_{E_l}(x_i^l), \delta_0)$ defined before lemma 12.2. We conclude that

$$P\big(\phi^{U_1,\dots,U_m}_{x_1,\dots,x_m}(\mathcal{C}_n) \in \mathcal{U}\big) \le P\Big( \bigcap_{1 \le l \le m} \bigcap_{i \in I(l)} \mathrm{Sep}_\theta(n, x_l + x_i^l, r_i^l, \nu_{E_l}(x_i^l), \delta_0) \Big)\,.$$

Since the balls $B(x_i^l, r_i^l)$, $i \in I(l)$, are disjoint for any $l$ in $\{1,\dots,m\}$ and since $(x_1,\dots,x_m)$ belongs to $T(U_1,\dots,U_m)$, then the balls

$$B(nx_l + nx_i^l, nr_i^l)\,, \quad i \in I(l)\,, \quad 1 \le l \le m\,,$$

are still disjoint and the events

$$\mathrm{Sep}_\theta(n, x_l + x_i^l, r_i^l, \nu_{E_l}(x_i^l), \delta_0)\,, \quad i \in I(l)\,, \quad 1 \le l \le m\,,$$

are independent. Since the model is invariant under integer translations, then for $l$ in $\{1,\dots,m\}$ and $i$ in $I(l)$,

$$P\big(\mathrm{Sep}_\theta(n, x_l + x_i^l, r_i^l, \nu_{E_l}(x_i^l), \delta_0)\big) = P\big(\mathrm{Sep}_\theta(n, x_i^l, r_i^l, \nu_{E_l}(x_i^l), \delta_0)\big)\,.$$

We therefore obtain

$$P\big(\phi^{U_1,\dots,U_m}_{x_1,\dots,x_m}(\mathcal{C}_n) \in \mathcal{U}\big) \le \prod_{1 \le l \le m} \prod_{i \in I(l)} P\big(\mathrm{Sep}_\theta(n, x_i^l, r_i^l, \nu_{E_l}(x_i^l), \delta_0)\big)\,.$$

This last bound is independent of $(x_1,\dots,x_m)$ in $T(U_1,\dots,U_m)$. By the choice of $\delta_0$ (see the interface lemmas 12.1 and 12.2) and of the balls $B(x_i^l, r_i^l)$,

$$\limsup_{n\to\infty} \frac{1}{n^{d-1}} \ln \sup\Big\{ P(\phi^{U_1,\dots,U_m}_{x_1,\dots,x_m}(\mathcal{C}_n) \in \mathcal{U}) : (x_1,\dots,x_m) \in T(U_1,\dots,U_m) \Big\}$$
$$\leq - \sum_{1\leq l\leq m} \sum_{i\in I(l)} \tau(\nu_{E_l}(x_i^l))\, \alpha_{d-1}(r_i^l)^{d-1}(1-c\sqrt{2\delta_0})$$
$$\leq - \sum_{1\leq l\leq m} \Big( \mathcal{I}(E_l,U_l)\,(1-\varepsilon) - \varepsilon \Big)$$
$$= -\mathcal{J}^m(\nu_1,\dots,\nu_m)\,(1-\varepsilon) + m\varepsilon\,.$$

This is the upper bound we were seeking. □

Lemmas 17.2, 17.3, 17.4 together yield in a standard fashion a uniform large deviation upper bound for $\mathcal{C}_n$ in the weak topology. Unfortunately, this upper bound is not enough for our purpose. We will need a stronger estimate on $\widetilde{\mathcal{C}}_n$. We define the variation $|\nu|$ of a measure $\nu \in \mathcal{M}(\mathbb{R}^d)$ by

$$|\nu| = \sup\,\big\{\, \nu(f) : f \in C_c(\mathbb{R}^d,\mathbb{R}),\, ||f||_\infty \leq 1 \,\big\}\,.$$

This is a norm on the space of the measures. We call the associated topology the variation topology. Let $U$ be an open bounded subset of $\mathbb{R}^d$ and let $E, F$ be two Borel subsets of $U$. Let $\varrho$, $\nu$ be the measures having densities $\theta 1_E$, $\theta 1_F$ with respect to the Lebesgue measure. We have then $|\varrho-\nu| = \theta\mathcal{L}^d(E\Delta F)$. Therefore the space

$$\mathcal{M}^\theta_{\mathrm{ac}}(U) = \big\{\, \nu \in \mathcal{M}(U) : \exists E \in \mathcal{B}(U) \quad \nu(dx) = \theta 1_E(x)\,\mathcal{L}^d(dx) \,\big\}$$

endowed with the variation topology is isometric to $\mathcal{B}(U)$ endowed with the $L^1$ topology. By theorem 14.5, the level sets of $\mathcal{J}$ restricted to $\mathcal{M}(U)$ are compact with respect to the variation topology. We endow now the product space $\mathcal{M}(U_1) \times \cdots \times \mathcal{M}(U_m)$ with the product of the variation topologies of each space $\mathcal{M}(U_1),\dots,\mathcal{M}(U_m)$.

**Corollary 17.5.** *For any Borel subset* $\mathbb{M}^m$ *of* $\mathcal{M}(U_1) \times \cdots \times \mathcal{M}(U_m)$ ,

$$\limsup_{n\to\infty} \frac{1}{n^{d-1}} \ln \sup\Big\{ P(\phi^{U_1,\dots,U_m}_{x_1,\dots,x_m}(\widetilde{\mathcal{C}}_n) \in \mathbb{M}^m) : (x_1,\dots,x_m) \in T(U_1,\dots,U_m) \Big\}$$
$$\leq -\inf\,\big\{\, \mathcal{J}^m(\nu_1,\dots,\nu_m) : (\nu_1,\dots,\nu_m) \in \overline{\mathbb{M}^m}^{var} \,\big\}\,,$$

*where* $\overline{\mathbb{M}^m}^{var}$ *is the closure of* $\mathbb{M}^m$ *in the variation topology.*

*Proof.* In fact, $\phi^{U_1,\dots,U_m}_{x_1,\dots,x_m}(\widetilde{\mathcal{C}}_n)$ takes its values in $\mathcal{M}^\theta_{\mathrm{ac}}(U_1) \times \cdots \times \mathcal{M}^\theta_{\mathrm{ac}}(U_m)$. By lemma 17.2, $\phi^{U_1,\dots,U_m}_{x_1,\dots,x_m}(\mathcal{C}_n)$ and $\phi^{U_1,\dots,U_m}_{x_1,\dots,x_m}(\widetilde{\mathcal{C}}_n)$ are exponentially contiguous in the weak topology. By lemma 17.4, $\phi^{U_1,\dots,U_m}_{x_1,\dots,x_m}(\mathcal{C}_n)$ satisfies the local large deviation upper bound in the weak topology. This implies that $\phi^{U_1,\dots,U_m}_{x_1,\dots,x_m}(\widetilde{\mathcal{C}}_n)$ satisfies the local large deviation upper bound in the weak topology, and therefore in the stronger variation topology. By lemma 17.3, $\phi^{U_1,\dots,U_m}_{x_1,\dots,x_m}(\widetilde{\mathcal{C}}_n)$ is exponentially tight in the variation topology. Moreover all these estimates are uniform with respect to $(x_1,\dots,x_m)$ in $T(U_1,\dots,U_m)$. By lemma 6.4 (in fact, a simpler version of it), we conclude that $\phi^{U_1,\dots,U_m}_{x_1,\dots,x_m}(\widetilde{\mathcal{C}}_n)$ satisfies the uniform large deviation upper bound in the variation topology. □

## 17.3 Conclusion of the proof

The final part of the proof relies on the set $C_n$ built in section 16.1. We first control the perimeter of $C_n$ under the conditioned measure $\widehat{P}$.

**Lemma 17.6.** *For $K$ large enough, there exists a positive constant $c$ such that*

$$\forall a>0 \qquad \limsup_{n\to\infty} \frac{1}{n^{d-1}} \ln \widehat{P}\big(\mathcal{P}(C_n)\geq a\big) \,\leq\, -ca\,.$$

*Proof.* In this proof, we use the notation of section 16.1. We consider first the event $\{\, s < \operatorname{diam} C(0) < \infty \,\}$. Let $i$ be an integer larger than $K$ and suppose that $C(0)$ has a diameter equal to $i$. Let $\underline{F}$ be the outer vertex boundary of the unbounded residual component of $\underline{\mathcal{C}}_n$. Then $\underline{F}$ is a $\mathbb{L}^{d,\infty}$–connected set of bad blocks, it surrounds the origin and $\operatorname{card}\underline{F}\geq i/K$. Thus

$$P(\operatorname{diam} C(0)=i) \,\leq\, \sum_{j\geq i/K}\sum_{\underline{F}} P(\forall \underline{x}\in \underline{F}\quad X(\underline{x})=0)\,,$$

where the second summation extends over all the subsets $\underline{F}$ of $\mathbb{Z}^d$ of cardinality $j$ which are $\mathbb{L}^{d,\infty}$–connected and surround the origin. The number of such sets is less than $jb^j$ for some constant $b$, whence, using the bound of lemma 9.9,

$$P(\operatorname{diam} C(0)=i) \,\leq\, \sum_{j\geq i/K} \exp\big(\ln j + j\ln b + j\,3^{-d}\ln P(X(\underline{0})=0)\big)\,.$$

Thus, for $K$ large enough, there exist two positive constants $b$ and $c$ such that for any positive $s$,

$$P(s\leq \operatorname{diam} C(0)<\infty) \,\leq\, b\exp(-cs)\,.$$

Let $a>0$. Proceeding as in lemma 16.1, but with the open ball $\overset{\circ}{B}(0,an^{d-2})$ instead of $O$ and $s=(n/K)^{d-1}(u/2d)$, we have

$$\begin{aligned}
&P\big(\mathcal{P}(C_n,\overset{\circ}{B}(0,an^{d-2}))\geq u\big)\\
&\quad\leq 2\sum_{j\geq s}\exp j\left(\frac{1}{\ln n}\ln \mathcal{L}^d\big(\overset{\circ}{B}(0,an^{d-2}+d)\big)+\ln b+\frac{1}{3^d}\ln P(X(\underline{0})=0)\right)\\
&\qquad\qquad\qquad\leq b'\exp\big(-c'n^{d-1}u\big)\,,
\end{aligned}$$

where the last inequality holds for some $b',c'>0$ and for $K$ large enough. Combining the two previous estimates, we get

$$\begin{aligned}
\widehat{P}(\mathcal{P}(C_n)\geq a) \,&\leq\, \widehat{P}(\operatorname{diam} C(0)\geq an^{d-1}/2) \,+\, \widehat{P}(\mathcal{P}(C_n,\overset{\circ}{B}(0,an^{d-2}))\geq a)\\
&\leq b\exp(-can^{d-1}/2)+b'\exp(-c'an^{d-1})
\end{aligned}$$

and the proof of the lemma is completed. □

Let $\mathbb{M}$ be a Borel subset of $\mathcal{M}(\mathbb{R}^d)$. Let $a, \delta$ be such that $0 < \delta \leq 1 \leq a$. We write

$$\widehat{P}(\mathcal{C}_n \in \mathbb{M}) \leq \widehat{P}(\mathcal{P}(C_n) \geq a) + \widehat{P}(\mathcal{E})$$

where $\mathcal{E}$ is the event

$$\Big\{ \mathcal{C}_n \in \mathbb{M},\, \mathcal{P}(C_n) < a \Big\}.$$

The first term is controlled with the help of lemma 17.6. Let us examine the second term $\widehat{P}(\mathcal{E})$. Suppose that $C(0)$ is finite and that $\mathcal{P}(C_n) < a$. By the isoperimetric inequality (see section 13.3), we have then $\mathcal{L}^d(C_n) \leq c_{\text{iso}} a^{\frac{d}{d-1}}$. Let $X$ be the random subset of $\mathbb{Z}^d$ defined by

$$X = \{ x \in \mathbb{Z}^d : \mathcal{L}^d(B(x,1) \cap C_n) \geq \delta \}.$$

Because $C_n$ is a union of disjoint boxes of diameter $1/n$, we have

$$\operatorname{diam} C_n \leq n^{d-2} \mathcal{P}(C_n) < a n^{d-2}.$$

Thus $X$ is included in the ball $B(0, 2an^{d-2})$. Since a point of $\mathbb{R}^d$ belongs to at most $2^d$ balls among the balls $B(x,1)$, $x \in \mathbb{Z}^d$, then

$$\delta |X| \leq \sum_{x \in X} \mathcal{L}^d(B(x,1) \cap C_n) \leq 2^d \mathcal{L}^d(C_n) \leq 2^d c_{\text{iso}} a^{\frac{d}{d-1}}$$

and therefore $|X| \leq 2^d c_{\text{iso}} a^{\frac{d}{d-1}} / \delta$. For $x$ in $\mathbb{Z}^d \setminus X$, by the isoperimetric inequality relative to the ball $B(x,1)$ (see the end of section 13.3), taking into account that $\mathcal{L}^d(C_n \cap B(x,1)) < \delta$, we have

$$\mathcal{L}^d(C_n \cap B(x,1)) \leq \delta^{\frac{1}{d}} b_{\text{iso}}^{\frac{d-1}{d}} \mathcal{P}(C_n, \overset{\circ}{B}(x,1)).$$

Summing this inequality over $x$ in $\mathbb{Z}^d \setminus X$,

$$\begin{aligned}
\mathcal{L}^d\Big(C_n \setminus \bigcup_{x \in X} B(x,1)\Big) &\leq \sum_{x \in \mathbb{Z}^d \setminus X} \mathcal{L}^d(C_n \cap B(x,1)) \\
&\leq \delta^{\frac{1}{d}} b_{\text{iso}}^{\frac{d-1}{d}} \sum_{x \in \mathbb{Z}^d \setminus X} \mathcal{P}(C_n, \overset{\circ}{B}(x,1)) = \delta^{\frac{1}{d}} b_{\text{iso}}^{\frac{d-1}{d}} \sum_{x \in \mathbb{Z}^d \setminus X} \mathcal{H}^{d-1}(\partial^* C_n \cap \overset{\circ}{B}(x,1)) \\
&\leq 2^d \delta^{\frac{1}{d}} b_{\text{iso}}^{\frac{d-1}{d}} \mathcal{H}^{d-1}(\partial^* C_n) = 2^d \delta^{\frac{1}{d}} b_{\text{iso}}^{\frac{d-1}{d}} \mathcal{P}(C_n) \leq 2^d a \delta^{\frac{1}{d}} b_{\text{iso}}^{\frac{d-1}{d}}.
\end{aligned}$$

Here we have used the fact that a point of $\mathbb{R}^d$ belongs to at most $2^d$ unit balls centered on $\mathbb{Z}^d$. We set

$$M = 2^d c_{\text{iso}} a^{\frac{d}{d-1}} / \delta, \qquad \eta = 2^d a \delta^{\frac{1}{d}} b_{\text{iso}}^{\frac{d-1}{d}}.$$

Whenever $\mathcal{E}$ occurs, we have therefore

$$X \subset B(0, 2an^{d-2}), \qquad |X| \le M, \qquad \mathcal{L}^d\Big(C_n \setminus \bigcup_{x \in X} B(x,1)\Big) \le \eta .$$

Let $E(X)$ be a subset of $X \times \{1, \dots, 3^{|X|}\}$ associated to $X$ as in lemma 17.1. We decompose the event $\mathcal{E}$ according to the possible values of the set $E(X)$:

$$\widehat{P}(\mathcal{E}) = \sum_{1 \le m \le M} \sum_{r_1, \dots, r_m} \sum_{y_1, \dots, y_m} \widehat{P}(\mathcal{E},\, E(X) = \{ (y_1, r_1), \dots, (y_m, r_m) \})$$

where the second summation extends over $r_1, \dots, r_m$ in $\{ 1, \dots, 3^M \}$ and the third summation extends over the points $y_1, \dots, y_m$ in $\mathbb{Z}^d \cap B(0, 2an^{d-2})$. The number of possible choices for $y_1, \dots, y_m$ is less than $\mathcal{L}^d(B(0, 2an^{d-2} + 2))^m$, which is a polynomial function in $n$. The number of possible choices for the first two sums is less than $3^{M(M+1)}$. We estimate now the term inside the sums. Let $\{ (y_1, r_1), \dots, (y_m, r_m) \}$ be a value for the random set $E(X)$ compatible with $\mathcal{E}$ (that is a value which occurs with positive probability). By the construction of $E(X)$, the balls $B(y_l, r_l + 1)$, $1 \le l \le m$, are pairwise disjoint so that

$$(y_1, \dots, y_m) \in T(\overset{\circ}{B}(0, r_1 + 1), \dots, \overset{\circ}{B}(0, r_m + 1)) .$$

Moreover the balls $B(y_i, r_i)$, $1 \le i \le m$, cover the balls $B(x, 1)$, $x \in X$, whence

$$\mathcal{L}^d(C_n \setminus B(y_1, r_1) \setminus \cdots \setminus B(y_m, r_m)) \le \eta .$$

Let $k \ge 1$ and let $f : \mathbb{R}^d \to \mathbb{R}^k$ be a uniformly continuous bounded function. For $\xi = (\xi_1, \dots, \xi_m) \in (\mathbb{R}^k)^m$, we define

$$|\xi|_1 = \sum_{1 \le i \le m} |\xi_i|_2 ,$$

where $|\xi_i|_2$ is the usual Euclidean norm of $\xi_i$ in $\mathbb{R}^k$. We have also

$$\big|\mathcal{C}_n(f) - \widetilde{\mathcal{C}}_n(f)\big|_2 \le \big|\phi^{U_1, \dots, U_m}_{y_1, \dots, y_m}(\mathcal{C}_n)(f) - \phi^{U_1, \dots, U_m}_{y_1, \dots, y_m}(\widetilde{\mathcal{C}}_n)(f)\big|_1 + 2\eta \|\, |f|_2 \|_\infty ,$$

where $\|\, |f|_2 \|_\infty = \sup_{x \in \mathbb{R}^d} |f(x)|_2$. We decompose the probability as follows:

$$\begin{aligned}
&\widehat{P}(\mathcal{E},\, E(X) = \{ (y_1, r_1), \dots, (y_m, r_m) \}) \le \\
&\widehat{P}(\mathcal{E},\, \big|\mathcal{C}_n(f) - \widetilde{\mathcal{C}}_n(f)\big|_2 < 3\eta \|\, |f|_2 \|_\infty,\, E(X) = \{ (y_1, r_1), \dots, (y_m, r_m) \}) + \\
&\qquad \widehat{P}(\big|\phi^{U_1, \dots, U_m}_{y_1, \dots, y_m}(\mathcal{C}_n)(f) - \phi^{U_1, \dots, U_m}_{y_1, \dots, y_m}(\widetilde{\mathcal{C}}_n)(f)\big|_1 \ge \eta \|\, |f|_2 \|_\infty) .
\end{aligned}$$

Let $\mathrm{M}^m_{\eta,f}(r_1, \dots, r_m)$ be the subset of $\mathcal{M}(\overset{\circ}{B}(0, r_1 + 1)) \times \cdots \times \mathcal{M}(\overset{\circ}{B}(0, r_m + 1))$ defined by

$$\begin{aligned}
\mathrm{M}^m_{\eta,f}(r_1, \dots, r_m) = \Big\{ &\phi^{\overset{\circ}{B}(0, r_1+1), \dots, \overset{\circ}{B}(0, r_m+1)}_{z_1, \dots, z_m}(\varrho) : \\
&(z_1, \dots, z_m) \in T(\overset{\circ}{B}(0, r_1 + 1), \dots, \overset{\circ}{B}(0, r_m + 1)),\ \varrho \in \mathcal{M}(\mathbb{R}^d), \\
&|\varrho|(\mathbb{R}^d \setminus \bigcup_{1 \le i \le m} \overset{\circ}{B}(z_i, r_i)) \le \eta, \quad \exists \nu \in \mathrm{M} \quad \big|\varrho(f) - \nu(f)\big|_2 < 3\eta \|\, |f|_2 \|_\infty \Big\} .
\end{aligned}$$

We write then

$$\widehat{P}(\mathcal{E},\,\big|\mathcal{C}_n(f)-\widetilde{\mathcal{C}}_n(f)\big|_2<3\eta\|\,|f|_2\|_\infty,\,E(X)=\{\,(y_1,r_1),\dots,(y_m,r_m)\,\})$$
$$\leq\,\widehat{P}\Big(\phi_{y_1,\dots,y_m}^{\overset{\circ}{B}(0,r_1+1),\dots,\overset{\circ}{B}(0,r_m+1)}(\widetilde{\mathcal{C}}_n)\in\mathbb{M}_{\eta,f}^m(r_1,\dots,r_m)\Big)\,.$$

But $\mathbb{M}_{\eta,f}^m(r_1,\dots,r_m)$ depends on $\eta$, $f$, $(r_1,\dots,r_m)$ only and not on $(y_1,\dots,y_m)$ in $T(\overset{\circ}{B}(0,r_1+1),\dots,\overset{\circ}{B}(0,r_m+1))$. Coming back to the innermost summation,

$$\sum_{y_1,\dots,y_m}\widehat{P}(\mathcal{E},\,E(X)=\{\,(y_1,r_1),\dots,(y_m,r_m)\,\})\,\leq\,\mathcal{L}^d(B(0,2an^{d-2}+2))^m$$
$$\times\Big(\sup_{y_1,\dots,y_m}\,\widehat{P}\Big(\phi_{y_1,\dots,y_m}^{\overset{\circ}{B}(0,r_1+1),\dots,\overset{\circ}{B}(0,r_m+1)}(\widetilde{\mathcal{C}}_n)\in\mathbb{M}_{\eta,f}^m(r_1,\dots,r_m)\Big)$$
$$+\,\sup_{y_1,\dots,y_m}\,\widehat{P}\big(\big|\phi_{y_1,\dots,y_m}^{U_1,\dots,U_m}(\mathcal{C}_n)(f)-\phi_{y_1,\dots,y_m}^{U_1,\dots,U_m}(\widetilde{\mathcal{C}}_n)(f)\big|_1\geq\eta\|\,|f|_2\|_\infty\big)\Big)$$

where the supremum is taken over $(y_1,\dots,y_m)$ in $T(\overset{\circ}{B}(0,r_1+1),\dots,\overset{\circ}{B}(0,r_m+1))$. This inequality, the upper bound of lemma 17.2 and the estimate of corollary 17.5 yield

$$\limsup_{n\to\infty}\frac{1}{n^{d-1}}\ln\sum_{y_1,\dots,y_m}\widehat{P}(\mathcal{E},\,E(X)=\{\,(y_1,r_1),\dots,(y_m,r_m)\,\})\,\leq$$
$$-\inf\,\big\{\,\mathcal{J}^m(\nu_1,\dots,\nu_m):(\nu_1,\dots,\nu_m)\in\overline{\mathbb{M}_{\eta,f}^m}^{var}(r_1,\dots,r_m)\,\big\}\,.$$

Since the number of terms involved in the first two sums is bounded by $3^{M(M+1)}$ which is independent of $n$, we conclude with the help of lemma 6.7 that

$$\limsup_{n\to\infty}\frac{1}{n^{d-1}}\ln\widehat{P}(\mathcal{E})\,\leq\,-\inf\,\Big\{\,\mathcal{J}^m(\nu_1,\dots,\nu_m):1\leq m\leq M,$$
$$r_1,\dots,r_m\in\{\,1,\dots,3^M\,\},\,(\nu_1,\dots,\nu_m)\in\overline{\mathbb{M}_{\eta,f}^m}^{var}(r_1,\dots,r_m)\Big\}\,.$$

It remains to evaluate this infimum. More precisely, we will relate the above infimum to

$$\inf\,\{\,\mathcal{J}(\varrho):\varrho(\mathbb{R}^d)<\infty,\,\exists\nu\in\mathbb{M}\,\;|\varrho(f)-\nu(f)|_2\leq 9\|\,|f|_2\|_\infty\eta\,\}\,.$$

Our strategy is the following. We pick $(\nu_1,\dots,\nu_m)$ an admissible candidate, i.e., $(\nu_1,\dots,\nu_m)$ belongs to $\overline{\mathbb{M}_{\eta,f}^m}^{var}(r_1,\dots,r_m)$ for some $r_1,\dots,r_m$. We approach it by an element of $\mathbb{M}_{\eta,f}^m(r_1,\dots,r_m)$, which itself comes from a measure $\varrho$ such that $|\varrho(f)-\nu(f)|_2<3\|\,|f|_2\|_\infty\eta$ for some $\nu\in\mathbb{M}$. The difficulty is that we have no information on the surface energy of $\varrho$ outside the balls $B(0,r_i+1)$, $1\leq i\leq m$. With the help of an adequate cutting procedure, we shall build a measure $\psi$ such that $\mathcal{J}(\psi)$ is close to $\mathcal{J}^m(\nu_1,\dots,\nu_m)$ and $|\psi(f)-\varrho(f)|_2$ is small as well.

So, let $m \in \{1,\dots,M\}$, let $r_1,\dots,r_m \in \{1,\dots,3^M\}$ and let $(\nu_1,\dots,\nu_m) \in \overline{\mathbb{M}^m_{\eta,f}}^{var}(r_1,\dots,r_m)$ such that $\mathcal{J}^m(\nu_1,\dots,\nu_m) < +\infty$. There exist $F_1,\dots,F_m$ Borel subsets of $B(0,r_1+1),\dots,B(0,r_m+1)$ such that $\theta 1_{F_1},\dots,\theta 1_{F_m}$ are the densities of $\nu_1,\dots,\nu_m$ with respect to the Lebesgue measure. By the definition of $\overline{\mathbb{M}^m_{\eta,f}}^{var}(r_1,\dots,r_m)$, there exist $\varrho \in \mathcal{M}(\mathbb{R}^d)$, $\nu \in \mathbb{M}$ and $(z_1,\dots,z_m) \in T(\overset{\circ}{B}(0,r_1+1),\dots,\overset{\circ}{B}(0,r_m+1))$ such that

$$\sum_{1\le i\le m} |\nu_i - \varrho|_{\overset{\circ}{B}(z_i,r_i+1)}(\cdot+z_i)| \le \eta\,,$$

$$|\varrho|(\mathbb{R}^d\setminus \bigcup_{1\le i\le m}\overset{\circ}{B}(z_i,r_i)) \le \eta\,,\quad \big|\varrho(f)-\nu(f)\big|_2 < 3\eta|||f|_2||_\infty\,.$$

This implies in particular that

$$\begin{aligned}\sum_{1\le i\le m}\Big|\int_{\overset{\circ}{B}(0,r_i+1)} f(z_i+x)\,d\nu_i(x) - \int_{\overset{\circ}{B}(z_i,r_i+1)} f(x)\,d\varrho(x)\Big|_2 \\ = \sum_{1\le i\le m}\Big|\int_{\overset{\circ}{B}(0,r_i+1)} f(z_i+x)\,d\big(\nu_i(x) - \varrho|_{\overset{\circ}{B}(z_i,r_i+1)}(x+z_i)\big)\Big|_2 \\ \le |||f|_2||_\infty \sum_{1\le i\le m} |\nu_i - \varrho|_{\overset{\circ}{B}(z_i,r_i+1)}(\cdot+z_i)| \le \eta|||f|_2||_\infty\,,\end{aligned}$$

$$\begin{aligned}\sum_{1\le i\le m}\nu_i\big(\overset{\circ}{B}(0,r_i+1)\setminus B(0,r_i)\big) \le \\ \sum_{1\le i\le m}|\nu_i - \varrho|_{\overset{\circ}{B}(z_i,r_i+1)}(\cdot+z_i)| + \sum_{1\le i\le m}|\varrho|\big(\overset{\circ}{B}(z_i,r_i+1)\setminus B(z_i,r_i)\big) \le 2\eta\,.\end{aligned}$$

By lemma 14.4, for $i$ in $\{1,\dots,m\}$, for $\mathcal{H}^1$ almost all $t$ in $]0,1[$,

$$\mathcal{I}(F_i\cap B(0,r_i+t)) \le \mathcal{I}(F_i,\overset{\circ}{B}(0,r_i+t)) + ||\tau||_\infty \mathcal{H}^{d-1}(F_i\cap\partial B(0,r_i+t))\,.$$

Let $T$ be the subset of $]0,1[$ where all the above inequalities hold simultaneously. Certainly $\mathcal{H}^1(T)=1$ and by integrating in polar coordinates,

$$\sum_{1\le i\le m}\mathcal{L}^d(F_i\cap B(0,r_i+1)\setminus B(0,r_i)) = \int_T \sum_{1\le i\le m}\mathcal{H}^{d-1}(F_i\cap\partial B(0,r_i+t))\,dt\,,$$

so that there exists $t$ in $T$ such that

$$\sum_{1\le i\le m}\mathcal{H}^{d-1}(F_i\cap\partial B(0,r_i+t)) \le \frac{2\eta}{\theta}\,.$$

Let $F$ be the set

$$F = \bigcup_{1\le i\le m}\big(z_i + (F_i\cap B(0,r_i+t))\big)$$

and let $\psi$ be the measure with density $\theta 1_F$ with respect to the Lebesgue measure. We have

$$\begin{aligned}
&\big|\psi(f)-\varrho(f)\big|_2 \leq \\
&\sum_{1\leq i\leq m}\Big|\int_{\overset{\circ}{B}(z_i,r_i)} f(x)\,d(\psi-\varrho)(x)\Big|_2 + \|\,|f|_2\|_\infty(\psi+|\varrho|)\Big(\mathbb{R}^d\setminus\bigcup_{1\leq i\leq m}\overset{\circ}{B}(z_i,r_i)\Big)\\
&\qquad\leq \sum_{1\leq i\leq m}\Big|\int_{\overset{\circ}{B}(0,r_i+1)} f(z_i+x)\,d\nu_i(x) - \int_{\overset{\circ}{B}(z_i,r_i+1)} f(x)\,d\varrho(x)\Big|_2\\
&\qquad\qquad + 2\|\,|f|_2\|_\infty(\psi+|\varrho|)\Big(\mathbb{R}^d\setminus\bigcup_{1\leq i\leq m}\overset{\circ}{B}(z_i,r_i)\Big) \leq 7\eta\|\,|f|_2\|_\infty,
\end{aligned}$$

whence

$$\big|\psi(f)-\nu(f)\big|_2 \leq \big|\psi(f)-\varrho(f)\big|_2 + \big|\varrho(f)-\nu(f)\big|_2 \leq 10\eta\|\,|f|_2\|_\infty.$$

Thanks to the choice of $t$, we have

$$\begin{aligned}
\mathcal{J}(\psi) &= \mathcal{I}(F) = \sum_{1\leq i\leq m}\mathcal{I}(F_i\cap B(0,r_i+t))\\
&\leq \sum_{1\leq i\leq m}\mathcal{I}(F_i,\overset{\circ}{B}(0,r_i+t)) + \|\tau\|_\infty\sum_{1\leq i\leq m}\mathcal{H}^{d-1}(F_i\cap\partial B(0,r_i+t))\\
&\qquad\leq \mathcal{J}^m(\nu_1,\dots,\nu_m)+\frac{2\eta}{\theta}\|\tau\|_\infty.
\end{aligned}$$

It follows that

$$\begin{aligned}
\inf\big\{\,\mathcal{J}(\psi):\psi(\mathbb{R}^d)<\infty,\ \exists\nu\in\mathbb{M}\quad |\psi(f)-\nu(f)|_2\leq 10\eta\|\,|f|_2\|_\infty\big\}\\
\leq \mathcal{J}^m(\nu_1,\dots,\nu_m)+\frac{2\eta}{\theta}\|\tau\|_\infty.
\end{aligned}$$

Passing to the infimum over $(\nu_1,\dots,\nu_m)$, $(r_1,\dots,r_m)$ and $m$,

$$\begin{aligned}
&\limsup_{n\to\infty}\frac{1}{n^{d-1}}\ln\widehat{P}(\mathcal{E}) \leq \frac{2\eta}{\theta}\|\tau\|_\infty\\
&\quad -\inf\big\{\,\mathcal{J}(\psi):\psi(\mathbb{R}^d)<\infty,\ \exists\nu\in\mathbb{M}\quad |\psi(f)-\nu(f)|_2\leq 10\eta\|\,|f|_2\|_\infty\big\}.
\end{aligned}$$

Coming back to the initial inequality,

$$\begin{aligned}
&\limsup_{n\to\infty}\frac{1}{n^{d-1}}\ln\widehat{P}(\mathcal{C}_n\in\mathbb{M}) \leq -\min\big(ca,\\
&\inf\big\{\,\mathcal{J}(\psi):\psi(\mathbb{R}^d)<\infty,\ \exists\nu\in\mathbb{M}\quad |\psi(f)-\nu(f)|_2\leq 10\eta\|\,|f|_2\|_\infty\big\}-\frac{2\eta}{\theta}\|\tau\|_\infty\big).
\end{aligned}$$

Recalling that $\eta = 2^d a\delta^{1/d} b_{\text{iso}}^{\frac{d-1}{d}}$, sending first $\delta$ to 0 and then $a$ to $\infty$, we get the enhanced upper bound we were seeking. □

## 17.4 Extension to FK percolation

All the results proved in chapters 16 and 17 hold also for the FK percolation model for $q \geq 1$ in the uniqueness region $]\widehat{p}_c, 1] \setminus \mathcal{U}(q)$, with the infinite volume FK measure $\Phi_\infty^{p,q}$ instead of $P$. Let us mention here the points which require some adaptation.

*Exponential tightness.* In the block estimates (lemmas 16.1, 17.6), the term $P(X(\underline{0}) = 0)$ should be replaced by

$$\max \left\{ \Phi(X(\underline{0}) = 0) : \Phi \in c\mathcal{FK}(p, q, B'(\underline{0})) \right\}.$$

*Local upper bound.* Instead of independence, one should use the decoupling lemma 10.10.

*Lower bound.* One should use a conditioning instead of independence (see the proof for a finite volume FK measure in section 18.4).

The most delicate point is the final part of the proof of the local estimate in the uniform large deviation upper bound (lemma 17.5). We need a decoupling which is uniform with respect to the positions of the sets $x_1+U_1, \dots, x_m+U_m$. Let us precise this point. We use the same notation than in section 17.2. For $l \in \{1 \cdots m\}$ and $x \in \mathbb{R}^d$, we define the event

$$\mathcal{E}_l(x) = \bigcap_{i \in I(l)} \mathrm{Sep}_\theta(n, x + x_i^l, r_i^l, \nu_{E_l}(x_i^l), \delta_0).$$

This event depends only on the edges included in $n(x_l + U_l)$. We have then

$$\begin{aligned}
\Phi_\infty(\phi_{x_1,\dots,x_m}^{U_1,\dots,U_m}(\mathcal{C}_n) \in \mathcal{U}) &\leq \Phi_\infty\Big( \bigcap_{1 \leq l \leq m} \mathcal{E}_l(x_l) \Big) \\
&= \Phi_\infty\Big( \Phi_\infty\Big( \bigcap_{1 \leq l \leq m} \mathcal{E}_l(x_l) \,\Big|\, \omega(e),\, e \in \mathbb{E}^d \setminus \mathbb{E}^d(nx_1 + nU_1) \Big)\Big) \\
&= \Phi_\infty\Big( \Phi_\infty\Big( \mathcal{E}_1(x_1) \,\Big|\, \omega(e),\, e \in \mathbb{E}^d \setminus \mathbb{E}^d(nx_1 + nU_1) \Big) \prod_{2 \leq l \leq m} 1_{\mathcal{E}_l(x_l)}(\omega) \Big) \\
&\leq \Big( \max_{\Phi \in c\mathcal{FK}(nx_1 + nU_1)} \Phi(\mathcal{E}_1(x_1)) \Big) \Phi_\infty\Big( \bigcap_{2 \leq l \leq m} \mathcal{E}_l(x_l) \Big).
\end{aligned}$$

By iterated conditioning, we get

$$\begin{aligned}
\Phi_\infty(\phi_{x_1,\dots,x_m}^{U_1,\dots,U_m}(\mathcal{C}_n) \in \mathcal{U}) &\leq \prod_{1 \leq l \leq m} \max_{\Phi \in c\mathcal{FK}(nx_l + nU_l)} \Phi(\mathcal{E}_l(x_l)) \\
&= \prod_{1 \leq l \leq m} \max_{\Phi \in c\mathcal{FK}(nU_l)} \Phi(\mathcal{E}_l(0)).
\end{aligned}$$

This last bound is independent of $(x_1, \dots, x_m)$ in $T(U_1, \dots, U_m)$. We apply next proposition 10.6 and we use the remark 10.8 to estimate, for $l \in \{1 \cdots m\}$,

$$\limsup_{n\to\infty} \frac{1}{n^{d-1}} \ln \Big( \max_{\Phi\in c\mathcal{FK}(nU_l)} \Phi\big(\mathcal{E}_l(0)\big)\Big)$$
$$\leq - \sum_{i\in I(l)} \tau(\nu_{E_l}(x_i^l))\, \alpha_{d-1}(r_i^l)^{d-1}\big(1 - c\sqrt{2\delta_0}\big)$$
$$\leq -\mathcal{I}(E_l, U_l)\,(1-\varepsilon) + \varepsilon\,.$$

The other parts of the proof of lemma 17.5 apply without changes.

# 18

# LDP for FK percolation

This chapter is devoted to the proof of the large deviation principle stated in theorem 7.5. Throughout this chapter, we fix $q \geq 1$ and $p > \widehat{p}_c$ such that $p \notin \mathcal{U}(q)$. We work here within the topological vector space $\mathcal{M}(Q)$ endowed with the weak topology, where

$$Q = [-1/2, 1/2]^d$$

is the $d$–dimensional unit cube. The collection consisting of the finite intersections of the sets

$$\big\{ \nu \in \mathcal{M}(Q) : |\nu(f)| \leq \eta \big\}, \quad \eta > 0, \quad f \in C_c(\mathbb{R}^d, \mathbb{R}),$$

is a basis of neighbourhoods of the origin. We first check that $\mathcal{J}$ is a good rate function. By theorem 14.5, the surface energy $\mathcal{I}$ restricted to the Borel subsets of $Q$ is a good rate function. Since the map

$$A \in (\mathcal{B}(Q), L^1) \mapsto \theta \, 1_A(x) \, d\mathcal{L}^d(x) \in \mathcal{M}(Q)$$

is continuous, then $\mathcal{J}$ is a good rate function (this is a consequence of the contraction principle theorem 6.8). Our next goal is to prove that the random measure $\mathcal{C}_n$ is $\mathcal{J}$–tight in the sense of definition 6.3: there exists a positive constant $c$ such that

$$\begin{aligned}
&\forall \eta, \lambda > 0, \quad \forall f \in C_c(\mathbb{R}^d, \mathbb{R}), \\
&\quad \limsup_{n \to \infty} \frac{1}{n^{d-1}} \ln \Phi^w_{\Lambda(n)} \big( \forall \nu \in \mathcal{J}^{-1}([0, \lambda]) \quad |\mathcal{C}_n(f) - \nu(f)| > \eta \big) \leq -c\lambda \,.
\end{aligned}$$

We then prove the local large deviation upper bound and finally the large deviation lower bound.

## 18.1 Coarse grained image

Let $n \in \mathbb{N}$ and let $K \in \mathbb{N}$. We work with the box $\Lambda(n)$ rescaled by a factor $K$, that is

$$\underline{\Lambda}(n) = \{\, \underline{x} \in \mathbb{Z}^d : B(\underline{x}) \cap \Lambda(n) \neq \emptyset \,\}.$$

For $\underline{x} \in \underline{\Lambda}(n)$, the block variable $X(\underline{x})$ is the indicator function of the event $R(B'(\underline{x}), K)$, i.e.,

• Inside the event block $B'(\underline{x})$ there is exactly one crossing cluster $C(\underline{x})$, it is the only cluster in $B'(\underline{x})$ having diameter larger than or equal to $K$ and it crosses every sub–box of $B'(\underline{x})$ with diameter larger than or equal to $K$.

In order to prove the exponential tightness, we build an auxiliary random measure $\widetilde{\mathcal{C}}_n$ which is exponentially contiguous to $\mathcal{C}_n$ and we provide suitable probabilistic estimates on the surface energy of $\widetilde{\mathcal{C}}_n$. Let $\underline{A}$ be a $\mathbb{L}^d$–connected subset of $\underline{\Lambda}(n)$. We recall that a residual $\mathbb{L}^{d,\infty}$–component $\underline{R}$ of $\underline{A}$ in $\underline{\Lambda}(n)$ is a $\mathbb{L}^{d,\infty}$–connected component of $\underline{\Lambda}(n) \setminus \underline{A}$. We define

$$\operatorname{fill} \underline{A} = \underline{A} \cup \bigcup \underline{R},$$

where the union runs over the residual $\mathbb{L}^{d,\infty}$–components $\underline{R}$ of $\underline{A}$ such that $\operatorname{diam}_\infty \underline{R} < \ln n$ and $\underline{R} \cap \partial^{in} \underline{\Lambda}(n) = \emptyset$. For a cluster $C$ of the configuration, we set

$$\underline{C} = \{\, \underline{x} \in \underline{\Lambda}(n) : B(\underline{x}) \cap C \neq \emptyset \,\}.$$

Since a cluster $C$ is $\mathbb{L}^d$–connected, so is $\underline{C}$. We say that a cluster $C$ is large if $\operatorname{diam}_\infty C \geq K \ln n$. Notice that if $C$ is a large cluster, then $\operatorname{diam}_\infty \underline{C} \geq \ln n$. For a large cluster $C$, we define

$$\widehat{\underline{C}} = \bigcup \operatorname{fill} \underline{A},$$

where the union runs over all the $\mathbb{L}^d$–connected components $\underline{A}$ of good blocks such that $\underline{C} \cap \underline{A} \neq \emptyset$ (or equivalently $\underline{A} \subset \underline{C}$). By definition, the $\mathbb{L}^d$ outer boundary $\partial^{out} \widehat{\underline{C}}$ of $\widehat{\underline{C}}$ consists of bad blocks whenever $\widehat{\underline{C}} \neq \emptyset$. In case $\widehat{\underline{C}} = \emptyset$, we define $\partial^{out} \widehat{\underline{C}}$ as $\underline{C}$ which again consists only of bad blocks.

**Lemma 18.1.** *If $C_1$, $C_2$ are two distinct large clusters, then $\widehat{\underline{C}}_1 \cap \widehat{\underline{C}}_2 = \emptyset$.*

*Proof.* Suppose that $\widehat{\underline{C}}_1 \cap \widehat{\underline{C}}_2 \neq \emptyset$. Then there exist $\underline{A}_1$, $\underline{A}_2$ two $\mathbb{L}^d$ components of good blocks such that

$$\underline{A}_1 \subset \underline{C}_1, \quad \underline{A}_2 \subset \underline{C}_2, \quad \operatorname{fill} \underline{A}_1 \cap \operatorname{fill} \underline{A}_2 \neq \emptyset.$$

Several cases arise:

• $\underline{A}_1 \cap \underline{A}_2 \neq \emptyset$. This would imply that $\underline{C}_1 \cap \underline{C}_2 \neq \emptyset$, which is impossible since $C_1 \cap C_2 = \emptyset$. From now on, we suppose that $\underline{A}_1 \cap \underline{A}_2 = \emptyset$.

• $\underline{A}_1 \cap \operatorname{fill} \underline{A}_2 \neq \emptyset$. Let $\underline{R}_2$ be a residual $\mathbb{L}^{d,\infty}$ component of $\underline{A}_2$ such that $\underline{R}_2 \subset \operatorname{fill} \underline{A}_2$ and $\underline{R}_2 \cap \underline{A}_1 \neq \emptyset$. Since $\underline{C}_1$ cannot fit into $\underline{R}_2$ for reasons of diameter, necessarily $\underline{C}_1 \cap \underline{A}_2 \neq \emptyset$, which contradicts the fact that $C_1 \cap C_2 = \emptyset$.

• $\underline{A}_2 \cap \operatorname{fill} \underline{A}_1 \neq \emptyset$. This case is similar to the previous case.

• $\underline{A}_1 \cap \operatorname{fill} \underline{A}_2 = \emptyset$ and $\underline{A}_2 \cap \operatorname{fill} \underline{A}_1 = \emptyset$. In this case, there exist $\underline{R}_1, \underline{R}_2$ residual $\mathbb{L}^{d,\infty}$ components of $\underline{A}_1, \underline{A}_2$ respectively such that $\underline{R}_1 \cap \underline{R}_2 \neq \emptyset$ and for $i = 1, 2,$

$$\operatorname{diam}_\infty \underline{R}_i < \ln n\,, \quad \underline{R}_i \cap \partial^{in} \underline{\Lambda}(n) = \emptyset\,.$$

Yet we have $\partial^{\,out,ext}\underline{R}_1 \subset \underline{A}_1$ and $\partial^{\,out,ext}\underline{R}_2 \subset \underline{A}_2$. Since $\underline{A}_1 \cap \operatorname{fill} \underline{A}_2 = \emptyset$, then $\partial^{\,out,ext}\underline{R}_1 \cap \underline{R}_2 = \emptyset$. Similarly $\partial^{\,out,ext}\underline{R}_2 \cap \underline{R}_1 = \emptyset$. The sets $\underline{R}_1$, $\underline{R}_2$ are $\mathbb{L}^{d,\infty}$ connected, not disjoint and neither of them contains a point of the external outer boundary of the other. Therefore these boundaries coincide, i.e., $\partial^{\,out,ext}\underline{R}_1 = \partial^{\,out,ext}\underline{R}_2$, which implies again that $\underline{A}_1 \cap \underline{A}_2 \neq \emptyset$ and $C_1 \cap C_2 \neq \emptyset$.
None of the above cases is compatible with the hypothesis $C_1 \cap C_2 = \emptyset$, therefore $\widehat{\underline{C}}_1 \cap \widehat{\underline{C}}_2 = \emptyset$. □

Finally, the measure $\widetilde{\mathcal{C}}_n$ is the measure with density $\theta\, 1_{C_n}$ with respect to the Lebesgue measure $\mathcal{L}^d$, where $C_n$ is the region

$$C_n = \Big(\Lambda(1) \setminus \Lambda(1 - \frac{5K}{n}\ln n)\Big) \cup \bigcup_{\substack{C \text{ large cluster} \\ C\cap\partial^{in}\Lambda(n)\neq\emptyset}} \ \bigcup_{\underline{x}\in\widehat{\underline{C}}} B_n(\underline{x})\,.$$

The first union runs over the large clusters $C$ which intersect the inner boundary of the box $\Lambda(n)$. We drop the parameter $K$ from the notation, yet the set $C_n$ and the measure $\widetilde{\mathcal{C}}_n$ depend on $K$.

Let $\widehat{\underline{F}}$ be the union of all the $\mathbb{L}^{d,\infty}$–connected components of bad blocks intersecting simultaneously the rescaled box $\underline{\Lambda}(n - 5K\ln n)$ and one of the sets $\partial^{\,out}\widehat{\underline{C}}$, where $C$ is a large cluster. We set

$$\widehat{F} = \bigcup_{\underline{x}\in\widehat{\underline{F}}} B_n(\underline{x})\,.$$

We have

$$\big\{\, x \in Q : \mathcal{C}_n(\{x\}) > 0 \,\big\} \setminus C_n \subset \widehat{F}\,.$$

Indeed, if $x \in \Lambda(1 - (5K/n)\ln n)$ is such that $\mathcal{C}_n(\{x\}) > 0$, then $x$ belongs to a large cluster $C$; if $x \notin C_n$, then the block containing $x$ is bad and its $\mathbb{L}^{d,\infty}$ component meets $\partial^{\,out}\widehat{\underline{C}}$.

**Lemma 18.2.** *There exists $K_0(d,p,q)$ such that, for $K \geq K_0$, there exist positive constants $b, c$ depending on $K, d, p, q$ such that*

$$\forall n \geq K \quad \forall u > 0 \qquad \Phi^w_{\Lambda(n)}\big(|\widehat{\underline{F}}| \geq u\big) \leq b\exp(-cu)\,.$$

*Proof.* Let $C$ be a large cluster and let $\underline{F}$ be an $\mathbb{L}^{d,\infty}$–connected component of bad blocks intersecting $\partial^{\,out}\widehat{\underline{C}}$ and the rescaled box $\underline{\Lambda}(n - 5K\ln n)$. We claim that $|\underline{F}| \geq \ln n$. To show this we consider only the case where $\widehat{\underline{C}} \neq \emptyset$ (the case $\widehat{\underline{C}} = \emptyset$ is straightforward because then $\underline{C} \subset \underline{F}$) and we assume that $|\underline{F}| < \ln n$. Necessarily $\underline{F} \cap \partial^{in}\underline{\Lambda}(n) = \emptyset$. Let $\underline{D} = \partial^{\,out,ext}_\infty \underline{F}$. Then $\underline{D}$ is a $\mathbb{L}^d$ connected set of good blocks surrounding $\underline{F}$, so that $\underline{F}$ is included in a

$\mathbb{L}^{d,\infty}$ residual component $\underline{R}$ of $\underline{D}$ satisfying $\mathrm{diam}_\infty \underline{R} < \ln n$. Let $\underline{A}$ be the $\mathbb{L}^d$–connected component of good blocks containing $\underline{D}$. The above properties imply that $\underline{R} \subset \mathrm{fill}\,\underline{A}$. Yet $\underline{F} \cap \partial^{out}\widehat{\underline{C}} \neq \emptyset$, thus there exist $\underline{x} \in \underline{F} \setminus \widehat{\underline{C}}$ and $\underline{y} \in \widehat{\underline{C}}$ such that $|\underline{x} - \underline{y}|_2 = 1$. If the block of $\underline{y}$ is bad, then it belongs to a residual component of $\underline{C}$ which has been added by the fill operator, hence $\underline{y}$ cannot be in $\partial^{in}\widehat{\underline{C}}$. Thus the block of $\underline{y}$ is good and $\underline{y} \notin \underline{F}$, so that $\widehat{\underline{C}} \cap \partial^{out}\underline{F} \neq \emptyset$. Two cases arise:

- $\underline{C} \cap \partial^{out}\underline{F} \neq \emptyset$. Since $\mathrm{diam}_\infty \underline{C} \geq \ln n > \mathrm{diam}_\infty \underline{F}$, necessarily $\underline{C} \cap \underline{D} \neq \emptyset$ and $\underline{R} \subset \widehat{\underline{C}}$, which is absurd.
- $\underline{C} \cap \partial^{out}\underline{F} = \emptyset$. We have then $\underline{C} \cap \underline{F} = \emptyset$ (otherwise we would have $\underline{C} \subset \underline{F}$, which is impossible for reasons of diameter). Moreover, there exists a $\mathbb{L}^d$ component of good blocks $\underline{A}'$ intersecting $\underline{C}$ and having a residual $\mathbb{L}^{d,\infty}$ component $\underline{R}'$ such that $\underline{R}' \subset \mathrm{fill}\,\underline{A}'$ and $\underline{R}' \cap \partial^{out}\underline{F} \neq \emptyset$. Yet $\underline{R}'$ is $\mathbb{L}^{d,\infty}$ connected, the blocks of $\partial^{out}_\infty \underline{R}'$ are good, therefore $\underline{F} \cap \partial^{out}_\infty \underline{R}' = \emptyset$. This implies that $\underline{F} \subset \underline{R}'$ and $\underline{F} \subset \widehat{\underline{C}}$, which is absurd.

Thus $|\underline{F}| \geq \ln n$, as claimed. We have now

$$|\widehat{\underline{F}}| \leq \Big|\{\, \underline{x} \in \underline{\Lambda}(n - 5K\ln n) : |\underline{C}(\underline{x})| \geq \ln n \,\}\Big|,$$

where $\underline{C}(\underline{x})$ denotes the $\mathbb{L}^{d,\infty}$ component of occupied sites containing $\underline{x}$ (a site $\underline{y}$ is occupied if $X(\underline{y}) = 0$). To estimate the right–hand side, we apply lemma 9.10 with $O = \underline{\Lambda}(n - 5K\ln n)$, $t = \ln n$, $s = u$, and we get

$$\Phi^w_{\Lambda(n)}\big(|\widehat{\underline{F}}| \geq u\big) \leq$$
$$2\sum_{j\geq u} \exp j\left(\frac{1}{\ln n}\ln \mathcal{V}\big(\underline{\Lambda}(n-5K\ln n), d\big) + \ln b + 3^{-d} \max_{\Phi \in c\mathcal{FK}(p,q,\Lambda(K))} \ln \Phi(X(0)=0)\right).$$

By lemma 9.4 and proposition 9.7, for $K$ sufficiently large, we obtain the required bound. □

**Corollary 18.3.** *There exists a constant $K_0(d,p,q)$ such that, for $K \geq K_0$, there exists a positive constant $c$ depending on $K, d, p, q$ such that*

$$\forall n \geq K \quad \forall u > 4d \qquad \limsup_{n\to\infty} \frac{1}{n^{d-1}} \ln \Phi^w_{\Lambda(n)}\big(\mathcal{P}(C_n) \geq u\big) \leq -cu\,.$$

*Proof.* By construction, the boundary of $C_n$ is located either on $\partial\Lambda(1) \cup \partial\Lambda(1 - (5K/n)\ln n)$ or on the faces of the blocks of $\widehat{\underline{F}}$. Thus

$$\begin{aligned}\mathcal{P}(C_n) &\leq \mathcal{H}^{d-1}(\partial\Lambda(1)) + \mathcal{H}^{d-1}(\partial\Lambda(1 - \frac{5K}{n}\ln n)) + 2d\Big(\frac{K}{n}\Big)^{d-1}|\widehat{\underline{F}}|\\ &\leq 4d + 2d\Big(\frac{K}{n}\Big)^{d-1}|\widehat{\underline{F}}|\,.\end{aligned}$$

The conclusion follows from lemma 18.2. □

## 18.2 Exponential contiguity

Let us fix $f \in C_c(\mathbb{R}^d, \mathbb{R})$. In order to estimate $|\mathcal{C}_n(f) - \widetilde{\mathcal{C}}_n(f)|$, we use another block argument with the scale $L = K \ln n$. We work with the lattice rescaled by a factor $L$. Let $\varepsilon > 0$. For $\underline{y} \in \mathbb{Z}^d$, the block variable $Y(\underline{y})$ is the indicator function of the event $RVW(\underline{B}(\underline{y}), B'(\underline{y}), L, \varepsilon)$ (see the definition before corollary 9.6). Since $f$ is continuous and has a compact support, it is uniformly continuous. We suppose that $L/n$ is less than 1 and small enough so that

$$\forall x, y \in \mathbb{R}^d \qquad |x - y| \leq \frac{L}{n} \quad \Rightarrow \quad |f(x) - f(y)| \leq \varepsilon ||f||_\infty .$$

Let

$$\underline{A} = \{\, \underline{y} \in \mathbb{Z}^d : B(\underline{y}) \cap \Lambda(n - 6L) \neq \emptyset \,\} .$$

Each block of size $L$ intersecting $\Lambda(n - 6L)$ is included in $\Lambda(n)$, thus

$$|\underline{A}| L^d \;\leq\; \mathcal{L}^d(\Lambda(n)) \;=\; n^d .$$

We evaluate $|\mathcal{C}_n(f) - \widetilde{\mathcal{C}}_n(f)|$ by decomposing it on blocks of size $L/n$ and using the uniform continuity of $f$:

$$\begin{aligned} |\mathcal{C}_n(f) - \widetilde{\mathcal{C}}_n(f)| &\leq \sum_{\underline{y} \in \underline{A}} \Big| \int_{B_n(\underline{y})} f \, d\mathcal{C}_n - \int_{B_n(\underline{y})} f \, d\widetilde{\mathcal{C}}_n \Big| + 2 \frac{||f||_\infty}{n^d} \mathcal{L}^d(\Lambda(n) \setminus \Lambda(n - 6L)) \\ &\leq ||f||_\infty \sum_{\underline{y} \in \underline{A}} \big| \mathcal{C}_n(B_n(\underline{y})) - \widetilde{\mathcal{C}}_n(B_n(\underline{y})) \big| + ||f||_\infty \big( \frac{12dL}{n} + 2\varepsilon \big) . \end{aligned}$$

Next we study the sum in the above inequality. If $\underline{y} \in \underline{A}$ is a block such that $\mathcal{C}_n(B_n(\underline{y})) = 0$ and $B_n(\underline{y}) \cap C_n = \emptyset$, then the corresponding term in the sum vanishes. We consider the other cases:

• $B_n(\underline{y}) \cap C_n = \emptyset$ and $\mathcal{C}_n(B_n(\underline{y})) \neq 0$. In this case, $B_n(\underline{y}) \subset \widehat{F}$.

So it remains only to consider the blocks such that $B_n(\underline{y}) \cap C_n \neq \emptyset$. Let $\underline{y} \in \underline{A}$ be such a block. We distinguish several cases.

• If $Y(\underline{y}) = 0$, then

$$\big| \mathcal{C}_n(B_n(\underline{y})) - \widetilde{\mathcal{C}}_n(B_n(\underline{y})) \big| \;\leq\; \frac{1}{n^d} |B(\underline{y})| 1_{Y(\underline{y}) = 0} .$$

Suppose next that $Y(\underline{y}) = 1$. Several subcases arise:

• $B_n(\underline{y}) \not\subset C_n$. Then we bound

$$\big| \mathcal{C}_n(B_n(\underline{y})) - \widetilde{\mathcal{C}}_n(B_n(\underline{y})) \big| \;\leq\; \frac{1}{n^d} |B(\underline{y})|$$

and we notice that since $B_n(\underline{y})$ meets $C_n$, it meets also the boundary of $C_n$. Thus

$$B_n(\underline{y}) \subset \{ x \in \mathbb{R}^d : \mathrm{d}_\infty(x, \partial C_n \cap \Lambda(1-3L/n)) \le \frac{L}{n} \}.$$

Moreover

$$\mathcal{L}^d\Big(\{ x \in \mathbb{R}^d : \mathrm{d}_\infty(x, \partial C_n \cap \Lambda(1-3L/n)) \le \frac{L}{n} \}\Big) \le |\partial^{in} nC_n| \Big(\frac{2L+2}{n}\Big)^d$$
$$\le \mathcal{P}(nC_n)\Big(\frac{3L}{n}\Big)^d \le \mathcal{P}(C_n)\frac{(3L)^d}{n}.$$

• $B_n(\underline{y}) \subset C_n$ and $\mathcal{C}_n(B_n(\underline{y})) = 0$. These conditions imply that the whole block $B_n(\underline{y})$ has been added to $C_n$ by the filling operation. Yet the regions which are added by the filling operation have a diameter at most $K(\ln n - 1)$, so this case cannot occur.

• $B_n(\underline{y}) \subset C_n$ and $\mathcal{C}_n(B_n(\underline{y})) > 0$. The definition of the block event associated to the variable $Y$ implies that

$$\big|\mathcal{C}_n(B_n(\underline{y})) - \widetilde{\mathcal{C}}_n(B_n(\underline{y}))\big| = \Big|\mathcal{C}_n(B_n(\underline{y})) - \frac{\theta}{n^d}|B(\underline{y})|\Big| \le \frac{\varepsilon}{n^d}|B(\underline{y})|.$$

Summing the previous inequalities over $\underline{y} \in \underline{A}$, we get

$$|\mathcal{C}_n(f) - \widetilde{\mathcal{C}}_n(f)| \le$$
$$||f||_\infty\Big(|\underline{\widehat{F}}|\frac{L^d}{n^d} + \frac{1}{|\underline{A}|}\sum_{\underline{y}\in\underline{A}} 1_{Y(\underline{y})=0} + \frac{(3L)^d}{n}\mathcal{P}(C_n) + 12d\frac{L}{n} + 3\varepsilon\Big).$$

Proposition 9.7 implies that for $L$ large enough the block process satisfies the condition stated at the beginning of section 9.4 (with $\delta = \varepsilon/2$). Using the estimates of lemmas 9.11, 18.2 and 18.3, we have for $K$ large enough

$$\Phi^w_{\Lambda(n)}\Big(|\mathcal{C}_n(f) - \widetilde{\mathcal{C}}_n(f)| \ge ||f||_\infty(6+3^d)\varepsilon\Big) \le$$
$$\Phi^w_{\Lambda(n)}\big(|\underline{\widehat{F}}|\frac{L^d}{n^d} \ge \varepsilon\big) + \Phi^w_{\Lambda(n)}\Big(\frac{1}{|\underline{A}|}\sum_{\underline{y}\in\underline{A}} 1_{Y(\underline{y})=0} \ge \varepsilon\Big) + \Phi^w_{\Lambda(n)}\Big(\frac{L^d}{n}\mathcal{P}(C_n) \ge \varepsilon\Big)$$
$$\le b\exp\big(-c\varepsilon\frac{n^d}{L^d}\big) + 3^d\exp\Big(-\Lambda^*\big(\varepsilon,\frac{\varepsilon}{2}\big)\Big\lfloor\frac{n-6L}{6L}\Big\rfloor^d\Big) + b\exp\Big(-c\varepsilon\frac{n^d}{L^d}\Big).$$

Therefore

$$\forall \varepsilon > 0 \qquad \limsup_{n\to\infty} \frac{1}{n^{d-1}} \ln \Phi^w_{\Lambda(n)}\big(|\mathcal{C}_n(f) - \widetilde{\mathcal{C}}_n(f)| > \varepsilon\big) = -\infty.$$

This estimate and corollary 18.3 imply the $\mathcal{J}$–tightness announced at the beginning of the section.

## 18.3 Local upper bound

The final ingredient needed to get the large deviation upper bound is the local large deviation upper bound, which we prove in this section.

**Lemma 18.4.** *Let $\nu \in \mathcal{M}(Q)$ be such that $\mathcal{J}(\nu) < \infty$. For every $\varepsilon > 0$, there exists a weak neighbourhood $\mathcal{U}$ of $\nu$ in $\mathcal{M}(Q)$ such that*

$$\limsup_{n\to\infty} \frac{1}{n^{d-1}} \ln \Phi^w_{\Lambda(n)}\big(\mathcal{C}_n \in \mathcal{U}\big) \le -(1-\varepsilon)\,\mathcal{J}(\nu)\,.$$

*Proof.* By definition of $\mathcal{J}$, since $\mathcal{J}(\nu) < \infty$, there exists a Borel subset $A$ of $Q$ such that $\nu$ is the measure with density $\theta\,1_{Q\setminus A}$ with respect to the Lebesgue measure and $\mathcal{J}(\nu) = \mathcal{I}(A)$. If $\mathcal{I}(A) = 0$, there is nothing to prove. Suppose that $\mathcal{I}(A) > 0$. Let $\varepsilon > 0$ and let us set $\varepsilon' = \varepsilon(1 + 1/\mathcal{I}(A))^{-1}$. Pick $\delta_0 \in \,]0,1[$ such that $c\sqrt{2\delta_0} < \varepsilon'$ where $c = c(p,d,\zeta)$ is the constant appearing in the interface lemma 12.1. Let $B(x_i, r_i)$, $i \in I \cup I^{\text{bd}}$, be a finite collection of disjoint balls associated with $A, \varepsilon'$ and $\delta_0/8$, as given in the covering lemma 14.7. For $i$ in $I \cup I^{\text{bd}}$, let $f_i, g_i$ be two continuous functions having compact support and taking values in $[0,1]$ such that

$$\begin{gathered}
\forall x \in \mathbb{R}^d \setminus \mathring{B}_-(x_i, r_i, -\nu_A(x_i)) \qquad f_i(x) = 0\,,\\
\forall x \in \overline{B}_+(x_i, r_i, -\nu_A(x_i)) \qquad g_i(x) = 1\,,\\
\Big(\frac{1}{2} - \frac{\delta_0}{8\theta}\Big)\alpha_d r_i^d \le \int f_i\, d\mathcal{L}^d \le \int g_i\, d\mathcal{L}^d \le \Big(\frac{1}{2} + \frac{\delta_0}{8\theta}\Big)\alpha_d r_i^d\,.
\end{gathered}$$

We have then

$$\begin{aligned}
\nu(f_i) &= \theta \int_{Q\setminus A} f_i\, d\mathcal{L}^d \ge \Big(\frac{\theta}{2} - \frac{\delta_0}{4}\Big)\alpha_d r_i^d\,,\\
\nu(g_i) &= \theta \int_{Q\setminus A} g_i\, d\mathcal{L}^d \le \Big(\frac{\theta}{2} + \frac{\delta_0}{4}\Big)\alpha_d r_i^d\,.
\end{aligned}$$

Let $\mathcal{U}$ be the weak neighbourhood of $\nu$ in $\mathcal{M}(Q)$ defined by

$$\begin{aligned}
\mathcal{U} = \Big\{ \varrho \in \mathcal{M}(Q) : \forall i \in I \quad & \varrho(f_i) > \nu(f_i) - \frac{\delta_0}{4}\alpha_d r_i^d\,,\\
& \forall i \in I \cup I^{\text{bd}} \quad \varrho(g_i) < \nu(g_i) + \frac{\delta_0}{4}\alpha_d r_i^d \Big\}.
\end{aligned}$$

Suppose that $\mathcal{C}_n \in \mathcal{U}$. Let us fix $i \in I$. The intersection with the ball $B(nx_i, nr_i)$ of the clusters connected to $\partial^{in}\Lambda(n)$ splits into a collection $\mathcal{C}(i)$ of $B(nx_i, nr_i)$–clusters which all intersect the boundary $\partial^{in}B(nx_i, nr_i)$. Setting

$$B_-(n,i) = B_-(nx_i, nr_i, -\nu_A(x_i))\,, \quad B_+(n,i) = B_+(nx_i, nr_i, -\nu_A(x_i))\,,$$

we have also

$$\mathcal{C}_n(f_i) \leq \sum_{C \in \mathcal{C}(i)} \big|C \cap \overset{\circ}{B}_-(n,i)\big| \leq \sum_{C \in \mathcal{C}(i)} \big|C \cap B_-(n,i)\big| ,$$
$$\mathcal{C}_n(g_i) \geq \sum_{C \in \mathcal{C}(i)} \big|C \cap \overline{B}_+(n,i)\big| \geq \sum_{C \in \mathcal{C}(i)} \big|C \cap B_+(n,i)\big| .$$

The very definition of the neighbourhood $\mathcal{U}$ and these inequalities yield that the event $\{\, \mathcal{C}_n \in \mathcal{U}\,\}$ is included in the event $\mathrm{Sep}_\theta(n, x_i, r_i, -\nu_A(x_i), \delta_0)$ defined before lemma 12.2. For $i \in I^{\mathrm{bd}}$, we obtain with a similar reasoning that the event $\{\, \mathcal{C}_n \in \mathcal{U}\,\}$ is included in the event $\mathrm{Sep}^{\mathrm{bd}}(n, x_i, r_i, -\nu_A(x_i), \delta_0)$ defined in section 12.2. We conclude that

$$\Phi^w_{\Lambda(n)}(\mathcal{C}_n \in \mathcal{U}) \leq$$
$$\Phi^w_{\Lambda(n)}\Big(\bigcap_{i\in I} \mathrm{Sep}_\theta(n, x_i, r_i, -\nu_A(x_i), \delta_0) \cap \bigcap_{i\in I^{\mathrm{bd}}} \mathrm{Sep}^{\mathrm{bd}}(n, x_i, r_i, -\nu_A(x_i), \delta_0)\Big) .$$

Yet the sets $B(x_i, r_i)$, $i \in I \cup I^{\mathrm{bd}}$, are compact and disjoint. Applying the interface lemma 12.1, lemma 12.2 and the decoupling lemma 10.10, we get

$$\limsup_{n\to\infty} \frac{1}{n^{d-1}} \ln \Phi^w_{\Lambda(n)}(\mathcal{C}_n \in \mathcal{U}) \leq - \sum_{i\in I\cup I^{\mathrm{bd}}} \alpha_{d-1} r_i^{d-1} \tau(-\nu_A(x_i)) \left(1 - c\sqrt{2\delta_0}\right)$$
$$\leq -\mathcal{I}(A)\,(1-\varepsilon') + \varepsilon' = -\mathcal{J}(\nu)(1-\varepsilon)$$

and we are done. □

## 18.4 Lower bound

We finish the proof of the large deviation principle with the large deviation lower bound.

**Lemma 18.5.** *Let $d \geq 3$ and let $\nu \in \mathcal{M}(Q)$. For any weak neighbourhood $\mathcal{U}$ of $\nu$ in $\mathcal{M}(Q)$, we have*

$$\liminf_{n\to\infty} \frac{1}{n^{d-1}} \ln \Phi^w_{\Lambda(n)}(\mathcal{C}_n \in \mathcal{U}) \geq -\mathcal{J}(\nu) .$$

*In two dimensions, the same lower bound holds for $\nu \in \mathcal{M}(Q)$ of the form $\nu(dx) = \theta 1_{Q\setminus A}\mathcal{L}^d(dx)$, where $A$ is an open connected subset of $Q$.*

*Proof.* If $\mathcal{J}(\nu) = +\infty$, there is nothing to prove. Let $\nu \in \mathcal{M}(Q)$ be such that $\mathcal{J}(\nu) < +\infty$. By definition of $\mathcal{J}$, there exists a Borel subset $A$ of $Q$ such that $\nu$ is the measure with density $\theta\, 1_{Q\setminus A}$ with respect to the Lebesgue measure and $\mathcal{J}(\nu) = \mathcal{I}(A)$. Let $\mathcal{U}$ be a weak neighbourhood of $\nu$ and let $\varepsilon > 0$. By corollary 14.12, there exists a polyhedral set $D$ such that $\overline{D} \subset \overset{\circ}{Q}$, the measure $\psi$ with density $\theta\, 1_{Q\setminus D}$ with respect to the Lebesgue measure belongs to $\mathcal{U}$ and moreover $\mathcal{I}(D) \leq \mathcal{I}(A) + \varepsilon$. We can even assume that the set $\overset{\circ}{D}$

is connected: in dimensions $d \geq 3$, we can connect together the components of $D$ with thin cylinders having negligible surface energy, while in dimensions $d = 2$, we suppose from the beginning that $A$ is open and connected, and these properties can be preserved when building the approximating polyhedral set $D$. By definition of a polyhedral element, the boundary $\partial D$ is the union of a finite number of $d-1$ dimensional sets $F_1, \dots, F_s$. For $j \in \{1 \cdots s\}$, let $\nu_j$ be a unit vector normal to $F_j$. We have

$$\sum_{1 \leq j \leq s} \mathcal{H}^{d-1}(F_j)\tau(\nu_j) \leq \mathcal{I}(A) + \varepsilon .$$

Moreover, for each $j$ in $\{1 \cdots s\}$, the relative boundary $\partial F_j$ has finite $d-2$ dimensional Hausdorff measure (we can achieve this by a slight perturbation of the polyhedral set if necessary).

Let $f \in C_c(\mathbb{R}^d, \mathbb{R})$. We are going to estimate the probability that $\mathcal{C}_n(f)$ and $\psi(f)$ are close. Let $\varepsilon > 0$. Since $f$ is continuous and has a compact support, it is uniformly continuous. Let $\delta \in ]0,1[$ be small enough so that

$$\mathcal{L}^d\big(\mathcal{V}(\partial D \cup \partial Q, 4d\delta)\big) \leq \varepsilon ,$$

$$\forall x, y \in \mathbb{R}^d \qquad |x-y| \leq \delta \quad \Rightarrow \quad |f(x) - f(y)| \leq \varepsilon ||f||_\infty .$$

We work with the lattice rescaled by a factor $L = \lfloor \delta n \rfloor$. For $\underline{y} \in \mathbb{Z}^d$, the block variable $Y(\underline{y})$ is the indicator function of the event $RVW(B(\underline{y}), B'(\underline{y}), L, \varepsilon)$ (see the definition before corollary 9.6). For $\delta$ small and $n$ large enough, there exists a connected set $\underline{E} \subset \mathbb{Z}^d$ such that

$$\mathrm{dist}\Big( \bigcup_{\underline{x} \in \underline{E}} B_n(\underline{x}), (\mathbb{R}^d \setminus Q) \cup D \Big) > 3d\delta , \quad \mathcal{L}^d\Big( (Q \setminus D) \setminus \bigcup_{\underline{x} \in \underline{E}} B_n(\underline{x}) \Big) \leq \varepsilon ,$$

and moreover $|\underline{E}| \leq c$ where $c = c(\delta)$ is a constant independent of $n$. Let $\gamma$ be a path of length less than $n$ in $\Lambda(n) \setminus \bigcup_{\underline{x} \in \underline{E}} B'(\underline{x})$ joining one vertex of $\partial^{in}\Lambda(n)$ to a vertex of one face $F$ of a block $B'(\underline{x}_0)$, where $\underline{x}_0 \in \underline{E}$ (we can for instance choose the leftmost face of one of the leftmost blocks of $\underline{E}$). Let

$$G = \Big\{ z \in \Lambda(n) \setminus \bigcup_{\underline{x} \in \underline{E}} B'(\underline{x}) : \exists y \in F \quad |z - y| = 1 \Big\} .$$

Let $T$ be the event

$$T = \{ \text{ the edges of } \gamma \text{ and those having one endpoint in } G \text{ are open } \} .$$

On the event $T$, the vertices of $F$ are connected together through the edges of $G$ and also connected to $\partial^{in}\Lambda(n)$ through the path $\gamma$. Let $\mathcal{E}$ be the intersection of the events

$$T , \qquad \{ Y(\underline{x}) = 1 \} , \quad \underline{x} \in \underline{E} , \qquad \mathrm{wall}\,(F_j, n) , \quad 1 \leq j \leq s .$$

See lemma 12.3 for the definition of the event "wall". Let us estimate the probability of $\mathcal{E}$:

$$\Phi^w_{\Lambda(n)}(\mathcal{E}) \geq \Phi^w_{\Lambda(n)}\Big(T \mid \bigcap_{\underline{x}\in\underline{E}}\{\,Y(\underline{x})=1\,\}\cap \bigcap_{1\leq j\leq s} \mathrm{wall}\,(F_j,n)\Big) \times$$
$$\Phi^w_{\Lambda(n)}\Big(\bigcap_{\underline{x}\in\underline{E}}\{\,Y(\underline{x})=1\,\} \mid \bigcap_{1\leq j\leq s} \mathrm{wall}\,(F_j,n)\Big)\, \Phi^w_{\Lambda(n)}\Big(\bigcap_{1\leq j\leq s} \mathrm{wall}\,(F_j,n)\Big)\,.$$

The number of edges involved in the event $T$ is at most $n+3d(\delta n)^{d-1}$ and these edges are distinct from the ones involved in the events $Y(\underline{x})$, $\underline{x}\in\underline{E}$ and $\mathrm{wall}\,(F_j,n)$, $1\leq j\leq s$, thus the conditional probability of $T$ is larger than

$$\exp\Big(\big(n+3d(\delta n)^{d-1}\big)\ln\Big(\frac{p}{p+q(1-p)}\Big)\Big)\,.$$

We deduce from lemmas 9.1, 9.3, corollary 9.6 and proposition 9.7 that the probability $\Phi^w_{\Lambda(n)}(Y(\underline{x})=1)$ goes to 1 as $n$ goes to $\infty$, uniformly over the boundary conditions on $B'(\underline{x})$. By the FKG inequality and lemma 12.3,

$$\liminf_{n\to\infty}\frac{1}{n^{d-1}}\ln\Phi^w_{\Lambda(n)}\Big(\bigcap_{1\leq j\leq s}\mathrm{wall}\,(F_j,n)\Big)\,\geq\, -\sum_{1\leq j\leq s}\mathcal{H}^{d-1}(F_j)\tau(\nu_j)\,.$$

Combining the previous inequalities, we get

$$\liminf_{n\to\infty}\frac{1}{n^{d-1}}\ln\Phi^w_{\Lambda(n)}(\mathcal{E})\,\geq\,-\mathcal{I}(A)-\varepsilon+3d\delta^{d-1}\ln\Big(\frac{p}{p+q(1-p)}\Big)\,.$$

Suppose that $\mathcal{E}$ occurs. Thanks to the event $T$ and the connectedness of $\underline{E}$, all the crossing clusters of the good blocks $B'(\underline{x})$, $\underline{x}\in\underline{E}$, are connected to $\partial^{in}\Lambda(n)$. Let

$$\underline{A}=\{\,\underline{y}\in\mathbb{Z}^d : B_n(\underline{y})\cap\mathrm{supp}\,f\neq\emptyset\,\}\,.$$

Since $\delta<1$, each block of size $L$ which intersects $n\mathrm{supp}\,f$ is included in $n\mathcal{V}(\mathrm{supp}\,f,d)$, thus

$$|\underline{A}|L^d\,\leq\, n^d\mathcal{L}^d(\mathcal{V}(\mathrm{supp}\,f,d))\,.$$

To evaluate $|\mathcal{C}_n(f)-\psi(f)|$, we decompose it on blocks of size $L/n$ and we use the uniform continuity of $f$:

$$|\mathcal{C}_n(f)-\psi(f)|\,\leq\,\sum_{\underline{y}\in\underline{A}}\Big|\int_{B_n(\underline{y})}f\,d\mathcal{C}_n-\int_{B_n(\underline{y})}f\,d\psi\Big|$$
$$\leq\,2\varepsilon||f||_\infty\mathcal{L}^d(\mathcal{V}(\mathrm{supp}\,f,d))+||f||_\infty\sum_{\underline{y}\in\underline{A}}\big|\mathcal{C}_n(B_n(\underline{y}))-\psi(B_n(\underline{y}))\big|\,.$$

Next we study the sum in the above inequality. Let $\underline{y}\in\underline{A}$. We distinguish several cases:

• If $\underline{y} \in \underline{E}$ then $B_n(\underline{y}) \subset Q \setminus D$ and

$$\left|\mathcal{C}_n(B_n(\underline{y})) - \psi(B_n(\underline{y}))\right| = \left|\mathcal{C}_n(B_n(\underline{y})) - \frac{\theta}{n^d}|B_n(\underline{y})|\right| \leq \frac{\varepsilon}{n^d}|B_n(\underline{y})|.$$

• If $\mathrm{dist}(B_n(\underline{y}), Q \setminus D) > 2d\delta$, then both terms $\mathcal{C}_n(B_n(\underline{y}))$ and $\psi(B_n(\underline{y}))$ are equal to 0. Indeed, the term $\mathcal{C}_n(B_n(\underline{y}))$ vanishes because the occurrence of the events wall $(F_j, n)$, $1 \leq j \leq s$, precludes the existence of an open path between $\partial^{in} \Lambda(n)$ and a site $x$ such that $\mathrm{dist}(x, Q \setminus D) > 2d\delta$.

• If $\mathrm{dist}(B_n(\underline{y}), Q \setminus D) \leq 2d\delta$ and $\underline{y} \notin \underline{E}$, then $B_n(\underline{y})$ is included in the set

$$\mathcal{V}(\partial D \cup \partial Q, 4d\delta) \cup \Big(Q \setminus D \setminus \bigcup_{\underline{x} \in \underline{E}} B_n(\underline{x})\Big)$$

whose Lebesgue measure is less than $2\varepsilon$.

Putting together the previous inequalities, we see that on the event $\mathcal{E}$,

$$|\mathcal{C}_n(f) - \psi(f)| \leq 5\varepsilon ||f||_\infty \mathcal{L}^d(\mathcal{V}(\mathrm{supp}\, f, d)).$$

These estimates can be carried out simultaneously for a finite number of functions $f_1, \dots, f_r \in C_c(\mathbb{R}^d, \mathbb{R})$: there exists $c > 0$ (depending on $f_1, \dots, f_r$) such that, for any $\varepsilon > 0$, for $\delta$ small enough, there exists an event $\mathcal{E}$ satisfying

$$\mathcal{E} \subset \Big\{ \forall i \in \{1 \cdots r\} \quad |\mathcal{C}_n(f_i) - \psi(f_i)| \leq c\varepsilon \Big\},$$

$$\liminf_{n \to \infty} \frac{1}{n^{d-1}} \ln \Phi^w_{\Lambda(n)}(\mathcal{E}) \geq -\mathcal{J}(\nu) - \varepsilon + 3d\delta^{d-1} \ln \Big(\frac{p}{p + q(1-p)}\Big).$$

Yet $\mathcal{U}$ is a weak neighbourhood of $\psi$, hence there exist $f_1, \dots, f_r \in C_c(\mathbb{R}^d, \mathbb{R})$ and $\varepsilon > 0$ such that

$$\Big\{ \varrho \in \mathcal{M}(Q) : \forall i \in \{1 \cdots r\} \quad |\varrho(f_i) - \psi(f_i)| \leq c\varepsilon \Big\} \subset \mathcal{U}.$$

Therefore, for any $\varepsilon > 0$ and for $\delta > 0$ small enough,

$$\liminf_{n \to \infty} \frac{1}{n^{d-1}} \ln \Phi^w_{\Lambda(n)}\big(\mathcal{C}_n \in \mathcal{U}\big) \geq -\mathcal{J}(\nu) - \varepsilon - c'\delta^{d-1},$$

where $c'$ is a positive constant depending on $p, q, d$ only. We conclude by sending successively $\delta$ and $\varepsilon$ to 0. □

# 19

# LDP for Ising

This chapter is devoted to the proof of the large deviation principle stated in theorem 7.1. We first check that $\mathcal{J}$ is a good rate function. By theorem 14.5, the surface energy $\mathcal{I}$ restricted to the Borel subsets of $Q$ is a good rate function. Since the map

$$A \in (\mathcal{B}(Q), L^1) \mapsto \big( -m^*\, 1_A(x) + m^* 1_{Q\setminus A}(x)\big)\, d\mathcal{L}^d(x) \in \mathcal{M}(Q)$$

is continuous, then $\mathcal{J}$ is a good rate function (this is a consequence of the contraction principle theorem 6.8). We work here within the topological vector space $\mathcal{M}(Q)$ endowed with the weak topology, where

$$Q = [-1/2, 1/2]^d$$

is the $d$–dimensional unit cube. The collection consisting of the finite intersections of the sets

$$\big\{\, \nu \in \mathcal{M}(Q) : |\nu(f)| \leq \eta \,\big\}, \quad \eta > 0, \quad f \in C_c(\mathbb{R}^d, \mathbb{R}),$$

is a basis of neighbourhoods of the origin. Our next goal is to prove that the random measure $\sigma_n$ is $\mathcal{J}$–tight in the sense of definition 6.3: there exists a positive constant $c$ such that

$$\begin{aligned}&\forall \eta, \lambda > 0, \quad \forall f \in C_c(\mathbb{R}^d, \mathbb{R}),\\ &\quad \limsup_{n\to\infty} \frac{1}{n^{d-1}} \ln \mu^+_{\Lambda(n),T}\big(\forall \nu \in \mathcal{J}^{-1}([0,\lambda]) \quad |\sigma_n(f) - \nu(f)| > \eta\big) \leq -c\lambda .\end{aligned}$$

We then prove the local large deviation upper bound and finally the large deviation lower bound.

## 19.1 Coarse grained image

The proof relies on the FK–Ising coupling measure $\mathbb{P}^+_n$ on edge–spin configurations in $\Lambda(n)$ with wired and plus boundary conditions (see section 4.3).

Let $n \in \mathbb{N}$ and let $K \in \mathbb{N}$. We work with the box $\Lambda(n)$ rescaled by a factor $K$, that is

$$\underline{\Lambda}(n) = \{\, \underline{x} \in \mathbb{Z}^d : B(\underline{x}) \cap \Lambda(n) \neq \emptyset \,\}.$$

For $\underline{x} \in \underline{\Lambda}(n)$, the block variable $X(\underline{x})$ is the indicator function of the event $R(B'(\underline{x}), K)$, i.e.,

• Inside the event block $B'(\underline{x})$ there is exactly one crossing cluster $C(\underline{x})$, it is the only cluster in $B'(\underline{x})$ having diameter larger than or equal to $K$ and it crosses every sub–box of $B'(\underline{x})$ with diameter larger than or equal to $K$.

In order to prove the exponential tightness, we build an auxiliary random measure $\widetilde{\sigma}_n$ which is exponentially contiguous to $\sigma_n$ and we provide suitable probabilistic estimates on the surface energy of $\widetilde{\sigma}_n$. Let $\underline{A}$ be a $\mathbb{L}^d$–connected subset of $\underline{\Lambda}(n)$. We recall that a residual $\mathbb{L}^{d,\infty}$–component $\underline{R}$ of $\underline{A}$ in $\underline{\Lambda}(n)$ is a $\mathbb{L}^{d,\infty}$–connected component of $\underline{\Lambda}(n) \setminus \underline{A}$. We define

$$\operatorname{fill} \underline{A} = \underline{A} \cup \bigcup \underline{R},$$

where the union runs over the residual $\mathbb{L}^{d,\infty}$–components $\underline{R}$ of $\underline{A}$ such that $\operatorname{diam}_\infty \underline{R} < \ln n$ and $\underline{R} \cap \partial^{in} \underline{\Lambda}(n) = \emptyset$. For a cluster $C$ of the configuration, we set

$$\underline{C} = \{\, \underline{x} \in \underline{\Lambda}(n) : B(\underline{x}) \cap C \neq \emptyset \,\}.$$

Since a cluster $C$ is $\mathbb{L}^d$–connected, so is $\underline{C}$. We say that a cluster $C$ is large if $\operatorname{diam}_\infty C \geq K \ln n$. Notice that if $C$ is a large cluster, then $\operatorname{diam}_\infty \underline{C} \geq \ln n$. For a large cluster $C$, we define

$$\widehat{\underline{C}} = \bigcup \operatorname{fill} \underline{A},$$

where the union runs over all the $\mathbb{L}^d$–connected components $\underline{A}$ of good blocks such that $\underline{C} \cap \underline{A} \neq \emptyset$. By lemma 18.1, if $C_1$, $C_2$ are two distinct large clusters, then $\widehat{\underline{C}}_1 \cap \widehat{\underline{C}}_2 = \emptyset$. Finally, the auxiliary measure $\widetilde{\sigma}_n$ is the measure with density

$$m^* 1_{P_n} - m^* 1_{M_n} + m^* 1_{Q \setminus M_n \setminus P_n}$$

with respect to the Lebesgue measure $\mathcal{L}^d$, where $M_n, P_n$ are the regions

$$P_n = \Big( \bigcup_{\substack{C \text{ large cluster} \\ \sigma(C)=+}} \bigcup_{\underline{x} \in \widehat{\underline{C}}} B_n(\underline{x}) \Big) \cup \Big( Q \setminus \Lambda\big(1 - \frac{5K}{n}\big) \Big),$$

$$M_n = \Big( \bigcup_{\substack{C \text{ large cluster} \\ \sigma(C)=-}} \bigcup_{\underline{x} \in \widehat{\underline{C}}} B_n(\underline{x}) \Big) \setminus \Big( Q \setminus \Lambda\big(1 - \frac{5K}{n}\big) \Big).$$

The union in the formula defining $P_n$ (respectively $M_n$) runs over the large clusters $C$ which are coloured positively (respectively negatively). The region $P_n$ corresponds to the plus phase while $M_n$ corresponds to the minus phase. The residual region $Q \setminus M_n \setminus P_n$ should typically be small. We drop the parameter $K$ from the notation, yet the sets $P_n, M_n$ and the measure $\widetilde{\sigma}_n$ depend on $K$.

**Corollary 19.1.** *There exists a constant $K_0(d,p,q)$ such that, for $K \geq K_0$, there exists a positive constant $c$ depending on $K,d,p,q$ such that*

$$\forall n \geq K \quad \forall u > 8d \qquad \limsup_{n\to\infty} \frac{1}{n^{d-1}} \ln \mathbb{P}_n^+\big(\mathcal{P}(P_n) + \mathcal{P}(M_n) \geq u\big) \leq -cu\,.$$

*Proof.* Let $\underline{\widehat{F}}$ be the union of all the $\mathbb{L}^{d,\infty}$-connected components of bad blocks intersecting simultaneously the rescaled box $\underline{\Lambda}(n - 5K\ln n)$ and one of the sets $\partial^{\,out}\underline{\widehat{C}}$, where $C$ is a large cluster. By definition, if $C$ is a large cluster, the $\mathbb{L}^d$ outer boundary $\partial^{\,out}\underline{\widehat{C}}$ of $\underline{\widehat{C}}$ consists of bad blocks whenever $\underline{\widehat{C}} \neq \emptyset$. In case $\underline{\widehat{C}} = \emptyset$, we define $\partial^{\,out}\underline{\widehat{C}}$ as $\underline{C}$ which again consists only of bad blocks. We set

$$\widehat{F} = \bigcup_{\underline{x}\in\underline{\widehat{F}}} B_n(\underline{x})\,.$$

By construction, the boundaries of $P_n$ and $M_n$ are located either on $\partial\Lambda(1) \cup \partial\Lambda(1-(5K/n)\ln n)$ or on the faces of the blocks of $\underline{\widehat{F}}$. Thus

$$\begin{aligned}\mathcal{P}(P_n)+\mathcal{P}(M_n) &\leq 2\mathcal{H}^{d-1}(\partial\Lambda(1)) + 2\mathcal{H}^{d-1}\Big(\partial\Lambda\big(1-\frac{5K}{n}\ln n\big)\Big) + 4d\Big(\frac{K}{n}\Big)^{d-1}|\underline{\widehat{F}}|\\ &\leq 8d + 4d\Big(\frac{K}{n}\Big)^{d-1}|\underline{\widehat{F}}|\,.\end{aligned}$$

The conclusion follows from lemma 18.2. □

## 19.2 Exponential contiguity

Let us fix $f \in C_c(\mathbb{R}^d, \mathbb{R})$. In order to estimate $|\sigma_n(f) - \widetilde{\sigma}_n(f)|$, we use another block argument with the scale $L = K\ln n$. We work with the lattice rescaled by a factor $L$. Let $\varepsilon > 0$. For $\underline{y} \in \mathbb{Z}^d$, the block variable $Y(\underline{y})$ is the indicator function of the event $RVW(B(\underline{y}), B'(\underline{y}), L, \varepsilon) \cap T(B(\underline{y}), \varepsilon)$ (see the definitions before corollary 9.6 and lemma 9.8). Since $f$ is continuous and has a compact support, it is uniformly continuous. We suppose that $L/n$ is less than 1 and small enough so that

$$\forall x, y \in \mathbb{R}^d \qquad |x-y| \leq \frac{L}{n} \quad\Rightarrow\quad |f(x) - f(y)| \leq \varepsilon ||f||_\infty\,.$$

Let

$$\underline{A} = \big\{\, \underline{y} \in \mathbb{Z}^d : B(\underline{y}) \cap \Lambda(n - 6L) \neq \emptyset \,\big\}\,.$$

Each block of size $L$ intersecting $\Lambda(n-6L)$ is included in $\Lambda(n)$, thus

$$|\underline{A}| L^d \leq \mathcal{L}^d(\Lambda(n)) = n^d\,.$$

We evaluate $|\sigma_n(f) - \widetilde{\sigma}_n(f)|$ by decomposing it on blocks of size $L/n$ and using the uniform continuity of $f$:

$$|\sigma_n(f)-\widetilde{\sigma}_n(f)| \leq \sum_{\underline{y}\in\underline{A}} \Big| \int_{B_n(\underline{y})} f\, d\sigma_n - \int_{B_n(\underline{y})} f\, d\widetilde{\sigma}_n \Big| + 2\frac{||f||_\infty}{n^d} \mathcal{L}^d\big(\Lambda(n)\setminus\Lambda(n-6L)\big)$$

$$\leq ||f||_\infty \sum_{\underline{y}\in\underline{A}} |\sigma_n(B_n(\underline{y})) - \widetilde{\sigma}_n(B_n(\underline{y}))| + ||f||_\infty \big(\frac{12dL}{n} + 2\varepsilon\big)\,.$$

Next we study the sum in the above inequality. Let $\underline{y} \in \underline{A}$.

- If $Y(\underline{y}) = 0$, then

$$\big|\sigma_n(B_n(\underline{y})) - \widetilde{\sigma}_n(B_n(\underline{y}))\big| \leq \frac{2}{n^d} |B_n(\underline{y})| 1_{Y(\underline{y})=0}\,.$$

Suppose next that $Y(\underline{y}) = 1$. Recall that $C(\underline{y})$ is the unique crossing cluster of $|B(\underline{y})|$. Denoting by $\sigma(C(\underline{y}))$ its colour, we decompose $\sigma_n(B_n(\underline{y}))$ as

$$\sigma_n(B_n(\underline{y})) = \frac{1}{n^d}\Big( \sum_{\substack{x\in B(\underline{y}) \\ x \not\longleftrightarrow \partial^{in} B(\underline{y})}} \sigma(x) + \sigma(C(\underline{y}))|C(\underline{y})| + \sum_{\substack{x\in B(\underline{y})\setminus C(\underline{y}) \\ x \longleftrightarrow \partial^{in} B(\underline{y})}} \sigma(x) \Big)\,.$$

Since $T(B(\underline{y}), \varepsilon)$ occurs, then the modulus of the first sum is less than $\varepsilon|B(\underline{y})|$. Since $RVW(B(\underline{y}), B'(\underline{y}), L, \varepsilon)$ occurs, then

$$(1-\varepsilon)\,\theta\,|B(\underline{y})| \leq |C(\underline{y})| \leq \big|\{ x \in B(\underline{y}) : x \longleftrightarrow \partial^{in} B(\underline{y}) \}\big| \leq (1+\varepsilon)\,\theta\,|B(\underline{y})|$$

whence

$$\Big| \sum_{\substack{x\in B(\underline{y})\setminus C(\underline{y}) \\ x \longleftrightarrow \partial^{in} B(\underline{y})}} \sigma(x) \Big| \leq (1+\varepsilon)\,\theta\,|B(\underline{y})| - |C(\underline{y})| \leq 2\varepsilon\,\theta\,|B(\underline{y})|\,.$$

In conclusion, recalling that $\theta = m^*$, we see that

$$\Big| |\sigma_n(B_n(\underline{y}))| - \frac{m^*}{n^d} |B_n(\underline{y})| \Big| \leq \frac{4\varepsilon}{n^d} |B_n(\underline{y})|\,.$$

Several subcases arise:

- $B_n(\underline{y}) \subset M_n$ and $\sigma_n(B_n(\underline{y})) < 0$. We have then

$$\big|\sigma_n(B_n(\underline{y})) - \widetilde{\sigma}_n(B_n(\underline{y}))\big| = \Big| - |\sigma_n(B_n(\underline{y}))| + \frac{m^*}{n^d} |B_n(\underline{y})| \Big| \leq \frac{4\varepsilon}{n^d} |B_n(\underline{y})|\,.$$

- $B_n(\underline{y}) \subset P_n$ and $\sigma_n(B_n(\underline{y})) > 0$. This case is symmetric to the previous one:

$$\big|\sigma_n(B_n(\underline{y})) - \widetilde{\sigma}_n(B_n(\underline{y}))\big| = \Big| |\sigma_n(B_n(\underline{y}))| - \frac{m^*}{n^d} |B_n(\underline{y})| \Big| \leq \frac{4\varepsilon}{n^d} |B_n(\underline{y})|\,.$$

- $B_n(\underline{y}) \cap M_n \neq \emptyset$ and $B_n(\underline{y}) \not\subset M_n$. Then $B_n(\underline{y})$ meets the boundary of $M_n$.
- $B_n(\underline{y}) \cap P_n \neq \emptyset$ and $B_n(\underline{y}) \not\subset P_n$. Then $B_n(\underline{y})$ meets the boundary of $P_n$.

In the last two cases, we bound

$$\Big|\sigma_n(B_n(\underline{y})) - \widetilde{\sigma}_n(B_n(\underline{y}))\Big| \leq 2\mathcal{L}^d(B_n(\underline{y}))$$

and we notice that

$$B_n(\underline{y}) \subset \Big\{ x \in \mathbb{R}^d : \mathrm{d}_\infty\big(x, (\partial P_n \cup \partial M_n) \cap \Lambda(1-3L/n)\big) \leq \frac{L}{n} \Big\}.$$

Moreover

$$\mathcal{L}^d\Big(\Big\{ x \in \mathbb{R}^d : \mathrm{d}_\infty\big(x, (\partial P_n \cup \partial M_n) \cap \Lambda(1-3L/n)\big) \leq \frac{L}{n} \Big\}\Big)$$
$$\leq \big(|\partial^{in} nP_n| + |\partial^{in} nM_n|\big)\Big(\frac{2L+2}{n}\Big)^d \leq \big(\mathcal{P}(M_n) + \mathcal{P}(P_n)\big)\frac{(3L)^d}{n}.$$

- $B_n(\underline{y}) \subset M_n$ and $\sigma_n(B_n(\underline{y})) > 0$;
- $B_n(\underline{y}) \subset P_n$ and $\sigma_n(B_n(\underline{y})) < 0$.

These conditions imply that the whole block $B_n(\underline{y})$ has been added to $M_n$ or $P_n$ by the filling operation. Indeed, since $Y(\underline{y}) = 1$, the crossing cluster $C'(\underline{y})$ of $B'(\underline{y})$ is a large cluster intersecting $B(\underline{y})$ and it is the only one. Since $T(B(\underline{y}), \varepsilon)$ occurs as well, the sign of $\sigma_n(B_n(\underline{y}))$ is given by the color of $C'(\underline{y})$. Suppose for instance that $\sigma_n(B_n(\underline{y})) > 0$. Then $C'(\underline{y})$ is colored $+$ and no large cluster colored $-$ intersects $B_n(\underline{y})$. For $B_n(\underline{y})$ to be included in $M_n$, there must exist a large cluster $C$ colored $-$ such that $B_n(\underline{y}) \subset \widehat{\underline{C}} \setminus \underline{C}$. Yet the regions which are added by the filling operation have a diameter at most $K(\ln n - 1)$, so this case cannot occur.

- $B_n(\underline{y}) \cap M_n = \emptyset$ and $B_n(\underline{y}) \cap P_n = \emptyset$. In this case, $B(\underline{y}) \subset \widehat{F}$.

Summing the previous inequalities over $\underline{y} \in \underline{A}$, we get

$$|\sigma_n(f) - \widetilde{\sigma}_n(f)| \leq$$
$$||f||_\infty\Big(\frac{2}{|\underline{A}|}\sum_{\underline{y}\in\underline{A}} 1_{Y(\underline{y})=0} + 2\cdot 3^d \frac{L^d}{n}(\mathcal{P}(M_n) + \mathcal{P}(P_n)) + \frac{L^d}{n^d}|\underline{\widehat{F}}| + 12d\frac{L}{n} + 6\varepsilon\Big).$$

Corollary 9.6, proposition 9.7 and lemma 9.8 imply that for $L$ large enough the block process $(Y(\underline{y}), \underline{y} \in \underline{A})$ satisfies the condition stated at the beginning of section 9.4 with $\delta = \varepsilon/4$. Let $n$ be large enough, so that $12d(L/n) < \varepsilon$. Using the estimates of lemmas 9.11, 18.2 and corollary 19.1, we obtain

$$\mathbb{P}_n^+\Big(|\sigma_n(f) - \widetilde{\sigma}_n(f)| \geq ||f||_\infty(9 + 2\cdot 3^d)\varepsilon\Big) \leq \mathbb{P}_n^+\Big(\frac{L^d}{n^d}|\underline{\widehat{F}}| \geq \varepsilon\Big)$$
$$+\mathbb{P}_n^+\Big(\frac{2}{|\underline{A}|}\sum_{\underline{y}\in\underline{A}} 1_{Y(\underline{y})=0} \geq \varepsilon\Big) + \mathbb{P}_n^+\Big(\frac{L^d}{n}(\mathcal{P}(M_n) + \mathcal{P}(P_n)) \geq \varepsilon\Big)$$
$$\leq b\exp\Big(-c\varepsilon\frac{n^d}{L^d}\Big) + 3^d\exp\Big(-\Lambda^*\big(\frac{\varepsilon}{2},\frac{\varepsilon}{4}\big)\Big\lfloor\frac{n-6L}{6L}\Big\rfloor^d\Big) + b\exp\Big(-c\varepsilon\frac{n^d}{L^d}\Big).$$

Therefore

$$\forall \varepsilon > 0 \qquad \limsup_{n\to\infty} \frac{1}{n^{d-1}} \ln \mathbb{P}_n^+\big(|\sigma_n(f) - \widetilde{\sigma}_n(f)| > \varepsilon\big) = -\infty .$$

This estimate and corollary 19.1 imply the $\mathcal{J}$–tightness announced at the beginning of the section.

## 19.3 Local upper bound

The final ingredient needed to get the large deviation upper bound is the local large deviation upper bound, which we prove in this section.

**Lemma 19.2.** *Let $\nu \in \mathcal{M}(Q)$ be such that $\mathcal{J}(\nu) < \infty$. For every $\varepsilon > 0$, there exists a weak neighbourhood $\mathcal{U}$ of $\nu$ in $\mathcal{M}(Q)$ such that*

$$\limsup_{n\to\infty} \frac{1}{n^{d-1}} \ln \mu^+_{\Lambda(n),T}\big(\sigma_n \in \mathcal{U}\big) \leq -(1-\varepsilon)\,\mathcal{J}(\nu) .$$

*Proof.* By definition of $\mathcal{J}$, since $\mathcal{J}(\nu) < \infty$, there exists a Borel subset $A$ of $Q$ such that $\nu$ is the measure with density $-m^*\,1_A + m^* 1_{Q\setminus A}$ with respect to the Lebesgue measure, and $\mathcal{J}(\nu) = \mathcal{I}(A) = \mathcal{I}(\mathbb{R}^d \setminus A)$. If $\mathcal{I}(A) = 0$, there is nothing to prove. Suppose that $\mathcal{I}(A) > 0$. Let $\varepsilon > 0$ and let us set $\varepsilon' = \varepsilon(1+1/\mathcal{I}(A))^{-1}$. Pick $\delta \in \,]0,1[$ such that $c\sqrt{10\delta} < \varepsilon'$ where $c = c(p,d,\zeta)$ is the constant appearing in the interface lemma 12.1. Let $B(x_i,r_i)$, $i \in I \cup I^{\mathrm{bd}}$, be a finite collection of disjoint balls associated with $A, \varepsilon'$ and $\delta/8$, as given in the covering lemma 14.7. For $i$ in $I \cup I^{\mathrm{bd}}$, let $f_i, g_i$ be two continuous functions having compact support and taking values in $[0,1]$ such that

$$\begin{gathered}
\forall x \in \mathbb{R}^d \setminus \overset{\circ}{B}_-(x_i,r_i,\nu_A(x_i)) \qquad f_i(x) = 0 ,\\
\forall x \in \overline{B}_+(x_i,r_i,\nu_A(x_i)) \qquad g_i(x) = 1 ,\\
\Big(\frac{1}{2} - \frac{\delta}{4m^*}\Big)\alpha_d r_i^d \leq \int f_i \, d\mathcal{L}^d , \qquad \int g_i \, d\mathcal{L}^d \leq \Big(\frac{1}{2} + \frac{\delta}{4m^*}\Big)\alpha_d r_i^d ,
\end{gathered}$$

and there exists $s_i > 0$ such that, if we set

$$\begin{aligned}
D^i_- &= \big\{\, y \in B_-(x_i,r_i,\nu_A(x_i)) : \mathrm{dist}(y, \mathbb{R}^d \setminus B_-(x_i,r_i,\nu_A(x_i))) \leq s_i \,\big\} ,\\
D^i_+ &= \big\{\, y \notin B_+(x_i,r_i,\nu_A(x_i)) : \mathrm{dist}(y, B_+(x_i,r_i,\nu_A(x_i))) \leq s_i \,\big\}
\end{aligned}$$

then we have

$$\begin{aligned}
\mathcal{L}^d(D^i_-) &\leq \frac{\delta}{8}\alpha_d r_i^d , \qquad \forall x \in B_-(x_i,r_i,\nu_A(x_i))) \setminus D^i_- \qquad f_i(x) = 1 ,\\
\mathcal{L}^d(D^i_+) &\leq \frac{\delta}{8}\alpha_d r_i^d , \qquad \forall x \in \mathbb{R}^d \setminus B_+(x_i,r_i,\nu_A(x_i)) \setminus D^i_+ \qquad g_i(x) = 0 .
\end{aligned}$$

These conditions imply that

$$\nu(f_i) = -m^* \int_A f_i \, d\mathcal{L}^d + m^* \int_{Q\setminus A} f_i \, d\mathcal{L}^d$$
$$\leq -m^* \int f_i \, d\mathcal{L}^d + 2\mathcal{L}^d\big(B_-(x_i, r_i, \nu_A(x_i)) \setminus A\big)$$
$$\leq (-m^* + \delta)\frac{1}{2}\alpha_d r_i^d \,,$$
$$\nu(g_i) = -m^* \int_A g_i \, d\mathcal{L}^d + m^* \int_{Q\setminus A} g_i \, d\mathcal{L}^d$$
$$\geq m^* \int g_i \, d\mathcal{L}^d - 2m^*\mathcal{L}^d\big(A \cap B_+(x_i, r_i, \nu_A(x_i))\big) - 2m^*\mathcal{L}^d(D_+^i)$$
$$\geq (m^* - \delta)\frac{1}{2}\alpha_d r_i^d \,.$$

Let $\mathcal{U}$ be the weak neighbourhood of $\nu$ in $\mathcal{M}(Q)$ defined by

$$\mathcal{U} = \Big\{ \varrho \in \mathcal{M}(Q) : \forall i \in I \quad \varrho(f_i) < \nu(f_i) + \frac{\delta}{2}\alpha_d r_i^d \,, \varrho(g_i) > \nu(g_i) - \frac{\delta}{2}\alpha_d r_i^d \Big\}.$$

We work with the lattice rescaled by a factor $L = \ln n$. Let $\delta > 0$. For $\underline{x} \in \mathbb{Z}^d$, the block variable $X(\underline{x})$ is the indicator function of the event

$$RVW(B(\underline{x}), B'(\underline{x}), L, \delta) \cap T(B(\underline{x}), \delta)$$

(see the definitions before corollary 9.6 and lemma 9.8). Let us fix $i \in I$. Let $*$ be a symbol representing either $-$ or $+$. We define

$$B_*(n,i) = B_*(nx_i, nr_i, \nu_A(x_i))\,, \quad \underline{B}_*(n,i) = \big\{ \underline{x} \in \mathbb{Z}^d : B(\underline{x}) \subset \overset{\circ}{B}_*(n,i) \big\}\,.$$

For $n$ large enough and $* = -$ or $* = +$, we have

$$\mathcal{L}^d\Big(B_*(n,i) \setminus \bigcup_{\underline{x} \in \underline{B}_*(n,i)} B(\underline{x})\Big) \leq \delta\mathcal{L}^d(B_*(n,i))\,,$$
$$\big|\mathbb{Z}^d \cap D_*^i\big| \leq \delta\mathcal{L}^d(B_*(n,i))\,.$$

We define $\mathcal{S}_*$ as the collection of the clusters which are included in one of the boxes $B(\underline{x})$, $\underline{x} \in \underline{B}_*(n,i)$, but which do not intersect the boundary of the box:

$$\mathcal{S}_* = \bigcup_{\underline{x} \in \underline{B}_*(n,i)} \big\{ C \text{ cluster in } B(\underline{x}) \text{ such that } C \cap \partial^{in} B(\underline{x}) = \emptyset \big\}\,.$$

Let $\underline{x} \in \underline{B}_*(n,i)$ be such that $X(\underline{x}) = 1$. Then

$$\sum_{C \in \mathcal{S}_*} |C \cap B(\underline{x})| \geq |B(\underline{x})| - \big|\big\{ x \in B(\underline{x}) : x \longleftrightarrow \partial^{in} B(\underline{x}) \big\}\big| \geq (1-\theta-\delta)|B(\underline{x})|\,.$$

Summing these inequalities, we get

$$\sum_{C\in\mathcal{S}_*} \big|C\cap B_*(n,i)\big| \geq (1-\theta-2\delta)\mathcal{L}^d\big(B_*(n,i)\big)\,.$$

We define also $\mathcal{C}$ as the collection of the $B(nx_i, nr_i)$–clusters which do not belong to $\mathcal{S}_- \cup \mathcal{S}_+$. For a cluster $C \in \mathcal{C}$, we denote by $\sigma(C)$ its colour. For $\underline{x} \in \underline{B}_-(n,i) \cup \underline{B}_+(n,i)$, we have

$$n^d\sigma_n(B(\underline{x})) = \sum_{C\in\mathcal{C}} \sigma(C)\big|C\cap B(\underline{x})\big| + \sum_{\substack{x\in B(\underline{x})\\ x \not\leftrightarrow \partial^{in} B(\underline{x})}} \sigma(x)\,.$$

Whenever $X(\underline{x}) = 1$, the modulus of the last sum is less than $\delta|B(\underline{x})|$. Suppose that, for $* = -$ and $* = +$,

$$L^d \sum_{\underline{x}\in\underline{B}_*(n,i)} 1_{X(\underline{x})=0} \leq \delta\mathcal{L}^d\big(B_*(n,i)\big)\,.$$

Summing the previous inequalities, we get

$$n^d\sigma_n(f_i) \geq \sum_{C\in\mathcal{C}} \sigma(C)\big|C\cap B_-(n,i)\big| - 5\delta\mathcal{L}^d\big(B_-(n,i)\big)\,,$$

$$n^d\sigma_n(g_i) \leq \sum_{C\in\mathcal{C}} \sigma(C)\big|C\cap B_+(n,i)\big| + 5\delta\mathcal{L}^d\big(B_+(n,i)\big)\,.$$

Let us denote by $\mathcal{C}_-$ (respectively $\mathcal{C}_+$) the collection of the negatively (respectively positively) coloured clusters of $\mathcal{C}$. The four collections $\mathcal{S}_-, \mathcal{S}_+, \mathcal{C}_-, \mathcal{C}_+$ are pairwise disjoint. Suppose in addition that $\sigma_n \in \mathcal{U}$. The very definition of the neighbourhood $\mathcal{U}$ and the previous inequalities yield that

$$\sum_{C\in\mathcal{C}_+} \big|C\cap B_-(n,i)\big| - \sum_{C\in\mathcal{C}_-\cup\mathcal{S}_-} \big|C\cap B_-(n,i)\big| \leq -(1-9\delta)\mathcal{L}^d\big(B_-(n,i)\big)\,,$$

$$\sum_{C\in\mathcal{C}_+\cup\mathcal{S}_+} \big|C\cap B_+(n,i)\big| - \sum_{C\in\mathcal{C}_-} \big|C\cap B_+(n,i)\big| \geq (1-9\delta)\mathcal{L}^d\big(B_+(n,i)\big)\,.$$

Let $n$ be large enough so that $\big|B_+(n,i)\cap\mathbb{Z}^d\big| \leq (1+\delta)\mathcal{L}^d\big(B_+(n,i)\big)$. These inequalities imply furthermore that

$$\sum_{C\in\mathcal{C}_-\cup\mathcal{S}_-} \big|C\cap B_-(n,i)\big| \geq (1-10\delta)\mathcal{L}^d\big(B_-(n,i)\big)\,,$$

$$\sum_{C\in\mathcal{C}_-\cup\mathcal{S}_-} \big|C\cap B_+(n,i)\big| = \sum_{C\in\mathcal{C}_-} \big|C\cap B_+(n,i)\big| \leq 10\delta\mathcal{L}^d\big(B_+(n,i)\big)\,,$$

so that the collection $\mathcal{C}_- \cup \mathcal{S}_-$ realizes the event $\mathrm{Sep}(n, x_i, r_i, \nu_A(x_i), 10\delta)$ defined before lemma 12.1. A similar reasoning holds for the indices $i \in I^{\mathrm{bd}}$: the event $\mathrm{Sep}^{\mathrm{bd}}(n, x_i, r_i, \nu_A(x_i), 10\delta)$ defined at the end of section 12.2 is realized. We conclude that

$$\mu^+_{\Lambda(n),T}(\sigma_n \in \mathcal{U}) \leq \sum_{*=-,+}\sum_{i\in I} \mathbb{P}^+_n\Big(L^d \sum_{\underline{x}\in \underline{B}_*(n,i)} 1_{X(\underline{x})=0} > \delta\mathcal{L}^d\big(B_*(n,i)\big)\Big)$$
$$+\Phi^w_{\Lambda(n)}\Big(\bigcap_{i\in I}\mathrm{Sep}(n,x_i,r_i,\nu_A(x_i),10\delta)\cap\bigcap_{i\in I^{\mathrm{bd}}}\mathrm{Sep}^{\mathrm{bd}}(n,x_i,r_i,\nu_A(x_i),10\delta)\Big).$$

Proposition 9.7 and lemma 9.8 imply that for $L$ large enough the block process satisfies the condition stated at the beginning of section 9.4, with $2^{-d-2}\delta$ instead of $\delta$. Applying lemma 9.11 with a rescaled box $\underline{\Lambda}(n,i)$ such that

$$\underline{B}_-(n,i)\cup\underline{B}_+(n,i)\subset\underline{\Lambda}(n,i)\,,\qquad 2^d\mathcal{L}^d\big(B(nx_i,nr_i)\big)\geq L^d\big|\underline{\Lambda}(n,i)\big|\,,$$

we obtain that, for $*=-,+$,

$$\mathbb{P}^+_n\Big(L^d\sum_{\underline{x}\in\underline{B}_*(n,i)}1_{X(\underline{x})=0}>\delta\mathcal{L}^d\big(B_*(n,i)\big)\Big)$$
$$\leq \mathbb{P}^+_n\Big(\sum_{\underline{x}\in\underline{\Lambda}(n,i)}1_{X(\underline{x})=0}>2^{-d-1}\delta|\underline{\Lambda}(n,i)|\Big)$$
$$\leq 3^d\exp\Big(-\Lambda^*(2^{-d-1}\delta,2^{-d-2}\delta)\,\Big\lfloor\frac{2nr_i}{6L}\Big\rfloor^d\Big).$$

Since $\Lambda^*(2^{-d-1}\delta,2^{-d-2}\delta)$ is positive, we conclude that

$$\limsup_{n\to\infty}\frac{1}{n^{d-1}}\ln\mathbb{P}^+_n\Big(L^d\sum_{\underline{x}\in\underline{B}_*(n,i)}1_{X(\underline{x})=0}>\delta\mathcal{L}^d\big(B_*(n,i)\big)\Big)=-\infty\,.$$

Next, the sets $B(x_i,r_i)$, $i\in I\cup I^{\mathrm{bd}}$, are compact and disjoint. Applying the interface lemma 12.1 and the decoupling lemma 10.10, we get

$$\limsup_{n\to\infty}\frac{1}{n^{d-1}}\ln\mu^+_{\Lambda(n),T}(\sigma_n\in\mathcal{U})\leq-\sum_{i\in I\cup I^{\mathrm{bd}}}\alpha_{d-1}r_i^{d-1}\tau(\nu_A(x_i))\,(1-c\sqrt{10\delta})$$
$$\leq-\mathcal{I}(A)\,(1-\varepsilon')+\varepsilon'=-\mathcal{J}(\nu)(1-\varepsilon)$$

and we are done. □

## 19.4 Lower bound

We finish the proof of the large deviation principle with the large deviation lower bound.

**Lemma 19.3.** *Let $\nu\in\mathcal{M}(Q)$. For any weak neighbourhood $\mathcal{U}$ of $\nu$ in $\mathcal{M}(Q)$, we have*

$$\liminf_{n\to\infty}\frac{1}{n^{d-1}}\ln\mu^+_{\Lambda(n),T}\big(\sigma_n\in\mathcal{U}\big)\geq-\mathcal{J}(\nu)\,.$$

*Proof.* If $\mathcal{J}(\nu) = +\infty$, there is nothing to prove. Let $\nu \in \mathcal{M}(Q)$ be such that $\mathcal{J}(\nu) < +\infty$. By definition of $\mathcal{J}$, there exists a Borel subset $A$ of $Q$ such that $\nu$ is the measure with density $-m^* 1_A + m^* 1_{Q \setminus A}$ with respect to the Lebesgue measure and $\mathcal{J}(\nu) = \mathcal{I}(A)$. Let $\mathcal{U}$ be a weak neighbourhood of $\nu$ and let $\varepsilon > 0$. By corollary 14.12, there exists a polyhedral set $D$ such that $\overline{D} \subset \overset{\circ}{Q}$, the measure $\psi$ with density $-m^* 1_D + m^* 1_{Q \setminus D}$ with respect to the Lebesgue measure belongs to $\mathcal{U}$ and moreover $\mathcal{I}(D) \leq \mathcal{I}(A) + \varepsilon$. By definition of a polyhedral element, the boundary $\partial D$ is the union of a finite number of $d-1$ dimensional sets $F_1, \dots, F_s$. For $j \in \{1 \cdots s\}$, let $\nu_j$ be a unit vector normal to $F_j$. We have

$$\sum_{1 \leq j \leq s} \mathcal{H}^{d-1}(F_j)\tau(\nu_j) \leq \mathcal{I}(A) + \varepsilon .$$

Moreover, for each $j$ in $\{1 \cdots s\}$, the relative boundary $\partial F_j$ has finite $d-2$ dimensional Hausdorff measure (we can achieve this by a slight perturbation of the polyhedral set if necessary).

Let $f \in C_c(\mathbb{R}^d, \mathbb{R})$. We are going to estimate the probability that $|\sigma_n(f) - \psi(f)|$ is small. Let $\varepsilon > 0$. Since $f$ is continuous and has a compact support, it is uniformly continuous. Let $\delta \in ]0,1[$ be small enough so that

$$\mathcal{L}^d\big(\mathcal{V}(\partial D \cup \partial Q, 4d\delta)\big) \leq \varepsilon ,$$
$$\forall x, y \in \mathbb{R}^d \qquad |x - y| \leq \delta \quad \Rightarrow \quad |f(x) - f(y)| \leq \varepsilon ||f||_\infty .$$

We work with the lattice rescaled by a factor $L = \lfloor \delta n \rfloor$. For $\underline{y} \in \mathbb{Z}^d$, the block variable $Y(\underline{y})$ is the indicator function of the event

$$RVW(B(\underline{y}), B'(\underline{y}), L, \varepsilon) \cap T(B(\underline{y}), \varepsilon)$$

(see the definitions before corollary 9.6 and lemma 9.8). For $\delta$ small and $n$ large enough, there exist two sets $\underline{E}$ and $\underline{F}$ such that

$$\mathrm{dist}\Big( \bigcup_{\underline{x} \in \underline{E}} B_n(\underline{x}), \mathbb{R}^d \setminus D \Big) \geq 3d\delta , \quad \mathcal{L}^d\Big( D \setminus \bigcup_{\underline{x} \in \underline{E}} B_n(\underline{x}) \Big) \leq \varepsilon ,$$
$$\mathrm{dist}\Big( \bigcup_{\underline{x} \in \underline{F}} B_n(\underline{x}), \mathbb{R}^d \setminus (Q \setminus D) \Big) \geq 3d\delta , \quad \mathcal{L}^d\Big( (Q \setminus D) \setminus \bigcup_{\underline{x} \in \underline{F}} B_n(\underline{x}) \Big) \leq \varepsilon ,$$

and moreover $|\underline{E}| + |\underline{F}| \leq c$ where $c = c(\delta)$ is a constant independent of $n$. Let $\mathcal{E}$ be the intersection of the events

$$\big\{ Y(\underline{x}) = 1 \big\} , \quad \underline{x} \in \underline{E} \cup \underline{F} , \qquad \mathrm{wall}\,(F_j, n) , \quad 1 \leq j \leq s .$$

See lemma 12.3 for the definition of the event "wall". Let us estimate the probability of $\mathcal{E}$:

$$\mathbb{P}_n^+(\mathcal{E}) \geq$$
$$\mathbb{P}_n^+\Big( \bigcap_{\underline{x} \in \underline{E} \cup \underline{F}} \{ Y(\underline{x}) = 1 \} \;\Big|\; \bigcap_{1 \leq j \leq s} \mathrm{wall}\,(F_j, n) \Big) \, \Phi^w_{\Lambda(n)}\Big( \bigcap_{1 \leq j \leq s} \mathrm{wall}\,(F_j, n) \Big) .$$

From proposition 9.7 and lemma 9.8, the probability $\mathbb{P}_n^+(Y(\underline{x})=1)$ goes to 1 as $n$ goes to $\infty$, uniformly over the boundary conditions on $B'(\underline{x})$. By the FKG inequality and lemma 12.3,

$$\liminf_{n\to\infty} \frac{1}{n^{d-1}} \ln \Phi^w_{\Lambda(n)}\Big( \bigcap_{1\le j\le s} \text{wall}\,(F_j,n)\Big) \geq - \sum_{1\le j\le s} \mathcal{H}^{d-1}(F_j)\tau(\nu_j)\,.$$

Combining the previous inequalities, we get

$$\liminf_{n\to\infty} \frac{1}{n^{d-1}} \ln \mathbb{P}_n^+(\mathcal{E}) \geq -\mathcal{I}(A) - \varepsilon\,.$$

Suppose that $\mathcal{E}$ occurs. Let $\underline{E}_i$, $i\in I$ (respectively $\underline{F}_i$, $i\in J$) be the $\mathbb{L}^d$–connected components of $\underline{E}$ (respectively $\underline{F}$). For $i\in I$, all the crossing clusters of the good blocks $B(\underline{x})$, $\underline{x}\in \underline{E}_i$, are connected together and belong to one big cluster, which we denote by $C^i_-$; similarly, for $i\in J$, all the crossing clusters of the good blocks $B(\underline{x})$, $\underline{x}\in\underline{F}_i$, belong to one big cluster $C^i_+$. Because of the events wall $(F_j,n)$, $1\le j\le s$, for any $i\in I$ and $k\in J$, the two clusters $C^i_-$ and $C^k_+$ are disjoint and moreover $C^i_-$ cannot be connected to $\partial^{in}\Lambda(n)$. Suppose in addition that the clusters $C^i_-$, $i\in I$, are coloured negatively and that the clusters $C^i_+$, $i\in J$, are coloured positively. Let

$$\underline{A} = \{\,\underline{y}\in\mathbb{Z}^d : B_n(\underline{y})\cap Q\neq\emptyset\,\}\,.$$

Since $\delta\le 1$, each block of size $L$ which intersects $nQ$ is included in $n\mathcal{V}(Q,d)$, thus $|\underline{A}|L^d \le n^d\mathcal{L}^d(\mathcal{V}(Q,d))$. We evaluate $|\sigma_n(f)-\psi(f)|$ by decomposing it on blocks of size $L/n$ and using the uniform continuity of $f$:

$$\begin{aligned}|\sigma_n(f)-\psi(f)| &\le \sum_{\underline{y}\in\underline{A}} \Big| \int_{B_n(\underline{y})} f\,d\sigma_n - \int_{B_n(\underline{y})} f\,d\psi \Big| \\ &\le 2\varepsilon\|f\|_\infty \mathcal{L}^d(\mathcal{V}(Q,d)) + \|f\|_\infty \sum_{\underline{y}\in\underline{A}} \big|\sigma_n(B_n(\underline{y}))-\psi(B_n(\underline{y}))\big|\,.\end{aligned}$$

Next we study the sum in the above inequality. Let $\underline{y}\in\underline{A}$. As in the proof of the exponential contiguity (section 19.2), we notice that, if $Y(\underline{y})=1$, then

$$\Big| |\sigma_n(B_n(\underline{y}))| - \frac{m^*}{n^d}|B_n(\underline{y})| \Big| \le \frac{4\varepsilon}{n^d}|B_n(\underline{y})|\,.$$

We distinguish several cases:

- If $\underline{y}\in\underline{E}$ then $B_n(\underline{y})\subset D$ and

$$\big|\sigma_n(B_n(\underline{y}))-\psi(B_n(\underline{y}))\big| = \Big| -|\sigma_n(B_n(\underline{y}))| + \frac{m^*}{n^d}|B_n(\underline{y})| \Big| \le \frac{4\varepsilon}{n^d}|B_n(\underline{y})|\,.$$

- If $\underline{y}\in\underline{F}$ then $B_n(\underline{y})\subset Q\setminus D$ and

$$\left|\sigma_n(B_n(\underline{y})) - \psi(B_n(\underline{y}))\right| = \left|\left|\sigma_n(B_n(\underline{y}))\right| - \frac{m^*}{n^d}\left|B_n(\underline{y})\right|\right| \leq \frac{4\varepsilon}{n^d}\left|B_n(\underline{y})\right|.$$

• If $\underline{y} \notin \underline{E} \cup \underline{F}$ then $B_n(\underline{y})$ is included in the set

$$\Big(D \setminus \bigcup_{\underline{x}\in\underline{E}} B_n(\underline{x})\Big) \cup \Big((Q\setminus D) \setminus \bigcup_{\underline{x}\in\underline{F}} B_n(\underline{x})\Big) \cup \mathcal{V}(\partial D \cup \partial Q, 4d\delta)$$

whose Lebesgue measure is less than $3\varepsilon$. Thus, if the event $\mathcal{E}$ occurs, if the clusters $C_-^i$, $i \in I$, are coloured negatively and if the clusters $C_+^i$, $i \in J$, are coloured positively, then we have

$$|\sigma_n(f) - \psi(f)| \leq 7\varepsilon ||f||_\infty \big(\mathcal{L}^d(\mathcal{V}(Q,d)) + 1\big).$$

These estimates can be carried out simultaneously for a finite number of functions $f_1, \dots, f_r \in C_c(\mathbb{R}^d, \mathbb{R})$: there exists $c > 0$ (depending on $f_1, \dots, f_r$) such that, for any $\varepsilon > 0$, there exists an event $\mathcal{E}$ satisfying

$$\mu^+_{\Lambda(n),T}\Big(\forall i \in \{1\cdots r\} \quad |\sigma_n(f_i) - \psi(f_i)| \leq c\varepsilon\Big) \geq 2^{-|I|-|J|}\mathbb{P}_n^+(\mathcal{E}),$$

$$\liminf_{n\to\infty} \frac{1}{n^{d-1}} \ln \mathbb{P}_n^+(\mathcal{E}) \geq -\mathcal{J}(\nu) - \varepsilon.$$

Here the factor $2^{-|I|-|J|}$ accounts for the probability that the clusters $C_-^i$, $i \in I$, $C_+^i$, $i \in J$, are coloured adequately. The cardinalities of $I$ and $J$ depend on $\varepsilon$, but not on $n$. Yet $\mathcal{U}$ is a weak neighbourhood of $\psi$, hence there exist $f_1, \dots, f_r \in C_c(\mathbb{R}^d, \mathbb{R})$ and $\varepsilon > 0$ such that

$$\Big\{\varrho \in \mathcal{M}(Q) : \forall i \in \{1\cdots r\} \quad |\varrho(f_i) - \psi(f_i)| \leq c\varepsilon\Big\} \subset \mathcal{U}.$$

Therefore

$$\liminf_{n\to\infty} \frac{1}{n^{d-1}} \ln \mu^+_{\Lambda(n),T}\big(\sigma_n \in \mathcal{U}\big) \geq -\mathcal{J}(\nu) - \varepsilon$$

and we conclude by sending $\varepsilon$ to 0. □

# References


1. M.A. Akcoglu and U. Krengel, *Ergodic theorems for superadditive processes*, J. Reine Angew. Math., vol. 323, 53–67 (1981)
2. M. Aizenman, J. Bricmont and J.L. Lebowitz, *Percolation of the minority spins in high-dimensional Ising models*, J. Stat. Phys., Vol. 49. 859–865 (1987)
3. M. Aizenman, J.T. Chayes, L. Chayes, J. Fröhlich and L. Russo, *On a sharp transition from area law to perimeter law in a system of random surfaces*, Commun. Math. Phys. 92, 19–69 (1983)
4. G. Alberti, G. Bellettini, M. Cassandro and E. Presutti, *Surface tension in Ising systems with Kac potentials*, J. Stat. Phys. 82, 743–796 (1996)
5. K.S. Alexander, *Stability of the Wulff minimum and fluctuations in shape for large finite clusters in two-dimensional percolation*, Probab. Theory Relat. Fields 91, 507–532 (1992)
6. K.S. Alexander, *Cube root boundary fluctuations for droplets in random cluster models* , Comm. Math. Phys. 224, no. 3, 733–781 (2001)
7. K.S. Alexander, M. Biskup and L. Chayes, *Colligative properties of solutions: I. Fixed concentrations*, Jour. Stat. Phys. 119, no. 3–4, 479–507 (2005)
8. K.S. Alexander, M. Biskup and L. Chayes, *Colligative properties of solutions: II. Vanishing concentrations*, Jour. Stat. Phys. 119, no. 3–4, 509–537 (2005)
9. K.S. Alexander, J.T. Chayes and L. Chayes, *The Wulff construction and asymptotics of the finite cluster distribution for two-dimensional Bernoulli percolation*, Comm. Math. Phys. 131, no. 1, 1–50 (1990)
10. K.S. Alexander and H. Uzun, *Lower bounds for boundary roughness for droplets in Bernoulli percolation*, Probab. Theory Relat. Fields 127, no. 1, 62–88 (2003)
11. L. Ambrosio, N. Fusco and D. Pallara, *Functions of bounded variation and free discontinuity problems*, Oxford Mathematical Monographs (2000)
12. L. Ambrosio, M. Novaga and E. Paolini, *Some regularity results for minimal crystals*, ESAIM Control Optim. Calc. Var. 8, 69–103 (2002)
13. P. Assouad and T. Quentin de Gromard, *Sur la dérivation des mesures dans* $\mathbb{R}^n$, Unpublished note (1998)
14. R.R. Bahadur and S.L. Zabell, *Large deviations of the sample mean in general vector spaces*, Ann. Prob. 7, no. 4, 587–621 (1979)
15. D. Barbato, *Tesi di Laurea*, Universita di Pisa (2002)
16. G. Bellettini, M. Cassandro and E. Presutti, *Constrained minima of nonlocal free energy functionals*, J. Stat. Phys. 84, no. 5–6, 1337–1349 (1996)

17. O. Benois, T. Bodineau, P. Buttà and E. Presutti, *On the validity of van der Waals theory of surface tension*, Markov Process. Rel. Fields 3, 175–198 (1997)
18. O. Benois, T. Bodineau and E. Presutti, *Large deviations in the van der Waals limit*, Stochastic Process. Appl. 75, 89–104 (1998)
19. A.S. Besicovitch, *A general form of the covering principle and relative differentiation of additive functions*, Proc. Cambridge Philos. Soc. 41, 103–110 (1945). Part II. Proc. Cambridge Philos. Soc. 42, 1–10 (1946)
20. M. Biskup, L. Chayes and R. Kotecky, *Comment on: "Theory of the evaporation/condensation transition of equilibrium droplets in finite volumes"* (2003)
21. M. Biskup, L. Chayes and R. Kotecky, *Critical region for droplet formation in the two-dimensional Ising model*, Comm. Math. Phys. 242, no. 1–2, 137–183 (2003)
22. M. Biskup, L. Chayes and R. Kotecky, *A proof of the Gibbs–Thomson formula in the droplet formation regime*, J. Stat. Phys. 116, no. 1–4, 175–203 (2004)
23. T. Bodineau, *The Wulff construction in three and more dimensions*, Comm. Math. Phys. 207, no. 1, 197–229 (1999)
24. T. Bodineau, *On the van der Waals theory of surface tension*, Inhomogeneous random systems (Cergy-Pontoise, 2001), Markov Process. Related Fields 8, no. 2, 319–338 (2002)
25. T. Bodineau, *Phase coexistence for the Kac-Ising models*, In and out of equilibrium (Mambucaba, 2000), Progr. Probab. 51, 75–111, Birkhäuser (2002)
26. T. Bodineau, *Slab percolation for the Ising model*, Probab. Theory Relat. Fields 132, no. 1, 83–118 (2005).
27. T. Bodineau, *Translation invariant Gibbs states for the Ising model*, Probab. Theory Relat. Fields (2005).
28. T. Bodineau, D. Ioffe and Y. Velenik, *Rigorous probabilistic analysis of equilibrium crystal shapes*, Probabilistic techniques in equilibrium and nonequilibrium statistical physics, J. Math. Phys. 41, no. 3, 1033–1098 (2000)
29. T. Bodineau, D. Ioffe and Y. Velenik, *Winterbottom construction for finite range ferromagnetic models: an $\mathbb{L}_1$-approach.*, J. Stat. Phys. 105, no. 1–2, 93–131 (2001)
30. T. Bodineau and F. Martinelli, *Some new results on the kinetic Ising model in a pure phase*, J. Stat. Phys. 109, no. 1–2, 207–235 (2002)
31. T. Bodineau, R. Schonmann and S. Shlosman, *3D crystal: how flat its flat facets are?*, Comm. Math. Phys. 255, no. 3, 747–766 (2005)
32. S.R. Broadbent and J.M. Hammersley, *Percolation processes I: Crystals and mazes*, Proceedings of the Cambridge Philosophical Society 53, 629–641 (1957).
33. J.E. Brothers and F. Morgan, *The isoperimetric theorem for general integrands*, Michigan Math. J. 41, no. 3, 419–431 (1994)
34. Y.D. Burago, V.A. Zalgaller, *Geometric inequalities*, Grundlehren der Mathematischen Wissenschaften 285, Springer-Verlag, Berlin (1988)
35. R. Caccioppoli, *Misura e integrazione sugli insiemi dimensionalmente orientati I, II*, Rend. Accad. Naz. Lincei, Cl. Sci. Fis. Mat. Nat., Ser. VIII, Vol. XII, fasc. 1,2 (gennaio–febbraio 1952), 3–11, 137–146
36. R. Caccioppoli, *Misura e integrazione sulle varietà parametriche I, II, III*, Rend. Accad. Naz. Lincei, Cl. Sci. Fis. Mat. Nat., Ser. VIII, Vol. XII, fasc. 3,4,6 (marzo, aprile, giugno 1952), 219–227, 365–373, 629–634
37. R. Cerf, *Large deviations of the finite cluster shape for two-dimensional percolation in the Hausdorff and $L^1$ metric*, J. Theoret. Probab. 13, no. 2, 491–517 (2000)

38. R. Cerf, *Large deviations for three dimensional supercritical percolation*, Astérisque 267 (2000)
39. R. Cerf, *The Hausdorff lower semicontinuous envelope of the length in the plane*, Ann. Scuola Norm. Sup. Pisa Cl. Sci. (5) I, no. 1, 33–71 (2002)
40. R. Cerf, *On Cramér theory in infinite dimensions*, preprint (2005)
41. R. Cerf and R. Kenyon, *The low temperature expansion of the Wulff crystal in the 3D Ising model*, Comm. Math. Phys. 222, no. 1, 147–179 (2001)
42. R. Cerf and R.J. Messikh, On the 2d Ising Wulff crystal near criticality, preprint.
43. R. Cerf and A. Pisztora, *On the Wulff crystal in the Ising model*, Ann. Probab. 28, no. 3, 947–1017 (2000)
44. R. Cerf and A. Pisztora, *Phase coexistence in Ising, Potts and percolation models*, Ann. Inst. H. Poincaré Probab. Statist. 37, no. 6, 643–724 (2001)
45. F. Cesi, G. Guadagni, F. Martinelli and R. Schonmann, *On the 2D stochastic Ising model in the phase coexistence region near the critical point*, J. Stat. Phys. 85, no. 1–2, 55–102 (1996)
46. J.T. Chayes, L. Chayes and R.H. Schonmann, *Exponential decay of connectivities in the two-dimensional Ising model*, J. Stat. Phys. 49, no. 3–4, 433–445 (1987)
47. F. Comets, *Grandes déviations pour des champs de Gibbs sur $\mathbb{Z}^d$*, C.R. Acad. Sci. Paris Ser. I Math. 303, no. 11, 511–513 (1986)
48. O. Couronné, The Wulff crystal for oriented percolation, preprint (2004)
49. O. Couronné, A large deviation result for the subcritical Bernoulli percolation, Ann. Fac. Sci. Toulouse Math. (6) 14, no. 2, 201–214 (2005)
50. O. Couronné, Poisson approximation for large finite clusters in the supercritical FK model, preprint (2004)
51. O. Couronné and R.J. Messikh, *Surface order large deviations for 2D FK percolation and Potts models*, Stochastic Process. Appl. 113, no. 1, 81–99 (2004)
52. M. Crandall, H. Ishii and P.L. Lions, *User's guide to viscosity solutions of second order partial differential equations*, Bull. Amer. Math. Soc. 27, no. 1, 1–67 (1992).
53. B. Dacorogna and C.–E. Pfister, *Wulff theorem and best constant in Sobolev inequality*, J. Math. Pures Appl. (9) 71, no. 2, 97–118 (1992).
54. E. De Giorgi, *Su una teoria generale della misura $(r-1)$-dimensionale in uno spazio ad $r$ dimensioni*, Ann. Mat. Pura Appl. IV Ser. 36, 191–213 (1954)
55. E. De Giorgi, *Nuovi teoremi relativi alle misure $(r-1)$-dimensionali in uno spazio ad $r$ dimensioni*, Ricerche Mat. 4, 95–113 (1955)
56. E. De Giorgi, *Sulla proprieta isoperimetrica dell'ipersfera, nella classe degli insiemi aventi frontiera orientata di misura finita*, Atti Accad. Naz. Lincei, Cl. Sci. Fis. Mat. Nat., VIII Ser. 5, 33–44 (1958)
57. E. De Giorgi, F. Colombini and L.C. Piccinini, *Frontiere orientate di misura minima e questioni collegate*, Scuola Normale Superiore di Pisa (1972)
58. J.–D. Deuschel and A. Pisztora, *Surface order large deviations for high-density percolation*, Probab. Theory Relat. Fields 104, no. 4, 467–482 (1996).
59. J.–D. Deuschel and D.W. Stroock, *Large deviations*, Academic Press (1989)
60. A. Dinghas, *Uber einen geometrischen Satz von Wulff für die Gleichgewichtsform von Kristallen*, Z. Kristallogr. 105, 304–314 (1944)
61. R.L. Dobrushin, *Gibbs state describing coexistence of phases for a three–dimensional Ising model*, Theor. Probability Appl. 17, no. 4, 582–600 (1973)
62. R.L. Dobrushin and O. Hryniv, *Fluctuations of the phase boundary in the 2D Ising ferromagnet*, Commun. Math. Phys. 189, no. 2, 395–445 (1997)

63. R.L. Dobrushin, R. Kotecký and S.B. Shlosman, *Wulff construction: a global shape from local interaction*, AMS translations series, Providence (Rhode Island) (1992)
64. R.L. Dobrushin and S.B. Shlosman, *Thermodynamic inequalities for the surface tension and the geometry of the Wulff construction*, Ideas and methods in quantum and statistical physics, S. Albeverio (ed.), Cambridge University Press 2, 461–483 (1992)
65. R.G. Edwards and A.D. Sokal, *Generalization of the Fortuin–Kasteleyn–Swendsen–Wang representation and Monte Carlo algorithm*, Phys. Rev. D (3) 38, no. 6, 2009–2012 (1988)
66. R.S. Ellis, *Entropy, Large Deviations and Statistical Mechanics*, Springer (1985)
67. L.C. Evans and R.F. Gariepy, *Measure theory and fine properties of functions*, Studies in Advanced Mathematics, Boca Raton: CRC Press (1992)
68. K.J. Falconer, *The geometry of fractal sets*, Cambridge Tracts in Mathematics 85, Cambridge Univ. Press (1986)
69. H. Federer, *Geometric measure theory*, Springer–Verlag (1969)
70. H. Föllmer and S. Orey, *Large deviations for the empirical field of a Gibbs measure*, Ann. Probab. 16, no. 3, 961–977 (1988)
71. H. Föllmer and M. Ort, *Large deviations and surface entropy for Markov fields*, Astérisque 157–158, 173–190 (1988)
72. I. Fonseca, *The Wulff theorem revisited*, Proc. R. Soc. Lond. Ser. A 432, no. 1884, 125–145 (1991)
73. I. Fonseca and S. Müller, *A uniqueness proof for the Wulff theorem*, Proc. R. Soc. Edinb. Sect. A 119, no. 1–2, 125–136 (1991)
74. C.M. Fortuin and P.W. Kasteleyn, *On the random-cluster model. I. Introduction and relation to other models.*, Physica 57, 536–564 (1972)
75. H.O. Georgii, O. Häggström and C. Maes, *The random geometry of equilibrium phases*, Phase transitions and critical phenomena, vol. 18, 1–142, Academic Press (2001).
76. E. Giusti, *Metodi diretti nel calcolo delle variazioni*, Unione Matematica Italiana, Bologna (1994).
77. E. Giusti, *Minimal surfaces and functions of bounded variation*, Birkhäuser (1984)
78. G.R. Grimmett, *Percolation*, Second edition. Grundlehren der Mathematischen Wissenschaften 321, Springer-Verlag, Berlin (1999)
79. G.R. Grimmett, *The stochastic random-cluster process and the uniqueness of random-cluster measures*, Ann. Probab. 23, no. 4, 1461–1510 (1995)
80. G.R. Grimmett, *The random-cluster model*, Probability on Discrete Structures, ed. H. Kesten, Encyclopedia of Mathematical Sciences, vol. 110, 73–123, Springer (2004)
81. G.R. Grimmett and J.M. Marstrand, *The supercritical phase of percolation is well behaved*, Proc. R. Soc. Lond. Ser. A 430, no. 1879, 439–457 (1990)
82. G.R. Grimmett and M.S.T. Piza, *Decay of correlations in random-cluster models*, Comm. Math. Phys. 189, no. 2, 465–480 (1997)
83. G. Gielis and G.R. Grimmett, *Rigidity of the interface in percolation and random cluster models*, J. Stat. Phys. 109, no. 1–2, 1–37 (2002)
84. B.L. Gurevich and G.E. Shilov, *Integral, measure and derivative: a unified approach*, Prentice–Hall (1966)
85. J. Hass and R. Schlafly, *Double bubbles minimize*, Ann. of Math. (2) 151, no. 2, 459–515 (2000)

86. R. Holley, *Remarks on the FKG inequalities*, Commun. Math. Phys. 36, 227–231 (1974)
87. J. Hass, M. Hutchings and R. Schlafly, *The double bubble conjecture*, Electron. Res. Announc. Amer. Math. Soc. 1, no. 3, 98–102 (electronic) (1995)
88. M. Hutchings, F. Morgan, M. Ritoré and A. Ros, *Proof of the double bubble conjecture*, Ann. of Math. (2) 155, no. 2, 459–489 (2002)
89. O. Hryniv, *On local behaviour of the phase separation line in the 2D Ising model*, Probab. Theory Relat. Fields 110, no. 1, 91–107 (1998)
90. D. Ioffe, *Large deviations for the 2D Ising model: a lower bound without cluster expansions*, J. Stat. Phys. 74, no. 1–2, 411–432 (1994)
91. D. Ioffe, *Exact large deviation bounds up to $T_c$ for the Ising model in two dimensions*, Probab. Theory Relat. Fields 102, no. 3, 313–330 (1995)
92. D. Ioffe and R. Schonmann, *Dobrushin–Kotecký–Shlosman Theorem up to the critical temperature*, Comm. Math. Phys. 199, no. 1, 117–167 (1998)
93. H. Kesten and Y. Zhang, *The probability of a large finite cluster in supercritical Bernoulli percolation*, Ann. Probab. 18, no. 2, 537–555 (1990)
94. O.E. Lanford, *Entropy and equilibrium states in classical statistical mechanics*, Statistical Mechanics and Mathematical Problems, Lecture Notes in Physics 20, 1–113 (1971)
95. J.L. Lebowitz, *Coexistence of phases in Ising ferromagnets*, J. Stat. Phys. 16, no. 6, 463–476 (1977)
96. J.L. Lebowitz and A. Martin-Löf, *On the uniqueness of the equilibrium state for Ising spin systems*, Comm. Math. Phys. 25, 276–282 (1972)
97. J.L. Lebowitz and C.E. Pfister, *Surface tension and phase coexistence*, Phys. Rev. Letters 46, no. 15, 1031–1033 (1981)
98. T.M. Liggett, R.H. Schonmann and A.M. Stacey, *Domination by product measures*, Ann. Probab. 25, no. 1, 71–95 (1997)
99. B.M. Mc Coy and T.T. Wu, *The Two Dimensional Ising Model*, Cambridge, MA: Harvard University Press (1973)
100. P. Mattila, *Geometry of sets and measures in Euclidean spaces*, Fractals and rectifiability. Cambridge Studies in Advanced Mathematics 44 (1995)
101. U. Massari and M. Miranda, *Minimal surfaces of codimension one*, North–Holland Mathematics Studies 91, Notas de Matematica 95 (1984)
102. A. Messager, S. Miracle–Solé and J. Ruiz, *Convexity properties of the surface tension and equilibrium crystals*, J. Stat. Phys. 67, no. 3–4, 449–470 (1992)
103. R.J. Messikh, *Asymptotiques de la tension de surface du modèle d'Ising 2-d près de la température critique*, Mémoire de DEA, Orsay.
104. R.J. Messikh, On the surface tension of the 2d Ising model near criticality, in preparation.
105. R.J. Messikh, Approximation of a Mumford–Shah functional using the Ising model: theory and numerics, in preparation.
106. S. Miracle–Solé, *Surface tension, step free energy, and facets in the equilibrium crystal*, J. Stat. Phys. 79, no. 1–2, 183–214 (1995)
107. C.M. Newman, *Disordered Ising systems and random cluster representations*, in Probability and Phase Transition, Kluwer, 247–260 (1994)
108. S. Olla, *Large deviations for Gibbs random fields*, Probab. Theory Relat. Fields 77, no. 3, 343–357 (1988)
109. C.–E. Pfister, *Large deviations and phase separation in the two-dimensional Ising model*, Helv. Phys. Acta 64, no. 7, 953–1054 (1991)

110. C.–E. Pfister, *Thermodynamical aspects of classical lattice systems*, In and out of equilibrium (Mambucaba, 2000), 393–472, Progr. Probab. 51, Birkhäuser Boston, Boston, MA (2002)
111. C.–E. Pfister and Y. Velenik, *Large deviations and continuum limit in the 2D Ising model*, Probab. Theory Relat. Fields 109, no. 4, 435–506 (1997)
112. A. Pisztora, *Surface order large deviations for Ising, Potts and percolation models*, Prob. Theor. Rel. Fields 104, 427–466 (1996)
113. R.T. Rockafellar, *Convex Analysis*, Princeton University Press (1970)
114. W. Rudin, *Real and complex analysis*, McGraw–Hill, 3rd ed. (1987)
115. W. Rudin, *Functional Analysis*, McGraw–Hill (1973)
116. R.H. Schonmann, *Second order large deviation estimates for ferromagnetic systems in the phase coexistence region*, Comm. Math. Phys. 112, no. 3, 409–422 (1987)
117. R.H. Schonmann, *Slow droplet–driven relaxation of stochastic Ising models in the vicinity of the phase coexistence region*, Comm. Math. Phys. 161, no. 1, 1–49 (1994)
118. R.H. Schonmann and S.B. Shlosman, *Constrained variational problem with applications to the Ising model*, J. Stat. Phys. 83, no. 5–6, 867–905 (1996)
119. R.H. Schonmann and S.B. Shlosman, *Complete analyticity for the 2D Ising model completed*, Commun. Math. Phys. 170, no. 2, 453–482 (1995)
120. R.H. Schonmann and S.B. Shlosman, *Wulff droplets and the metastable relaxation of kinetic Ising models*, Commun. Math. Phys. 194, no. 2, 389–462 (1998)
121. T. Seppäläinen, *Entropy for translation-invariant random-cluster measures*, Ann. Probab. 26, no. 3, 1139–1178 (1998)
122. R.T. Smythe, *Multiparameter subadditive processes*, Ann. Probability 4, no. 5, 772–782 (1976)
123. J.M. Steele, *Probability theory and combinatorial optimization*, SIAM (1997)
124. J.E. Taylor, *Crystalline variational problems*, Bull. Am. Math. Soc. 84, no. 4, 568–588 (1978)
125. J.E. Taylor, *Existence and structure of solutions to a class of nonelliptic variational problems*, Symposia Mathematica 14, no. 4, 499–508 (1974)
126. J.E. Taylor, *Unique structure of solutions to a class of nonelliptic variational problems*, Proc. Symp. pure Math. 27, 419–427 (1975)
127. Y. Velenik, Phase separation as a large deviations problem: a microscopic derivation of surface thermodynamics in some 2D spin systems, PhD Thesis, EPFL, Lausanne (1997)
128. A. Visintin, *Models of phase transitions*, Progress in Nonlinear Differential Equations and their Applications 28, Birkhäuser Boston (1996)
129. A. Visintin, *Generalized coarea formula and fractal sets*, Japan J. Indust. Appl. Math. 8, no. 2, 175–201 (1991)
130. W.L. Winterbottom, *Equilibrium shape of a small particle in contact with a foreign substrate*, Acta Metallurgica 15, 303–310 (1967)
131. G. Wulff, *Zur Frage der Geschwindigkeit des Wachstums und der Auflösung der Kristallflächen*, Z. Kristallogr. 34, 449–530 (1901)
132. B. Younovitch, *Sur la dérivation des fonctions absolument additives d'ensemble*, C. R. (Doklady) Acad. Sci. URSS, n. Ser. 30, 112–114 (1941)
133. W.P. Ziemer, *Weakly differentiable functions. Sobolev spaces and functions of bounded variation*, Graduate texts in Mathematics 120, Springer–Verlag (1989)

# Index

# List of participants

**Lecturers**

| | |
|---|---|
| CERF Raphaël | Univ. Paris-Sud, Orsay, F |
| LYONS Terry | Univ. Oxford, UK |
| SLADE Gordon | Univ. British Columbia, Vancouver, Canada |

**Participants**

| | |
|---|---|
| ASSELAH Amine | Univ. Provence, Marseille, F |
| AUTRET Solenn | Univ. Paul Sabatier, Toulouse, F |
| BAILLEUL Ismaël | Univ. Paris-Sud, Orsay, F |
| BAUDOIN Fabrice | Univ. Paul Sabatier, Toulouse, F |
| BEGYN Arnaud | Univ. Paul Sabatier, Toulouse, F |
| BEN-ARI Iddo | Technion Inst. Technology, Haifa, Israel |
| BERARD Jean | Univ. Lyon 1, F |
| BLACHE Fabrice | Univ. Blaise Pascal, Clermont-Ferrand, F |
| BROMAN Erik | Chalmers Univ. Techn., Gothenburg, Sweden |
| BROUTTELANDE Christophe | Univ. Paul Sabatier, Toulouse, F |
| BRYC Wlodzimierz | Univ. Cincinnati, USA |
| CARUANA Michael | Univ. Oxford, UK |
| CHIVORET Sebastien | Univ. Michigan, Ann Arbor, USA |
| COUPIER David | Univ. Paris 5, F |
| CROYDON David | Univ. Oxford, UK |
| DE CARVALHO BEZERRA S. | Univ. Henri Poincaré, Nancy, F |
| DE TILIERE Béatrice | Univ. Paris-Sud, Orsay, F |
| DELARUE François | Univ. Paris 7, F |
| DEVAUX Vincent | Univ. Rouen, F |
| DUNLOP François | Univ. Cergy-Pontoise, F |
| DUQUESNE Thomas | Univ. Paris-Sud, Orsay, F |
| FERAL Delphine | Univ. Paul Sabatier, Toulouse, F |
| FILLIGER Roger | EPFL, Lausanne, Switzerland |
| GARET Olivier | Univ. Orléans, F |
| GAUTIER Eric | Univ. Rennes 1 & INSEE, F |

| | |
|---|---|
| GOERGEN Laurent | ETH Zurich, Switzerland |
| GOUERE Jean-Baptiste | Univ. Claude Bernard, Lyon, F |
| GOURCY Mathieu | Univ. Blaise Pascal, Clermont-Ferrand, F |
| HOLROYD Alexander | Univ. British Columbia, Vancouver, Canada |
| ISHIKAWA Yasushi | Univ. Ehime, Matsuyama, Japan |
| JAKUBOWICZ Jérémie | ENS Cachan, F |
| JOULIN Aldéric | Univ. La Rochelle, F |
| KASPRZYK Arkadiusz | Univ. Wroclaw, Poland |
| KOVCHEGOV Yevgeniy | UCLA, Los Angeles, USA |
| KURT Noemi | Univ. Zurich, Switzerland |
| LACAUX Céline | Univ. Paul Sabatier, Toulouse, F |
| LACHAUD Béatrice | Univ. Paris 5, F |
| LE GALL Jean-François | ENS Paris, F |
| LE JAN Yves | Univ. Paris-Sud, Orsay, F |
| LEI Liangzhen | Univ. Blaise Pascal, Clermont-Ferrand, F |
| LEVY Thierry | ENS Paris, F |
| LUCZAK Malwina | London School of Economics, UK |
| MARCHAND Régine | Univ. Henri Poincaré, Nancy, F |
| MARDIN Arif | TELECOM-INT, Evry, F |
| MARTIN James | Univ. Paris 7, F |
| MARTY Renaud | Univ. Paul Sabatier, Toulouse, F |
| MERLE Mathieu | ENS Paris, F |
| MESSIKH Reda Jürg | Univ. Paris-Sud, Orsay, F |
| MOCIOALCA Oana | Purdue Univ., West Lafayette, USA |
| NIEDERHAUSEN Meike | Purdue Univ., West Lafayette, USA |
| NINOMIYA Syoiti | Tokyo Instit. Technology, Japan |
| NUALART David | Univ. Barcelona, Spain |
| PICARD Jean | Univ. Blaise Pascal, Clermont-Ferrand, F |
| PUDLO Pierre | Univ. Claude Bernard, Lyon, F |
| RIVIERE Olivier | Univ. Paris 5, F |
| ROUSSET Mathias | Univ. Paul Sabatier, Toulouse, F |
| ROUX Daniel | Univ. Blaise Pascal, Clermont-Ferrand, F |
| SAINT LOUBERT BIE Erwan | Univ. Blaise Pascal, Clermont-Ferrand, F |
| SAVONA Catherine | Univ. Blaise Pascal, Clermont-Ferrand, F |
| SERLET Laurent | Univ. Paris 5, F |
| THOMANN Philipp | Univ. Zurich, Switzerland |
| TORRECILLA Iván | Univ. Barcelona, Spain |
| TRASHORRAS Jose | Univ. Warwick, Coventry, UK |
| TURNER Amanda | Univ. Cambridge, UK |
| TYKESSON Johan | Chalmers Univ. Techn., Göteborg, Sweden |
| VIGNAUD Yvon | CPT, Marseille, F |
| WEILL Mathilde | ENS Paris, F |
| WINKEL Matthias | Univ. Oxford, UK |
| YU Yuhua | Purdue Univ., West Lafayette, USA |

# List of short lectures

| | |
|---|---|
| Amine Asselah | Hitting time for independent random walks on $Z^d$ |
| Fabrice Baudoin | The tangent space to a hypoelliptic diffusion and applications |
| Arnaud Begyn | On the second order quadratic variations for processes with Gaussian increments |
| Iddo Ben-Ari | The spectral shift of the principal eigenvalue of the Laplacian in bounded domains in the presence of small obstacles |
| Jean Bérard | Asymptotic results for a simple model of trail-following behaviour |
| Fabrice Blache | Harmonic mappings, martingales and BSDE on manifolds |
| Christophe Brouttelande | On Sobolev-type inequalities on Riemannian manifolds |
| Sebastien Chivoret | Properties of multiple stochastic integrals with respect to fractional Brownian motion |
| David Coupier | Technics of random graphs theory applied to image analysis |
| François Delarue | Decoupling a forward backward SDE |
| Thomas Duquesne | Probabilistic and fractal aspects of Lévy trees |
| Delphine Feral | On large deviations for the spectral measure of both continuous and discrete Coulomb gas |
| Roger Filliger | Diffusion mediated transport |
| Olivier Garet | Asymptotic shape for the chemical distance and first-passage percolation on a Bernoulli cluster |

| | |
|---|---|
| Eric Gautier | Large deviations for stochastic nonlinear Schrödinger equations and applications |
| Jean-Baptiste Gouéré | Probabilistic characterization of quasicrystals |
| Alexander Holroyd | A stable marriage of Poisson and Lebesgue |
| Aldéric Joulin | Functional inequalities for continuous time Markov chains: a modified curvature criterion |
| Yevgeniy Kovchegov | Multi-particle systems with reinforcements |
| Céline Lacaux | Atypic examples of locally self-similar fields |
| Thierry Lévy | Large deviations for the Yang-Mills measure |
| Malwina Luczak | On the maximum queue length in the supermarket model |
| Régine Marchand | Coexistence in a simple competition model |
| James Martin | Heavy tails in last-passage percolation |
| Renaud Marty | Asymptotic behavior of a nonlinear Schrödinger equation in a random medium via a splitting method |
| Mathieu Merle | Local behaviour of the occupation time measure of super-Brownian motion |
| Reda Messikh | Large deviations in the supercritical vicinity of $p_c$ |
| Oana Mocioalca | Skorohod integration and stochastic calculus beyond the fractional Brownian scale |
| David Nualart | Rough path analysis via fractional calculus |
| Pierre Pudlo | Precise estimates of large deviations of finite Markov chains with applications to biological sequence analysis |
| Mathias Rousset | Interacting particle systems approximations of Feynman-Kac formulae, applications to nonlinear filtering |
| Laurent Serlet | Super-Brownian motion conditioned to its total mass |
| José Trashorras | Fluctuations of the free energy in the high temperature Hopfield model |
| Yvon Vignaud | Magnetostriction |
| Matthias Winkel | Growth of the Brownian forest |

# Lecture Notes in Mathematics

For information about earlier volumes
please contact your bookseller or Springer
LNM Online archive: springerlink.com

Vol. 1681: G. J. Wirsching, The Dynamical System Generated by the 3n+1 Function (1998)
Vol. 1682: H.-D. Alber, Materials with Memory (1998)
Vol. 1683: A. Pomp, The Boundary-Domain Integral Method for Elliptic Systems (1998)
Vol. 1684: C. A. Berenstein, P. F. Ebenfelt, S. G. Gindikin, S. Helgason, A. E. Tumanov, Integral Geometry, Radon Transforms and Complex Analysis. Firenze, 1996. Editors: E. Casadio Tarabusi, M. A. Picardello, G. Zampieri (1998)
Vol. 1685: S. König, A. Zimmermann, Derived Equivalences for Group Rings (1998)
Vol. 1686: J. Azéma, M. Émery, M. Ledoux, M. Yor (Eds.), Séminaire de Probabilités XXXII (1998)
Vol. 1687: F. Bornemann, Homogenization in Time of Singularly Perturbed Mechanical Systems (1998)
Vol. 1688: S. Assing, W. Schmidt, Continuous Strong Markov Processes in Dimension One (1998)
Vol. 1689: W. Fulton, P. Pragacz, Schubert Varieties and Degeneracy Loci (1998)
Vol. 1690: M. T. Barlow, D. Nualart, Lectures on Probability Theory and Statistics. Editor: P. Bernard (1998)
Vol. 1691: R. Bezrukavnikov, M. Finkelberg, V. Schechtman, Factorizable Sheaves and Quantum Groups (1998)
Vol. 1692: T. M. W. Eyre, Quantum Stochastic Calculus and Representations of Lie Superalgebras (1998)
Vol. 1694: A. Braides, Approximation of Free-Discontinuity Problems (1998)
Vol. 1695: D. J. Hartfiel, Markov Set-Chains (1998)
Vol. 1696: E. Bouscaren (Ed.): Model Theory and Algebraic Geometry (1998)
Vol. 1697: B. Cockburn, C. Johnson, C.-W. Shu, E. Tadmor, Advanced Numerical Approximation of Nonlinear Hyperbolic Equations. Cetraro, Italy, 1997. Editor: A. Quarteroni (1998)
Vol. 1698: M. Bhattacharjee, D. Macpherson, R. G. Möller, P. Neumann, Notes on Infinite Permutation Groups (1998)
Vol. 1699: A. Inoue,Tomita-Takesaki Theory in Algebras of Unbounded Operators (1998)
Vol. 1700: W. A. Woyczyński, Burgers-KPZ Turbulence (1998)
Vol. 1701: Ti-Jun Xiao, J. Liang, The Cauchy Problem of Higher Order Abstract Differential Equations (1998)
Vol. 1702: J. Ma, J. Yong, Forward-Backward Stochastic Differential Equations and Their Applications (1999)
Vol. 1703: R. M. Dudley, R. Norvaiša, Differentiability of Six Operators on Nonsmooth Functions and p-Variation (1999)
Vol. 1704: H. Tamanoi, Elliptic Genera and Vertex Operator Super-Algebras (1999)
Vol. 1705: I. Nikolaev, E. Zhuzhoma, Flows in 2-dimensional Manifolds (1999)
Vol. 1706: S. Yu. Pilyugin, Shadowing in Dynamical Systems (1999)
Vol. 1707: R. Pytlak, Numerical Methods for Optimal Control Problems with State Constraints (1999)
Vol. 1708: K. Zuo, Representations of Fundamental Groups of Algebraic Varieties (1999)
Vol. 1709: J. Azéma, M. Émery, M. Ledoux, M. Yor (Eds.), Séminaire de Probabilités XXXIII (1999)
Vol. 1710: M. Koecher, The Minnesota Notes on Jordan Algebras and Their Applications (1999)
Vol. 1711: W. Ricker, Operator Algebras Generated by Commuting Projećtions: A Vector Measure Approach (1999)
Vol. 1712: N. Schwartz, J. J. Madden, Semi-algebraic Function Rings and Reflectors of Partially Ordered Rings (1999)
Vol. 1713: F. Bethuel, G. Huisken, S. Müller, K. Steffen, Calculus of Variations and Geometric Evolution Problems. Cetraro, 1996. Editors: S. Hildebrandt, M. Struwe (1999)
Vol. 1714: O. Diekmann, R. Durrett, K. P. Hadeler, P. K. Maini, H. L. Smith, Mathematics Inspired by Biology. Martina Franca, 1997. Editors: V. Capasso, O. Diekmann (1999)
Vol. 1715: N. V. Krylov, M. Röckner, J. Zabczyk, Stochastic PDE's and Kolmogorov Equations in Infinite Dimensions. Cetraro, 1998. Editor: G. Da Prato (1999)
Vol. 1716: J. Coates, R. Greenberg, K. A. Ribet, K. Rubin, Arithmetic Theory of Elliptic Curves. Cetraro, 1997. Editor: C. Viola (1999)
Vol. 1717: J. Bertoin, F. Martinelli, Y. Peres, Lectures on Probability Theory and Statistics. Saint-Flour, 1997. Editor: P. Bernard (1999)
Vol. 1718: A. Eberle, Uniqueness and Non-Uniqueness of Semigroups Generated by Singular Diffusion Operators (1999)
Vol. 1719: K. R. Meyer, Periodic Solutions of the N-Body Problem (1999)
Vol. 1720: D. Elworthy, Y. Le Jan, X-M. Li, On the Geometry of Diffusion Operators and Stochastic Flows (1999)
Vol. 1721: A. Iarrobino, V. Kanev, Power Sums, Gorenstein Algebras, and Determinantal Loci (1999)
Vol. 1722: R. McCutcheon, Elemental Methods in Ergodic Ramsey Theory (1999)
Vol. 1723: J. P. Croisille, C. Lebeau, Diffraction by an Immersed Elastic Wedge (1999)
Vol. 1724: V. N. Kolokoltsov, Semiclassical Analysis for Diffusions and Stochastic Processes (2000)
Vol. 1725: D. A. Wolf-Gladrow, Lattice-Gas Cellular Automata and Lattice Boltzmann Models (2000)
Vol. 1726: V. Marić, Regular Variation and Differential Equations (2000)
Vol. 1727: P. Kravanja M. Van Barel, Computing the Zeros of Analytic Functions (2000)
Vol. 1728: K. Gatermann Computer Algebra Methods for Equivariant Dynamical Systems (2000)
Vol. 1729: J. Azéma, M. Émery, M. Ledoux, M. Yor (Eds.)

Séminaire de Probabilités XXXIV (2000)
Vol. 1730: S. Graf, H. Luschgy, Foundations of Quantization for Probability Distributions (2000)
Vol. 1731: T. Hsu, Quilts: Central Extensions, Braid Actions, and Finite Groups (2000)
Vol. 1732: K. Keller, Invariant Factors, Julia Equivalences and the (Abstract) Mandelbrot Set (2000)
Vol. 1733: K. Ritter, Average-Case Analysis of Numerical Problems (2000)
Vol. 1734: M. Espedal, A. Fasano, A. Mikelić, Filtration in Porous Media and Industrial Applications. Cetraro 1998. Editor: A. Fasano. 2000.
Vol. 1735: D. Yafaev, Scattering Theory: Some Old and New Problems (2000)
Vol. 1736: B. O. Turesson, Nonlinear Potential Theory and Weighted Sobolev Spaces (2000)
Vol. 1737: S. Wakabayashi, Classical Microlocal Analysis in the Space of Hyperfunctions (2000)
Vol. 1738: M. Émery, A. Nemirovski, D. Voiculescu, Lectures on Probability Theory and Statistics (2000)
Vol. 1739: R. Burkard, P. Deuflhard, A. Jameson, J.-L. Lions, G. Strang, Computational Mathematics Driven by Industrial Problems. Martina Franca, 1999. Editors: V. Capasso, H. Engl, J. Periaux (2000)
Vol. 1740: B. Kawohl, O. Pironneau, L. Tartar, J.-P. Zolesio, Optimal Shape Design. Tróia, Portugal 1999. Editors: A. Cellina, A. Ornelas (2000)
Vol. 1741: E. Lombardi, Oscillatory Integrals and Phenomena Beyond all Algebraic Orders (2000)
Vol. 1742: A. Unterberger, Quantization and Nonholomorphic Modular Forms (2000)
Vol. 1743: L. Habermann, Riemannian Metrics of Constant Mass and Moduli Spaces of Conformal Structures (2000)
Vol. 1744: M. Kunze, Non-Smooth Dynamical Systems (2000)
Vol. 1745: V. D. Milman, G. Schechtman (Eds.), Geometric Aspects of Functional Analysis. Israel Seminar 1999-2000 (2000)
Vol. 1746: A. Degtyarev, I. Itenberg, V. Kharlamov, Real Enriques Surfaces (2000)
Vol. 1747: L. W. Christensen, Gorenstein Dimensions (2000)
Vol. 1748: M. Ruzicka, Electrorheological Fluids: Modeling and Mathematical Theory (2001)
Vol. 1749: M. Fuchs, G. Seregin, Variational Methods for Problems from Plasticity Theory and for Generalized Newtonian Fluids (2001)
Vol. 1750: B. Conrad, Grothendieck Duality and Base Change (2001)
Vol. 1751: N. J. Cutland, Loeb Measures in Practice: Recent Advances (2001)
Vol. 1752: Y. V. Nesterenko, P. Philippon, Introduction to Algebraic Independence Theory (2001)
Vol. 1753: A. I. Bobenko, U. Eitner, Painlevé Equations in the Differential Geometry of Surfaces (2001)
Vol. 1754: W. Bertram, The Geometry of Jordan and Lie Structures (2001)
Vol. 1755: J. Azéma, M. Émery, M. Ledoux, M. Yor (Eds.), Séminaire de Probabilités XXXV (2001)
Vol. 1756: P. E. Zhidkov, Korteweg de Vries and Nonlinear Schrödinger Equations: Qualitative Theory (2001)
Vol. 1757: R. R. Phelps, Lectures on Choquet's Theorem (2001)
Vol. 1758: N. Monod, Continuous Bounded Cohomology of Locally Compact Groups (2001)
Vol. 1759: Y. Abe, K. Kopfermann, Toroidal Groups (2001)
Vol. 1760: D. Filipović, Consistency Problems for Heath-Jarrow-Morton Interest Rate Models (2001)
Vol. 1761: C. Adelmann, The Decomposition of Primes in Torsion Point Fields (2001)
Vol. 1762: S. Cerrai, Second Order PDE's in Finite and Infinite Dimension (2001)
Vol. 1763: J.-L. Loday, A. Frabetti, F. Chapoton, F. Goichot, Dialgebras and Related Operads (2001)
Vol. 1764: A. Cannas da Silva, Lectures on Symplectic Geometry (2001)
Vol. 1765: T. Kerler, V. V. Lyubashenko, Non-Semisimple Topological Quantum Field Theories for 3-Manifolds with Corners (2001)
Vol. 1766: H. Hennion, L. Hervé, Limit Theorems for Markov Chains and Stochastic Properties of Dynamical Systems by Quasi-Compactness (2001)
Vol. 1767: J. Xiao, Holomorphic Q Classes (2001)
Vol. 1768: M.J. Pflaum, Analytic and Geometric Study of Stratified Spaces (2001)
Vol. 1769: M. Alberich-Carramiñana, Geometry of the Plane Cremona Maps (2002)
Vol. 1770: H. Gluesing-Luerssen, Linear Delay-Differential Systems with Commensurate Delays: An Algebraic Approach (2002)
Vol. 1771: M. Émery, M. Yor (Eds.), Séminaire de Probabilités 1967-1980. A Selection in Martingale Theory (2002)
Vol. 1772: F. Burstall, D. Ferus, K. Leschke, F. Pedit, U. Pinkall, Conformal Geometry of Surfaces in $S^4$ (2002)
Vol. 1773: Z. Arad, M. Muzychuk, Standard Integral Table Algebras Generated by a Non-real Element of Small Degree (2002)
Vol. 1774: V. Runde, Lectures on Amenability (2002)
Vol. 1775: W. H. Meeks, A. Ros, H. Rosenberg, The Global Theory of Minimal Surfaces in Flat Spaces. Martina Franca 1999. Editor: G. P. Pirola (2002)
Vol. 1776: K. Behrend, C. Gomez, V. Tarasov, G. Tian, Quantum Comohology. Cetraro 1997. Editors: P. de Bartolomeis, B. Dubrovin, C. Reina (2002)
Vol. 1777: E. García-Río, D. N. Kupeli, R. Vázquez-Lorenzo, Osserman Manifolds in Semi-Riemannian Geometry (2002)
Vol. 1778: H. Kiechle, Theory of K-Loops (2002)
Vol. 1779: I. Chueshov, Monotone Random Systems (2002)
Vol. 1780: J. H. Bruinier, Borcherds Products on O(2,1) and Chern Classes of Heegner Divisors (2002)
Vol. 1781: E. Bolthausen, E. Perkins, A. van der Vaart, Lectures on Probability Theory and Statistics. Ecole d' Eté de Probabilités de Saint-Flour XXIX-1999. Editor: P. Bernard (2002)
Vol. 1782: C.-H. Chu, A. T.-M. Lau, Harmonic Functions on Groups and Fourier Algebras (2002)
Vol. 1783: L. Grüne, Asymptotic Behavior of Dynamical and Control Systems under Perturbation and Discretization (2002)
Vol. 1784: L.H. Eliasson, S. B. Kuksin, S. Marmi, J.-C. Yoccoz, Dynamical Systems and Small Divisors. Cetraro, Italy 1998. Editors: S. Marmi, J.-C. Yoccoz (2002)
Vol. 1785: J. Arias de Reyna, Pointwise Convergence of Fourier Series (2002)
Vol. 1786: S. D. Cutkosky, Monomialization of Morphisms from 3-Folds to Surfaces (2002)
Vol. 1787: S. Caenepeel, G. Militaru, S. Zhu, Frobenius and Separable Functors for Generalized Module Categories and Nonlinear Equations (2002)

Vol. 1788: A. Vasil'ev, Moduli of Families of Curves for Conformal and Quasiconformal Mappings (2002)
Vol. 1789: Y. Sommerhäuser, Yetter-Drinfel'd Hopf algebras over groups of prime order (2002)
Vol. 1790: X. Zhan, Matrix Inequalities (2002)
Vol. 1791: M. Knebusch, D. Zhang, Manis Valuations and Prüfer Extensions I: A new Chapter in Commutative Algebra (2002)
Vol. 1792: D. D. Ang, R. Gorenflo, V. K. Le, D. D. Trong, Moment Theory and Some Inverse Problems in Potential Theory and Heat Conduction (2002)
Vol. 1793: J. Cortés Monforte, Geometric, Control and Numerical Aspects of Nonholonomic Systems (2002)
Vol. 1794: N. Pytheas Fogg, Substitution in Dynamics, Arithmetics and Combinatorics. Editors: V. Berthé, S. Ferenczi, C. Mauduit, A. Siegel (2002)
Vol. 1795: H. Li, Filtered-Graded Transfer in Using Noncommutative Gröbner Bases (2002)
Vol. 1796: J.M. Melenk, hp-Finite Element Methods for Singular Perturbations (2002)
Vol. 1797: B. Schmidt, Characters and Cyclotomic Fields in Finite Geometry (2002)
Vol. 1798: W.M. Oliva, Geometric Mechanics (2002)
Vol. 1799: H. Pajot, Analytic Capacity, Rectifiability, Menger Curvature and the Cauchy Integral (2002)
Vol. 1800: O. Gabber, L. Ramero, Almost Ring Theory (2003)
Vol. 1801: J. Azéma, M. Émery, M. Ledoux, M. Yor (Eds.), Séminaire de Probabilités XXXVI (2003)
Vol. 1802: V. Capasso, E. Merzbach, B.G. Ivanoff, M. Dozzi, R. Dalang, T. Mountford, Topics in Spatial Stochastic Processes. Martina Franca, Italy 2001. Editor: E. Merzbach (2003)
Vol. 1803: G. Dolzmann, Variational Methods for Crystalline Microstructure – Analysis and Computation (2003)
Vol. 1804: I. Cherednik, Ya. Markov, R. Howe, G. Lusztig, Iwahori-Hecke Algebras and their Representation Theory. Martina Franca, Italy 1999. Editors: V. Baldoni, D. Barbasch (2003)
Vol. 1805: F. Cao, Geometric Curve Evolution and Image Processing (2003)
Vol. 1806: H. Broer, I. Hoveijn. G. Lunther, G. Vegter, Bifurcations in Hamiltonian Systems. Computing Singularities by Gröbner Bases (2003)
Vol. 1807: V. D. Milman, G. Schechtman (Eds.), Geometric Aspects of Functional Analysis. Israel Seminar 2000-2002 (2003)
Vol. 1808: W. Schindler, Measures with Symmetry Properties (2003)
Vol. 1809: O. Steinbach, Stability Estimates for Hybrid Coupled Domain Decomposition Methods (2003)
Vol. 1810: J. Wengenroth, Derived Functors in Functional Analysis (2003)
Vol. 1811: J. Stevens, Deformations of Singularities (2003)
Vol. 1812: L. Ambrosio, K. Deckelnick, G. Dziuk, M. Mimura, V. A. Solonnikov, H. M. Soner, Mathematical Aspects of Evolving Interfaces. Madeira, Funchal, Portugal 2000. Editors: P. Colli, J. F. Rodrigues (2003)
Vol. 1813: L. Ambrosio, L. A. Caffarelli, Y. Brenier, G. Buttazzo, C. Villani, Optimal Transportation and its Applications. Martina Franca, Italy 2001. Editors: L. A. Caffarelli, S. Salsa (2003)
Vol. 1814: P. Bank, F. Baudoin, H. Föllmer, L.C.G. Rogers, M. Soner, N. Touzi, Paris-Princeton Lectures on Mathematical Finance 2002 (2003)
Vol. 1815: A. M. Vershik (Ed.), Asymptotic Combinatorics with Applications to Mathematical Physics. St. Petersburg, Russia 2001 (2003)
Vol. 1816: S. Albeverio, W. Schachermayer, M. Talagrand, Lectures on Probability Theory and Statistics. Ecole d'Eté de Probabilités de Saint-Flour XXX-2000. Editor: P. Bernard (2003)
Vol. 1817: E. Koelink, W. Van Assche(Eds.), Orthogonal Polynomials and Special Functions. Leuven 2002 (2003)
Vol. 1818: M. Bildhauer, Convex Variational Problems with Linear, nearly Linear and/or Anisotropic Growth Conditions (2003)
Vol. 1819: D. Masser, Yu. V. Nesterenko, H. P. Schlickewei, W. M. Schmidt, M. Waldschmidt, Diophantine Approximation. Cetraro, Italy 2000. Editors: F. Amoroso, U. Zannier (2003)
Vol. 1820: F. Hiai, H. Kosaki, Means of Hilbert Space Operators (2003)
Vol. 1821: S. Teufel, Adiabatic Perturbation Theory in Quantum Dynamics (2003)
Vol. 1822: S.-N. Chow, R. Conti, R. Johnson, J. Mallet-Paret, R. Nussbaum, Dynamical Systems. Cetraro, Italy 2000. Editors: J. W. Macki, P. Zecca (2003)
Vol. 1823: A. M. Anile, W. Allegretto, C. Ringhofer, Mathematical Problems in Semiconductor Physics. Cetraro, Italy 1998. Editor: A. M. Anile (2003)
Vol. 1824: J. A. Navarro González, J. B. Sancho de Salas, $\mathscr{C}^\infty$ – Differentiable Spaces (2003)
Vol. 1825: J. H. Bramble, A. Cohen, W. Dahmen, Multiscale Problems and Methods in Numerical Simulations, Martina Franca, Italy 2001. Editor: C. Canuto (2003)
Vol. 1826: K. Dohmen, Improved Bonferroni Inequalities via Abstract Tubes. Inequalities and Identities of Inclusion-Exclusion Type. VIII, 113 p, 2003.
Vol. 1827: K. M. Pilgrim, Combinations of Complex Dynamical Systems. IX, 118 p, 2003.
Vol. 1828: D. J. Green, Gröbner Bases and the Computation of Group Cohomology. XII, 138 p, 2003.
Vol. 1829: E. Altman, B. Gaujal, A. Hordijk, Discrete-Event Control of Stochastic Networks: Multimodularity and Regularity. XIV, 313 p, 2003.
Vol. 1830: M. I. Gil', Operator Functions and Localization of Spectra. XIV, 256 p, 2003.
Vol. 1831: A. Connes, J. Cuntz, E. Guentner, N. Higson, J. E. Kaminker, Noncommutative Geometry, Martina Franca, Italy 2002. Editors: S. Doplicher, L. Longo (2004)
Vol. 1832: J. Azéma, M. Émery, M. Ledoux, M. Yor (Eds.), Séminaire de Probabilités XXXVII (2003)
Vol. 1833: D.-Q. Jiang, M. Qian, M.-P. Qian, Mathematical Theory of Nonequilibrium Steady States. On the Frontier of Probability and Dynamical Systems. IX, 280 p, 2004.
Vol. 1834: Yo. Yomdin, G. Comte, Tame Geometry with Application in Smooth Analysis. VIII, 186 p, 2004.
Vol. 1835: O.T. Izhboldin, B. Kahn, N.A. Karpenko, A. Vishik, Geometric Methods in the Algebraic Theory of Quadratic Forms. Summer School, Lens, 2000. Editor: J.-P. Tignol (2004)
Vol. 1836: C. Năstăsescu, F. Van Oystaeyen, Methods of Graded Rings. XIII, 304 p, 2004.
Vol. 1837: S. Tavaré, O. Zeitouni, Lectures on Probability Theory and Statistics. Ecole d'Eté de Probabilités de Saint-Flour XXXI-2001. Editor: J. Picard (2004)
Vol. 1838: A.J. Ganesh, N.W. O'Connell, D.J. Wischik, Big Queues. XII, 254 p, 2004.
Vol. 1839: R. Gohm, Noncommutative Stationary Processes. VIII, 170 p, 2004.

Vol. 1840: B. Tsirelson, W. Werner, Lectures on Probability Theory and Statistics. Ecole d'Eté de Probabilités de Saint-Flour XXXII-2002. Editor: J. Picard (2004)
Vol. 1841: W. Reichel, Uniqueness Theorems for Variational Problems by the Method of Transformation Groups (2004)
Vol. 1842: T. Johnsen, A.L. Knutsen, K3 Projective Models in Scrolls (2004)
Vol. 1843: B. Jefferies, Spectral Properties of Noncommuting Operators (2004)
Vol. 1844: K.F. Siburg, The Principle of Least Action in Geometry and Dynamics (2004)
Vol. 1845: Min Ho Lee, Mixed Automorphic Forms, Torus Bundles, and Jacobi Forms (2004)
Vol. 1846: H. Ammari, H. Kang, Reconstruction of Small Inhomogeneities from Boundary Measurements (2004)
Vol. 1847: T.R. Bielecki, T. Björk, M. Jeanblanc, M. Rutkowski, J.A. Scheinkman, W. Xiong, Paris-Princeton Lectures on Mathematical Finance 2003 (2004)
Vol. 1848: M. Abate, J. E. Fornaess, X. Huang, J. P. Rosay, A. Tumanov, Real Methods in Complex and CR Geometry, Martina Franca, Italy 2002. Editors: D. Zaitsev, G. Zampieri (2004)
Vol. 1849: Martin L. Brown, Heegner Modules and Elliptic Curves (2004)
Vol. 1850: V. D. Milman, G. Schechtman (Eds.), Geometric Aspects of Functional Analysis. Israel Seminar 2002-2003 (2004)
Vol. 1851: O. Catoni, Statistical Learning Theory and Stochastic Optimization (2004)
Vol. 1852: A.S. Kechris, B.D. Miller, Topics in Orbit Equivalence (2004)
Vol. 1853: Ch. Favre, M. Jonsson, The Valuative Tree (2004)
Vol. 1854: O. Saeki, Topology of Singular Fibers of Differential Maps (2004)
Vol. 1855: G. Da Prato, P.C. Kunstmann, I. Lasiecka, A. Lunardi, R. Schnaubelt, L. Weis, Functional Analytic Methods for Evolution Equations. Editors: M. Iannelli, R. Nagel, S. Piazzera (2004)
Vol. 1856: K. Back, T.R. Bielecki, C. Hipp, S. Peng, W. Schachermayer, Stochastic Methods in Finance, Bressanone/Brixen, Italy, 2003. Editors: M. Fritelli, W. Runggaldier (2004)
Vol. 1857: M. Émery, M. Ledoux, M. Yor (Eds.), Séminaire de Probabilités XXXVIII (2005)
Vol. 1858: A.S. Cherny, H.-J. Engelbert, Singular Stochastic Differential Equations (2005)
Vol. 1859: E. Letellier, Fourier Transforms of Invariant Functions on Finite Reductive Lie Algebras (2005)
Vol. 1860: A. Borisyuk, G.B. Ermentrout, A. Friedman, D. Terman, Tutorials in Mathematical Biosciences I. Mathematical Neurosciences (2005)
Vol. 1861: G. Benettin, J. Henrard, S. Kuksin, Hamiltonian Dynamics – Theory and Applications, Cetraro, Italy, 1999. Editor: A. Giorgilli (2005)
Vol. 1862: B. Helffer, F. Nier, Hypoelliptic Estimates and Spectral Theory for Fokker-Planck Operators and Witten Laplacians (2005)
Vol. 1863: H. Fürh, Abstract Harmonic Analysis of Continuous Wavelet Transforms (2005)
Vol. 1864: K. Efstathiou, Metamorphoses of Hamiltonian Systems with Symmetries (2005)
Vol. 1865: D. Applebaum, B.V. R. Bhat, J. Kustermans, J. M. Lindsay, Quantum Independent Increment Processes I. From Classical Probability to Quantum Stochastic Calculus. Editors: M. Schürmann, U. Franz (2005)
Vol. 1866: O.E. Barndorff-Nielsen, U. Franz, R. Gohm, B. Kümmerer, S. Thorbjønsen, Quantum Independent Increment Processes II. Structure of Quantum Lévy Processes, Classical Probability, and Physics. Editors: M. Schürmann, U. Franz, (2005)
Vol. 1867: J. Sneyd (Ed.), Tutorials in Mathematical Biosciences II. Mathematical Modeling of Calcium Dynamics and Signal Transduction. (2005)
Vol. 1868: J. Jorgenson, S. Lang, $Pos_n(R)$ and Eisenstein Sereies. (2005)
Vol. 1869: A. Dembo, T. Funaki, Lectures on Probability Theory and Statistics. Ecole d'Eté de Probabilités de Saint-Flour XXXIII-2003. Editor: J. Picard (2005)
Vol. 1870: V.I. Gurariy, W. Lusky, Geometry of Müntz Spaces and Related Questions. (2005)
Vol. 1871: P. Constantin, G. Gallavotti, A.V. Kazhikhov, Y. Meyer, S. Ukai, Mathematical Foundation of Turbulent Viscous Flows, Martina Franca, Italy, 2003. Editors: M. Cannone, T. Miyakawa (2006)
Vol. 1872: A. Friedman (Ed.), Tutorials in Mathematical Biosciences III. Cell Cycle, Proliferation, and Cancer (2006)
Vol. 1873: R. Mansuy, M. Yor, Random Times and Enlargements of Filtrations in a Brownian Setting (2006)
Vol. 1874: M. Yor, M. Émery (Eds.), In Memoriam Paul-André Meyer - Séminaire de Probabilités XXXIX (2006)
Vol. 1875: J. Pitman, Combinatorial Stochastic Processes. Ecole d'Eté de Probabilités de Saint-Flour XXXII-2002. Editor: J. Picard (2006)
Vol. 1876: H. Herrlich, Axiom of Choice (2006)
Vol. 1877: J. Steuding, Value Distributions of $L$-Functions (2006)
Vol. 1878: R. Cerf, The Wulff Crystal in Ising and Percolation Models, Ecole d'Eté de Probabilités de Saint-Flour XXXIV-2004. Editor: Jean Picard (2006)
Vol. 1879: G. Slade, The Lace Expansion and its Applications, Ecole d'Eté de Probabilités de Saint-Flour XXXIV-2004. Editor: Jean Picard (2006)
Vol. 1880: S. Attal, A. Joye, C.-A. Pillet, Open Quantum Systems I, The Hamiltonian Approach (2006)
Vol. 1881: S. Attal, A. Joye, C.-A. Pillet, Open Quantum Systems II, The Markovian Approach (2006)
Vol. 1882: S. Attal, A. Joye, C.-A. Pillet, Open Quantum Systems III, Recent Developments (2006)
Vol. 1883: W. Van Assche, F. Marcellàn (Eds.), Orthogonal Polynomials and Special Functions, Computation and Application (2006)
Vol. 1884: N. Hayashi, E.I. Kaikina, P.I. Naumkin, I.A. Shishmarev, Asymptotics for Dissipative Nonlinear Equations (2006)
Vol. 1885: A. Telcs, The Art of Random Walks (2006)

## Recent Reprints and New Editions

Vol. 1471: M. Courtieu, A.A. Panchishkin, Non-Archimedean L-Functions and Arithmetical Siegel Modular Forms. – Second Edition (2003)
Vol. 1618: G. Pisier, Similarity Problems and Completely Bounded Maps. 1995 – Second, Expanded Edition (2001)
Vol. 1629: J.D. Moore, Lectures on Seiberg-Witten Invariants. 1997 – Second Edition (2001)
Vol. 1638: P. Vanhaecke, Integrable Systems in the realm of Algebraic Geometry. 1996 – Second Edition (2001)
Vol. 1702: J. Ma, J. Yong, Forward-Backward Stochastic Differential Equations and their Applications. 1999. – Corrected 3rd printing (2005)